T0300991

BIOMIMETIC ROBOTICS

This book is written as an initial course in robotics. It is ideal for the study of unmanned aerial or underwater vehicles, a topic on which few books exist. It presents the fundamentals of robotics, from an aerospace perspective, by considering only the field of robot mechanisms. For an aerospace engineer, three-dimensional and parallel mechanisms – flight simulation, unmanned aerial vehicles, and space robotics – take on an added significance. Biomimetic robot mechanisms are fundamental to manipulators and walking, mobile, and flying robots. As a distinguishing feature, this book gives a unified and integrated treatment of biomimetic robot mechanisms. It is ideal preparation for the next robotics module: practical robot-control design. Although the book focuses on principles, computational procedures are also given due importance. Students are encouraged to use computational tools to solve the examples in the exercises. The author has also included some additional topics beyond his course coverage for the enthusiastic reader to explore.

Ranjan Vepa has been with Queen Mary and Westfield College in the Department of Engineering since 1984, after receiving his Ph.D. from Stanford University. Dr. Vepa's research covers, in addition to biomimetic robotics, aspects of the design, analysis, simulation, and implementation of avionics and avionics-related systems, including nonlinear aerospace vehicle kinematics, dynamics, vibration, control, and filtering. He has published numerous journal and conference papers and has contributed to a textbook titled *Application of Artificial Intelligence in Process Control*. He is active in learning-based design education and the utilization of modern technology in engineering education. Dr. Vepa is a Chartered Engineer, a Member of the Royal Aeronautical Society (MRAeS) London, and a Fellow of the Higher Education Academy.

Biomimetic Robotics

MECHANISMS AND CONTROL

Ranjan Vepa
Queen Mary, University of London

CAMBRIDGE
UNIVERSITY PRESS

CAMBRIDGE
UNIVERSITY PRESS

Shaftesbury Road, Cambridge CB2 8EA, United Kingdom

One Liberty Plaza, 20th Floor, New York, NY 10006, USA

477 Williamstown Road, Port Melbourne, VIC 3207, Australia

314–321, 3rd Floor, Plot 3, Splendor Forum, Jasola District Centre, New Delhi – 110025, India

103 Penang Road, #05–06/07, Visioncrest Commercial, Singapore 238467

Cambridge University Press is part of Cambridge University Press & Assessment,
a department of the University of Cambridge.

We share the University's mission to contribute to society through the pursuit of
education, learning and research at the highest international levels of excellence.

www.cambridge.org
Information on this title: www.cambridge.org/9780521895941

First published 2009

A catalogue record for this publication is available from the British Library

Library of Congress Cataloging-in-Publication data
Vepa, Ranjan.
Biomimetic robotics : mechanisms and control / Ranjan Vepa.
   p. cm.
Includes bibliographical references and index.
ISBN 978-0-521-89594-1 (hardback)
1. Robotics. 2. Biomimetics. I. Title.
TJ211.V47 2009
629.8′93 – dc22   2008024401

ISBN   978-0-521-89594-1   Hardback

To my father, Narasimha Row, and mother, Annapurna

Contents

Preface   *page* xiii

Acronyms   xv

1. **The Robot** **1**

    1.1 Robotics: An Introduction   1
    1.2 Robot-Manipulator Fundamentals and Components   5
    1.3 From Kinematic Pairs to the Kinematics of Mechanisms   12
    1.4 Novel Mechanisms   13
        1.4.1 Rack-and-Pinion Mechanism   14
        1.4.2 Pawl-and-Ratchet Mechanism   14
        1.4.3 Pantograph   15
        1.4.4 Quick-Return Mechanisms   15
        1.4.5 Ackermann Steering Gear   16
        1.4.6 Sun and Planet Epicyclic Gear Train   17
        1.4.7 Universal Joints   17
    1.5 Spatial Mechanisms and Manipulators   18
    1.6 Meet Professor da Vinci the Surgeon, PUMA, and SCARA   20
    1.7 Back to the Future   23
    EXERCISES   24

2. **Biomimetic Mechanisms** **25**

    2.1 Introduction   25
    2.2 Principles of Legged Locomotion   27
        2.2.1 Inchworm Locomotion   29
        2.2.2 Walking Machines   30
        2.2.3 Autonomous Footstep Planning   31
    2.3 Imitating Animals   31
        2.3.1 Principles of Bird Flight   33
        2.3.2 Mechanisms Based on Bird Flight   34
        2.3.3 Swimming Like a Fish   37

2.4 Biomimetic Sensors and Actuators   39
 2.4.1 Action Potentials   43
 2.4.2 Measurement and Control of Cellular Action Potentials   46
 2.4.3 Bionic Limbs: Interfacing Artificial Limbs to Living Cells 47
 2.4.4 Artificial Muscles: Flexible Muscular Motors   51
 2.4.5 Prosthetic Control of Artificial Muscles   53
2.5 Applications in Computer-Aided Surgery and Manufacture   55
 2.5.1 Steady Hands: Active Tremor Compensation   56
 2.5.2 Design of Scalable Robotic Surgical Devices   58
 2.5.3 Robotic Needle Placement and Two-Hand Suturing   60
EXERCISES   61

3. **Homogeneous Transformations and Screw Motions** **62**
3.1 General Rigid Motions in Two Dimensions   62
 3.1.1 Instantaneous Centers of Rotation   64
3.2 Rigid Body Motions in Three Dimensions: Definition of *Pose*   64
 3.2.1 Homogeneous Coordinates: Transformations of Position
  and Orientation   65
3.3 General Motions of Rigid Frames in Three Dimensions: Frames
 with *Pose*   66
 3.3.1 The Denavit–Hartenberg Decomposition   66
 3.3.2 Instantaneous Axis of Screw Motion   67
 3.3.3. A Screw from a Twist   69
EXERCISES   70

4. **Direct Kinematics of Serial Robot Manipulators** **74**
4.1 Definition of Direct or Forward Kinematics   74
4.2 The Denavit–Hartenberg Convention   74
4.3 Planar Anthropomorphic Manipulators   76
4.4 Planar Nonanthropomorphic Manipulators   78
4.5 Kinematics of Wrists   80
4.6 Direct Kinematics of Two Industrial Manipulators   81
EXERCISES   86

5. **Manipulators with Multiple Postures and Compositions** **89**
5.1 Inverse Kinematics of Robot Manipulators   89
 5.1.1 The Nature of Inverse Kinematics: Postures   91
 5.1.2 Some Practical Examples   95
5.2 Parallel Manipulators: Compositions   99
 5.2.1 Parallel Spatial Manipulators: The Stewart Platform   101
5.3 Workspace of a Manipulator   105
EXERCISES   107

6. **Grasping: Mechanics and Constraints** **111**

  6.1 Forces and Moments 111
  6.2 Definition of a Wrench 112
  6.3 Mechanics of Gripping 112
  6.4 Transformation of Forces and Moments 114
  6.5 Compliance 115
      6.5.1 Passive and Active Compliance 116
      6.5.2 Constraints: Natural and Artificial 116
      6.5.3 Hybrid Control 117
  EXERCISES 118

7. **Jacobians** **120**

  7.1 Differential Motion 120
      7.1.1 Velocity Kinematics 123
      7.1.2 Translational Velocities and Acceleration 124
      7.1.3 Angular Velocities 127
  7.2 Definition of a Screw Vector: Instantaneous Screws 127
      7.2.1 Duality with the Wrench 129
      7.2.2 Transformation of a Compliant Body Wrench 130
  7.3 The Jacobian and the Inverse Jacobian 131
      7.3.1 The Mobility Criterion: Overconstrained Mechanisms 133
      7.3.2 Singularities: Physical Interpretation 134
      7.3.3 Manipulability: Putting Redundant Mechanisms to Work 136
      7.3.4 Computing the Inverse Kinematics: The Lyapunov
            Approach 137
  EXERCISES 140

8. **Newtonian, Eulerian, and Lagrangian Dynamics** **142**

  8.1 Newtonian and Eulerian Mechanics 142
      8.1.1 Kinetics of Screw Motion: The Newton–Euler Equations 145
      8.1.2 Moments of Inertia 146
      8.1.3 Dynamics of a Link's Moment of Inertia 147
      8.1.4 Recursive Form of the Newton–Euler Equations 149
  8.2 Lagrangian Dynamics of Manipulators 152
      8.2.1 Forward and Inverse Dynamics 154
  8.3 The Principle of Virtual Work 156
  EXERCISES 158

9. **Path Planning, Obstacle Avoidance, and Navigation** **164**

  9.1 Fundamentals of Trajectory Following 164
      9.1.1 Path Planning: Trajectory Generation 165
      9.1.2 Splines, Bézier Curves, and Bernstein Polynomials 167
  9.2 Dynamic Path Planning 172

9.3   Obstacle Avoidance   174
9.4   Inertial Measuring and Principles of Position and
      Orientation Fixing   180
      9.4.1 Gyro-Free Inertial Measuring Units   188
      9.4.2 Error Dynamics of Position and Orientation   189
      EXERCISES   193

10. **Hamiltonian Systems and Feedback Linearization** **198**

      10.1 Dynamical Systems of the Liouville Type   198
           10.1.1 Hamilton's Equations of Motion   199
           10.1.2 Passivity of Hamiltonian Dynamics   202
           10.1.3 Hamilton's Principle   203
      10.2 Contact Transformation   204
           10.2.1 Hamilton–Jacobi Theory   205
           10.2.2 Significance of the Hamiltonian Representations 206
      10.3 Canonical Representations of the Dynamics   207
           10.3.1 Lie Algebras   208
           10.3.2 Feedback Linearization   210
           10.3.3 Partial State–Feedback Linearization   213
           10.3.4 Involutive Transformations   214
      10.4 Applications of Feedback Linearization   215
      10.5 Optimal Control of Hamiltonian and Near-Hamiltonian Systems 223
      10.6 Dynamics of Nonholonomic Systems   225
           10.6.1 The Bicycle   228
      EXERCISES   236

11. **Robot Control** **242**

      11.1 Introduction   242
           11.1.1 Adaptive and Model-Based Control   242
           11.1.2 Taxonomies of Control Strategies   252
           11.1.3 Human-Centered Control Methods   252
           11.1.4 Robot-Control Tasks   257
           11.1.5 Robot-Control Implementations   258
           11.1.6 Controller Partitioning and Feedforward   259
           11.1.7 Independent Joint Control   260
      11.2 HAL, Do You Understand JAVA?   261
      11.3 Robot Sensing and Perception   263
      EXERCISES   269

12. **Biomimetic Motive Propulsion** **272**

      12.1 Introduction   272
      12.2 Dynamics and Balance of Walking Biped Robots   272
           12.2.1 Dynamic Model for Walking   272
           12.2.2 Dynamic Balance during Walking: The Zero-Moment Point 277

12.2.3 Half-Model for a Quadruped Robot: Dynamics and Control   279

12.3 Modeling Bird Flight: Robot Manipulators in Free Flight   281

12.3.1 Dynamics of a Free-Flying Space Robot   282

12.3.2 Controlling a Free-Flying Space Robot   284

12.4 Flapping Propulsion of Aerial Vehicles   285

12.4.1 Unsteady Aerodynamics of an Aerofoil   287

12.4.2 Generation of Thrust   294

12.4.3 Controlled Flapping for Flight Vehicles   299

12.5 Underwater Propulsion and Its Control   301

EXERCISES   304

Answers to Selected Exercises **309**

Appendix: Attitude and Quaternions **317**

Bibliography   335

Index   339

Preface

This book is intended for a first course in a robotics sequence. There are indeed a large number of books in the field of robotics, but for the engineer or student interested in unmanned aerial or underwater vehicles there are very few books. From the point of view of an aerospace engineer (flight simulation, unmanned aerial vehicles, and space robotics) three-dimensional and parallel mechanisms take on an added significance.

The fundamentals of robotics, from an aerospace perspective, can be very well presented by considering just the field of robot mechanisms. Biomimetic robot mechanisms are fundamental to manipulators and to walking, mobile, and flying robots. A distinguishing feature of the book is that a unified and integrated treatment of biomimetic robot mechanisms is offered.

This book is intended to prepare a student for the next logical module in a course on robotics: practical robot-control design. Throughout the book, in every chapter, are introduced the most important and recent developments relevant to the subject matter. Although the book's primary focus is on understanding principles, computational procedures are also given due importance as a matter of course. Students are encouraged to use whatever computational tools they consider appropriate to solve the examples in the exercises.

When writing this text, I included the topics in my course and some more for the enthusiastic reader with the initiative for self-learning.

Thus, Chapter 1 gives an overview of robotics, with a focus on generic robotic mechanisms, whereas Chapter 2 is about biomimetic mechanisms. Chapter 3 draws attention to the integrated representation of displacement and rotation. Chapters 4 and 5 are about direct and inverse kinematics. Chapter 6 is about the mechanics of grasping, and Chapter 7 provides a gentle introduction to Jacobians. Dynamics of mechanisms is introduced in Chapter 8, and path planning and trajectory following are discussed in Chapter 9. Chapter 10 is dedicated to the techniques of transforming the dynamics to simpler forms that are useful for such tasks as control synthesis. Chapter 10 also considers the ever-important problem of the dynamics of nonholonomic systems with the archetypal example of the bicycle.

Chapter 11 on robot control and Chapter 12 on biomimetic motive propulsion are unique elements of this book. Most books dive straight into the mathematics of robot-control synthesis. Here the student is introduced to important concepts in the preceding chapters and is thus better prepared. Additionally, the problem of robot control is inherently different from other applications of control engineering in that it is patently nonlinear. Thus, Chapter 11 introduces the strategies for robot control from a broad application perspective, keeping in mind the real-world and nonideal situations faced on a routine basis. The book concludes with a unique and lively chapter on biomimetic approaches to thrust generation in robots and robot vehicles.

All chapters feature exercises designed to reinforce the chapter content and ensure the student has sufficient background to address the subsequent chapters. The appendix provides a basic introduction to attitude representation, dynamics, and control.

Any general text draws heavily on the work of countless engineers and scientists. It is not possible to include references to the whole body of published work in the field. Yet those books and papers that were consulted in the preparation of the manuscript are referenced, along with selected texts and papers that are appropriate for the reader to obtain more detail and breadth.

I would like to thank my colleagues in the Department of Engineering at Queen Mary, for their support in this endeavor. I would also like to express my special thanks to Peter Gordon, Senior Editor, Aeronautical and Mechanical Engineering, Cambridge University Press, New York, for his enthusiastic support for this project.

I would like to thank my wife, Sudha, for her love, understanding, and patience. Her encouragement was a principal factor that provided the motivation to complete the project. Finally, I must add that my interest in the kinematics of mechanisms was inherited from my late father many years ago. It was, in fact, he who brought to my attention the work of Alexander B. W. Kennedy and of Denavit and Hartenberg, as well as countless other related aspects, a long time ago. To him I owe my deepest debt of gratitude. This book is dedicated to his memory.

R. Vepa
London, UK, 2008

Acronyms

ATP	adenosine triphosphate
BCF	body or caudal fin
CNS	central nervous system
CP	caudate and putamen
CVG	Coriolis vibratory gyroscope
DCM	direction cosine matrix
DH	Denavit–Hartenberg
DOF	degree of freedom
DTG	dynamically tuned gyroscope
DVT	deep vein thrombosis
ECG	electrocardiogram
EEG	electroencephalogram
EMF	electromotive force
EMG	electromyogram
ENG	electroneurogram
FET	field-effect transistor
FOG	fiber-optic gyroscope
GOD	glucose oxidase
GPpe	globus pallidus pars externa
IMU	inertial measuring unit
INS	inertial navigation system
I/O	input/output
IRS	inertial reference system
ISFET	ion-selective field-effect transistor
IUPAC	International Union of Pure and Applied Chemistry
LASER	light amplification by stimulated emission of radiation
LED	light-emitting diode
MPF	median or paired fin
MRAC	model reference adaptive control
MRI	magnetic resonance image
OS	operating system

OSI	open system interconnection
PD	proportional-derivative
PI	proportional-integral
PUMA	Programmable Universal Machine for Assembly
PVDF	poly-vinylidene fluoride trifluoroethylene
QCM	quartz crystal microbalance
RF	radio frequency
RLG	ring-laser gyroscope
ROV	remotely operated vehicle
RTOS	real-time operating system
SCARA	Selectively Compliant Assembly Robot Arm
SMA	shape memory alloy
SNpc	substantia nigra pars compacta
STN	subthalamic nucleus
TCP	tool center point
UAV	unmanned aerial vehicle
UT	Universal Time
VDT	virtual device table
VOR	vestibulo-ocular reflex

1

The Robot

1.1 Robotics: An Introduction

The concept of a robot as we know it today evolved over many years. In fact, its origins could be traced to ancient Greece well before the time when Archimedes invented the screw pump. Leonardo da Vinci (1452–1519) made far-reaching contributions to the field of robotics with his pioneering research into the brain that led him to make discoveries in neuroanatomy and neurophysiology. He provided physical explanations of how the brain processes visual and other sensory inputs and invented a number of ingenious machines. His flying devices, although not practicable, embodied sound principles of aerodynamics, and a toy built to bring to fruition Leonardo's drawing inspired the Wright brothers in building their own flying machine, which was successfully flown in 1903. The word robot itself seems to have first appeared in 1921 in Karel Capek's play, *Rossum's Universal Robots*, and originated from the Slavic languages. In many of these languages the word robot is quite common as it stands for *worker*. It is derived from the Czech word *robitit*, which implies drudgery. Indeed, robots were conceived as machines capable of repetitive tasks requiring a lower intelligence than that of humans. Yet today robots are thought to be capable of possessing intelligence, and the term is probably inappropriate. Nevertheless it is in use. The term *robotics* was probably first coined in science fiction work published around the 1950s by Isaac Asimov, who also enunciated his three laws of robotics. It was from Asimov's work that the concept of emulating humans emerged. A typical robot arm based on the anatomy of humans is illustrated in Figure 1.1.

The evolution of a robot has a long and fascinating history, and this is traced in Table 1.1. Apart from the down-to-earth inventors, a number of high-flying thinkers contributed to its development by imagining the unimaginable. Although there were many such philosophers, some of the recent thinkers who contributed to the evolution of robotics are identified in Table 1.2. In fact, the late Werner von Braun, the legendary head of the U.S. space agency, NASA, used Arthur C. Clarke's book, *Exploration of Space* (1951), to persuade President John F. Kennedy to commit the United States to going to the Moon. According to Arthur C. Clarke, Cyrano de

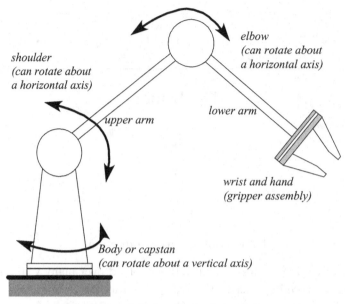

Figure 1.1. A typical robot arm.

Bergerac must be credited both for first employing the rocket in space travel and for inventing the ramjet in the 1650s. These inventions, however, became a reality 300 years later.

What then is a robot? An industrial robot in its simplest form is a programmable multifunctional manipulator designed to move materials, parts, tools, or specialized devices through a sequence of programmed motions to perform a variety of tasks. Some of the earliest developments in the field of robotics were concerned with carrying out tasks in environments that were either patently dangerous for humans to enter or simply inaccessible. A classic example is the handling of nuclear fuel rods in the environment of a nuclear reactor. Another, the American program to land on the Moon, was supported in no small measure by developments in robotics, particularly the Apollo Lunar Module. It was a remarkable vehicle that allowed the astronauts to descend softly onto the lunar surface, carry out a number of experiments on it, and return to the command and service modules in lunar orbit. Examples of robots today include the Space Shuttle's satellite manipulator, the Martian rover, unmanned aerial vehicles (UAVs), and autonomously operated underwater vehicles for ocean exploration. A broad taxonomy of robots is introduced in Figure 1.2. Although a true robot would be an intelligent, autonomous mobile entity, capable of interacting with its environment in a meaningful, adaptive, and intelligent manner, no such entities exist at the present time.

An intelligent autonomous vehicle does indeed possess a number of capabilities not found in a common machine, however ingenious it may be. Yet such intelligent machines are only evolving at the present time. Considering the impressive range of human achievements in the last century, it is fair to say that the development of

Table 1.1. *Evolution of robots [remotely operated vehicles (ROVs), unmanned aerial vehicles (UAVs)]: The chronology*

Year	Invention	Inventor
The pre-1900 period		
200 B.C.E.	Screw pump	Archimedes
1440 C.E.	Printing press	Johannes Gutenberg
1589	Knitting machine	William Lee
1636	Micrometer	William Gascoigne
1642	Calculating machine	Blaise Pascal
1725	Stereotyping	William Ged
1733	The flying shuttle	John Kay
1765	Steam engine	James Watt
1783	Parachute	Louis Lenormand
1783	Hot air balloon	Montgolfier Brothers
1785	Power loom	Edward Cartwright
1793	Cotton gin	Eli Whitney
1800	Lathe	Henry Maudslay
1804	Steam locomotive	Richard Trevithick
1816	Bicycle	Karl von Sauerbronn
1823	Digital calculating machine	Charles Babbage
1831	Dynamo	Michael Faraday
1834	Reaping machine	Cyrus McCormick
1846	Sewing machine	Elias Howe
1852	Gyroscope	Leon Foucault
1852	Passenger lift	Elisha Otis
1858	Washing machine	William Hamilton
1859	Internal combustion engine	Jean-Joseph-Etienne Lenoir
1867	Typewriter	Christopher Scholes
1868	Motorized bicycle	Michaux Brothers
1876	Carpet sweeper	Melville Bissell
1885	Motorcycle	Edward Butler
1885	Motor car engine	Gottlieb Daimle/Karl Benz
1886	Electric fan	Schuyler Wheeler
1892	Diesel engine	Rudolf Diesel
1895	Wireless	Guglielmo Marconi
1898	Submarine	John P. Holland
The post-1900 period		
1901	Vacuum cleaner	Hubert Booth
1903	Airplane	Wright Brothers
1911	Combine harvester	Benjamin Holt
1926	Rocket	Robert H. Goddard
1929	Electronic television	Vladimir Zworykin
1944	Automatic digital computer	Howard Aiken
1949	Transistor	Team of American engineers
1955	Hovercraft	Christopher Cockerell
1957	Satellite	Team of Russian engineers
1959	Microchip	Team of American engineers
1962	Industrial robot manipulator	George Devol, Unimation
1960s	Automated assembly line	Team of American engineers
1971	Microprocessor	Team of American engineers
1981	Space Shuttle	Team of American engineers

4 *The Robot*

Table 1.2. *Dreamers, thinkers (eccentrics) who imagined the unimaginable*

Year	Thinker/writer	Contribution
1510	Leonardo da Vinci	Conceptual painting of a helicopter
1648	Cyrano de Bergerac	*Histoire comique des Estats et Empires de la Lune* (book)
1831	Mary Wollstonecraft-Shelley	*Frankenstein or, The Modern Prometheus* (book)
1864	Jules Verne	*From Earth to the Moon, Journey to the Centre of the Earth, Twenty Thousand Leagues Under the Sea* (books)
1901	H. G. Wells	*The Time Machine, The First Men on the Moon* (books)
1921	Karel Capek	*Rossum's Universal Robots* (play)
1930	Edgar Rice Burroughs	The Martian Tales (a series of 11 books published between 1911 and 1930)
1945	Arthur C. Clarke	Concept of geostationary communication satellites
1948	Norbert Weiner	*Cybernetics or Control and Communication in the Animal* (book)
1950	Isaac Asimov	*I Robot* (book)
1954	Jules Verne	*Twenty Thousand Leagues Under the Sea* (Disney film)
1955	Walt Elias Disney	Robots for entertainment in the *Disneyland* theme park
1968	Arthur C. Clarke, Stanley Kubrick	*2001: A Space Odyssey* (book, film), HAL (computer)
1977	George Lucas	*Star Wars* (films), R2D2, C3P0 (robots).

these machines may be effected in the not-so-distant future. For our part, the study of industrial robotics is best initiated by considering the simplest of these creatures: the robot manipulator. The manipulator is indeed the basic element in a walking robot, which in turn could be considered to be a primary component of an intelligent autonomous robot.

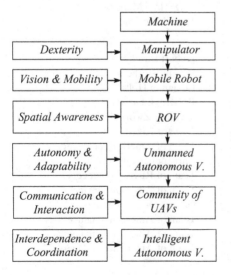

Figure 1.2. Taxonomy of robots.

Table 1.3. *Types of links*

Type of link	Illustration
Nullary	
Unary	
Binary	
Ternary	
Quaternary	
N-ary	

1.2 Robot-Manipulator Fundamentals and Components

Manipulation involves not only the motion, location, orientation, and geometry of a real physical object but also the procedures for changing these attributes as well as the physical design of a mechanical system that can effectively perform the manipulation functions. A robot manipulator is essentially a kinematic system that is an assemblage of a number of kinematic components. Although these components could in general be flexible, we shall restrict our attention to rigid components. The basic rigid-component elements are referred to as *kinematic links* or simply links. The basic interconnections between these links are referred to as *kinematic joints* or simply joints.

The simplest of kinematic arrangements is made up of two links. Thus we may consider two rigid bodies or links, one of which is free to move in three-dimensional space. Six independent parameters are required for completely specifying the motion of the free body in space, relative to the fixed body. The relative motion itself may be uniquely expressed in terms of the relative position of the center of gravity of the body in Cartesian space in the surge (forward and backward), sway (sideward), and heave (up and down) together with the orientation of the body. The orientation of the body is defined by its *attitude* and specified as a sequence of three successive rotations about the surge, sway, and heave directions in terms of a roll, a pitch, and a yaw angle, respectively. The six independent parameters or degrees of freedom (DOFs), as they are referred to by dynamicists, of the rigid body are the displacements of the center of gravity of the rigid body in the surge, sway, and heave directions and the sequence of roll, pitch, and yaw rotations about these directions.

We may classify the links on the basis of the number of joints used to interconnect them to each other, as illustrated in Table 1.3. Thus a *nullary* link is a rigid body that is not connected to any other link, and a *unary* link is one that is connected to just one other link by a single joint. A *binary* link is one that is connected to two other links by two distinct and separate joints. In a similar manner we define a *ternary* link as one that is connected to three other links at three distinct points on it by separate joints and a *quaternary* link as one that is connected to four other links at four distinct points on it by separate joints. In general, an *N-ary* link has *N* independent joints associated with it. Each link could be of arbitrary shape and the joints not only located at arbitrary points but also of arbitrary shape.

A *joint* itself is then defined as a connection between links that allows certain relative motion between links. At the outset such joints as welds, rigid connections, or bonded joints are excluded. Joints may then be classified, irrespective of their nature, based on the number of links associated with each. A binary joint connects two links together, *ternary* joints connect three links together, and in general an *N-ary* joint connects *N* links together.

In defining an *N*-ary link we associate *N* joints with it. It is sometimes more convenient to define an *N*-ary link as one that is connected to *N* other links. The two definitions are equivalent when every joint is a binary joint. A *kinematic system* is then a collection of kinematic links connected together by an associated set of kinematic joints. A typical robot manipulator is such an assemblage or *kinematic chain*, transmitting a definite motion. When one element of such a kinematic chain is fixed, the kinematic chain is a *mechanism*. From the point of view of robot manipulators, the objective is to investigate and synthesize kinematic chains to produce mechanisms that can transmit, constrain, or control various relative motions. This is the notion of *kinematic structure*.

The kinematic structure of a robot manipulator must satisfy two essential requirements:

1. First, it must have the ability to displace objects in three dimensions, operate within a finite workspace, have a reasonable amount of dexterity in avoiding obstacles, be able to reach into confined spaces and thus be able to arbitrarily alter an object's position and orientation from one to another;
2. Second, once positioned, the manipulator must have the ability to grasp with optimum force arbitrary or custom objects and hold them stationary irrespective of the other forces and moments they are subjected to.

Considering the first of these requirements, the primary design problem reduces to one of kinematic design. In general, to operate in three-dimensional space, a kinematic mechanism must possess at least six DOFs in order to function without any constraints. The DOFs are removed as one applies *kinematic constraints* to the assemblage.

To understand the kinematic design problem, the synthesis of a component system in a constrained kinematic mechanism so that it has a finite number of relative DOFs may first be considered. To ensure that the complete sets of feasible designs are finite in number, we need to consider the constraints that must be imposed. First, we consider the number of connected *planar* systems that can be put together from purely topological considerations. Considering four different links, we may identify

Table 1.4. *Four-link kinematic systems and their interchange graphs*

Four-link kinematic system	Interchange graph

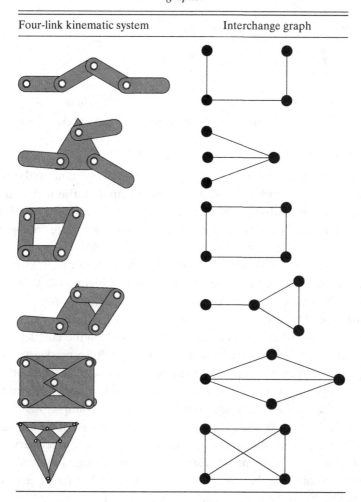

two "open-loop" arrangements, two "single-loop" arrangements, and two multiloop arrangements, as shown in Table 1.4. Also shown is a graph-theoretic representation of the kinematic structure in the form of an *interchange graph* in which a link in the physical mechanism is modeled as a node and a physical binary joint as an edge. One of the open-loop arrangements has three branches, and one of the closed-loop arrangements includes a branch in addition to the loop. Only one of these arrangements has all its links connected serially, whereas another has all links connected in a single loop. The former is a classic serial manipulator structure shown in Figure 1.3, which we will consider extensively in later sections. Also shown in the figure is an end-effector in the form of wrist and gripper. Without the wrist the manipulator would at most have the capability to approach a certain object from a single angle. The wrist gives the manipulator the mechanical flexibility to perform real tasks. Yet there are still a number of other limitations as the range of movement of each joint is

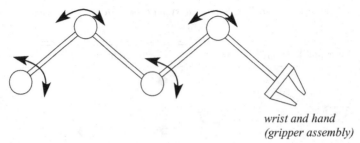

wrist and hand
(gripper assembly)

Figure 1.3. A typical robot serial manipulator including an end-effector (wrist-and-gripper assembly).

limited. As a result, the *workspace*, which is the space of positions and orientations reachable by the robot's end-effector, is still severely limited. But within this limited workspace it is the wrist that gives the end-effector the capability to point toward an arbitrary direction. The most common types of wrists are quite analogous to the human wrist, with three or more spatial (nonplanar) DOFs. The latter single-loop chain is the classic four-link kinematic chain, and when one of its links is fixed, one obtains the well-known four-bar mechanism shown in Figure 1.4. Link *OA* is the driving link or the *crank*, and link *AB* is the coupler link or *coupler*.

The path traced by a point on the coupler link *BA* is referred to as a *coupler curve*. Generally moving the point to different locations on the coupler would generate different coupler curves. A basic task is to be able to predict the geometry of the coupler curve generated by the point on the coupler. Coupler curves can be surprisingly complex. The corresponding design problem consists of synthesizing a mechanism for generating a specific coupler curve (for example, the shape of an aerofoil or a ship's hull). In the generic case, a coupler curve of a four-bar mechanism is an algebraic curve of degree six. Different shapes of coupler curves that can arise from four-bar and other mechanisms have been cataloged by several authors. It is interesting to observe that there exist at least two other four-bar mechanisms that can generate the same coupler curve as the one generated by a four-bar mechanism.

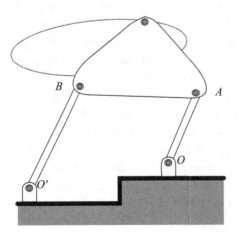

Figure 1.4. A four-bar linkage, with an example of a coupler curve generated by a point on the coupler.

Table 1.5. *(a) Kinematic chains with L links and J joints (shaded links are ternary links); (b) kinematic mechanisms derived from the chains in (a)*

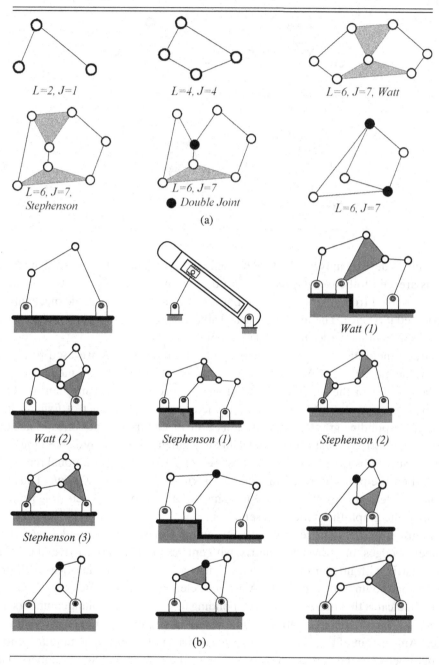

(a)

(b)

The basic four-link kinematic chain and the chains made more complex by the further addition of links are illustrated in Table 1.5(a). To transmit motion, a practical mechanism may be obtained by fixing one of the links in a kinematic chain. Thus from an *n*-link chain one may derive as many as *n* mechanisms. Some of the

Table 1.6. *A seven-link kinematic chain and*
its velocity graphs

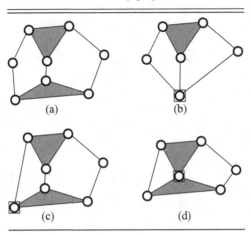

mechanisms derived may be identical to each other in a kinematic sense. Such mechanisms are said to be *isomorphic*. To find the number of distinct mechanisms that may be derived from a kinematic chain, it is essential that isomorphic mechanisms be counted just once. The mechanisms in Table 1.5(b) are obtained by fixing one of the links in the kinematic chains in Table 1.5(a).

Graph-theoretic representations of kinematic mechanisms are used to detect isomorphic mechanisms. Although these representations are by no means unique, they are useful in the analysis of the kinematic geometry of a mechanism. One method, for example, is based on clearly incorporating the effect of fixing a link in the corresponding graph representation of the mechanisms. Thus one obtains the graphical pattern by substituting a fixed link in a kinematic chain by a direct pin connection. Such substituted pin connections are indicated with an additional rectangle overlaid on the pin. The resulting diagram is known as a *velocity graph*. Thus the problem of detecting isomorphic mechanisms is reduced to one of detecting isomorphisms in the graph-theoretic representations.

A graph is characterized by a certain number of vertices or nodes, a certain number of edges, the adjacency matrix, with entries of ones and zeros, and the distance matrix with integer entries. The matrices are both square matrices with one row and one column for each node. A nonzero element of the former, the adjacency matrix, indicates that the vertices corresponding to the row and column numbers are connected. It records information about every edge except when there are parallel edges. An element of the latter, the distance matrix, indicates the minimum number of edges between the vertices corresponding to the row and column numbers. The distance matrix may be computed from the adjacency matrix. Thus two graphs may be considered to be isomorphic, provided the numbers of nodes and edges and the adjacency matrices are identical.

Considering, for example, the seven-link chain in Table 1.6(a), there are seven possible theoretical mechanisms one could associate it with. However, considering the velocity graphs, there are only three possible nonisomorphic velocity graphs.

Table 1.7. *Some of the mechanisms formed when a link is replaced with a slider*

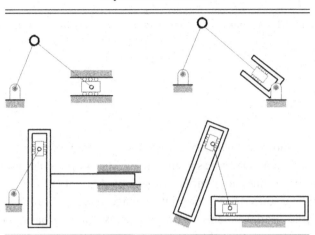

The three nonisomorphic velocity graphs are also shown in Table 1.6, thus indicating that only three nonisomorphic mechanisms may be derived from the seven-link kinematic chain in Table 1.6(a).

Often one may fix an alternative link in the same kinematic chain and thus obtain a different nonisomorphic mechanism. This process is referred to as the process of *kinematic inversion*. The slider–crank mechanism in Table 1.7 is obtained by the replacement of one of the links in the four-bar mechanism with a slider. The joint between the fixed link and the slider is a sliding joint. Alternatively the slider may be a collar that is pinned to the fixed link and permitted to slide over the coupler. Other mechanisms obtained by this process of kinematic inversion or by replacing a link with a slider are shown in Table 1.7.

When a mechanism is designed to have a single relative DOF, its motion must be completely determined by prescribing a single input variable. For design purposes, it is essential that all the distinct systems possible and their structures are identified so an optimum choice may be made. The general solution to the preceding problem results in an infinite number of possibilities. Yet, when certain restrictions are imposed, the number of possibilities reduces to a finite set. If one restricts one's attention to planar kinematic systems with a maximum of eight components or links, so that all components move in parallel planes, and imposes the following four constraints, the number of possibilities reduces to only 19 essential distinct systems:

1. all kinematic joints are of the same type (for example a pin joint);
2. the system should be kinematically closed loop so disconnecting any one joint will not separate the system into two parts;
3. the system should not have a separating component whose removal would reduce the system to two independent and uncoupled subsystems;
4. no connected subsystem consisting of two or more components should have zero DOFs (i.e., immobile).

Table 1.8. *Number of closed-loop kinematic chains with eight or fewer links*

Joints	Binary links	Ternary links	Quaternary links	Chains
4	4	0	0	1
7	4	2	0	2
10	4	4	0	9
10	5	2	1	5
10	6	0	2	2

When these conditions, which are not really restrictive from a practical standpoint, are imposed on a kinematic system, the number of possible systems satisfying all of the constraints is finite in number (Table 1.8). A number of properties of these subsystems can be deduced from their interchange graphs. Moreover, it is possible to synthesize large spatial kinematic systems by assembling basic modules whose interchange graphs are not only known but also satisfy certain specific properties.

1.3 From Kinematic Pairs to the Kinematics of Mechanisms

So far in our consideration of kinematic systems, we have restricted our attention to a single type of joint, whatever it may be. In practice, however, there are several types of binary joints of all shapes, sizes, and designs. One of the simplest joints is the revolute joint, which interconnects the two links in such a way that the free link is constrained to rotate only about a single axis. Each revolute joint introduces five constraints on a connecting link and thus permits it with one additional rotational DOF. It is one of three one-DOF-lower kinematic pairs or binary joints, characterized by a surface contact (identified by the German kinematician Franz Reuleaux in 1876), two of which are essentially nonrevolute with only a translational DOF. Reuleaux also identified three other lower kinematic pairs, characterized by surface contact binary joints and representing basic practical kinematic joints.

Franz Reuleaux, who is regarded as the founder of modern kinematics and as one of the forerunners of modern design theory of machines, was born in Eschweiler (near Aachen), Germany, on September 30, 1829. Reuleaux, who taught first at the Swiss Federal Polytechnic Institute in Zurich and then at the Technische Hochschule at Charlottenburg near Berlin, wrote a seminal text on theoretical kinematics that was translated into English in 1876 by another famous kinematician, Professor Alexander B.W. Kennedy of Great Britain. In the book the notion of a *kinematic pair* was first defined. When one element of a mechanism is constrained to move in a definite way by virtue of its connection to another, then the two form a kinematic pair.

Higher kinematic pairs are characterized by line contact. The ball-and-socket joint, a spherical pair, is characterized by three rotational DOFs, the planar pair by one rotational and two translational DOFs. and the cylindrical pair by one translational and one rotational DOF. The spherical pair introduces three translational constraints on the motion of the rigid body, thereby implying that only three additional parameters are required for describing the motion of the body linked to a free body in space by such a joint. The planar pair also introduces three constraints whereas the cylindrical pair introduces four constraints. All other lower kinematic pairs introduce five constraints on the motion of the rigid body so only one additional independent parameter is required for describing the relative motion of the rigid body linked to another by such a pair. However, the joints with maximal practical import are the one-DOF binary joints, the revolute joint with one rotational DOF, and the screw and parallel joints with one translational DOF each, which could be actuated by in-parallel rotating (the first two) and translational motors, respectively. Whereas the revolute permits a rotation only through an angle and the prismatic permits translation only through a distance, the screw allows a screw displacement involving a simultaneous coupled translation only through a distance and rotation through an angle that are related by the pitch of the screw.

Inspired by a former professor at Karlsruhe, F. J. Redtenbacher, Franz Reuleaux embarked on a systematic method to synthesize kinematic mechanisms. To this end, he created in Berlin a collection of over 800 models of mechanisms based on his book. These were cataloged and classified by another German engineer, Gustav Voigt. Apart from the lower and higher kinematic pairs and simple kinematic chains, they include a variety of pump mechanisms, screw mechanisms, slider–crank mechanisms, escapement mechanisms, cam mechanisms, chamber and chamber-wheel mechanisms, ratchet mechanisms, planetary gear trains, jointed couplings, straight-line mechanisms, guide mechanisms, coupling mechanisms, gyroscopic mechanisms, and a host of other kinematic mechanisms.

An industrial robot manipulator may be modeled as an open-loop articulated chain with several rigid links connected in series by revolute or prismatic joints, driven by suitable actuators, Motion of the joints results in the motion of the links that position and orient the manipulator's end-effector. The kinematics of the mechanism facilitates the analytical description of the position and orientation of the end-effector with reference to a fixed reference coordinate system in terms of relations between the joint coordinates. Two fundamental questions are of interest. First, it is desirable to obtain the position and orientation of the end-effector in the reference coordinate system in terms of the joint coordinates. This is known as the *direct* or *forward kinematics* of the manipulator. Second, it is often essential to be able to express the joint coordinates in terms of the position and orientation of the end-effector in the reference coordinate system. This is known as *inverse kinematics*. In most robotic applications it is relatively easy to find the one, usually the forward kinematics, but it is the other, the inverse kinematics, that must be known because the task is usually stated in the reference coordinates and the variables that could be controlled are the joint coordinates.

1.4 Novel Mechanisms

Although James Watt is credited with the invention of the steam engine, his singular contribution was not so much the synthesis of a coal-fired steam boiler to drive the

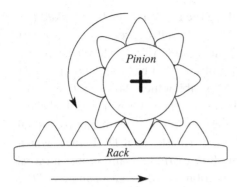

Figure 1.5. Rack-and-pinion mechanism.

piston in a cylinder by properly operating a set of pressure release valves; rather, it was the successful application of the slider–crank mechanism to convert the recip-rocating linear motion of the piston to a sustained and continuous rotary motion. It was indeed one of a novel and practical set of inventions that triggered the industrial revolution in Europe. Considering Gustav Voigt's collection of kinematic mecha-nisms, one could identify a basic set of novel mechanisms that serves as building blocks of other mechanisms and machines. A sample of these mechanisms is briefly discussed in this section.

1.4.1 Rack-and-Pinion Mechanism

The rack-and-pinion mechanism illustrated in Figure 1.5 is a common mechanism used to convert circular motion of the toothed pinion into the linear motion of the rack that is designed to continuously mate with the pinion.

1.4.2 Pawl-and-Ratchet Mechanism

When rotary motion is not required to be continuous or if it is causing the motion in one direction only while preventing it in the other, the pawl and ratchet in Fig-ure 1.6 is a suitable mechanism. The toothed wheel known as the ratchet may be rotated clockwise without any difficulty as the finger known as the pawl merely slips over the ratchet. However, when rotated in the counterclockwise direction, the pawl obstructs the ratchet, and rotation in this direction is not permitted.

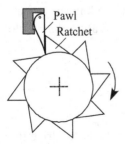

Figure 1.6. Pawl-and-ratchet mechanism.

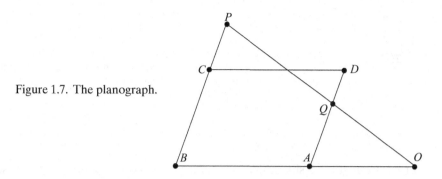

Figure 1.7. The planograph.

1.4.3 Pantograph

The pantograph was used by James Watt in his steam engine to exactly enlarge the reciprocating motion of the piston in the cylinder. One form of the pantograph, the planograph, is illustrated in Figure 1.7. The links are pin jointed at *A*, *B*, *C*, and *D*. The links *AB* and *DC* are parallel to each other, and the links *AD* and *BC* are parallel to each other. The link *BA* is extended to the fixed pin *O*. The point *Q* on the link *AD* and the point *P* on the extension of the link *BC* both lie on a straight line that passes through *O*. It can be shown that under these circumstances the point *P* will reproduce the motion of point *Q* to an enlarged scale; alternatively, point *Q* will reproduce the motion of the point *P* to a reduced scale.

1.4.4 Quick-Return Mechanisms

It is often necessary to convert rotary motion into reciprocating motion, and the slider–crank mechanism, driven by the crank, is commonly used. Moreover, what is often necessary in practice is for the forward stroke of the reciprocating motion to be slow and controlled because of the need for the mechanism to do work during this stroke. On the return stroke no work is done, and therefore it is desirable that it be completed as quickly as is possible. Thus the reciprocating motion is necessarily asymmetric.

A suitable mechanism to meet these requirements is the crank and slotted lever quick-return mechanism. In this mechanism, instead of the connecting rod being driven directly by the crank pin, it is attached to one end of a slotted link, which in turn is driven by the crank. In Figure 1.8 the crank pin *O* rotates at uniform speed about the center of rotation, *C*, and as it slides over the link *PQ*, it causes this link

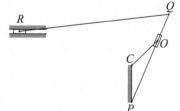

Figure 1.8. The crank and slotted lever quick-return mechanism.

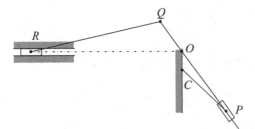

Figure 1.9. Whitworth quick-return
mechanism.

to rock from side to side about the pivot at *P*. The connecting rod *QR*, which is
attached to the rocker *PQ* at *Q*, transmits the rocking motion to the slider at *R*.
The forward stroke begins with *OC* being perpendicular to *PQ*, with the slider at
O located to the right of *C* and rotating counterclockwise. The forward stroke is
completed when *OC* is again perpendicular to *PQ*, with the slider at *O* located to
the left of *C*. During the forward stroke the slider at *O* traces the arc of a circle,
which in turn subtends an angle greater than 180° at the center of rotation *C*. Thus
the forward stroke takes a relatively longer time to complete within one cycle of
rotation of the crank pin *O*, in comparison with that of the return stroke.

   Another type of quick-return mechanism is the Whitworth quick-return mechanism, as illustrated in Figure 1.9. In this mechanism, the crank pin *P* also slides over a
link, although in this case the link does not merely rock but completes a full rotation
about a center, *O*, that is different from the center of rotation, *C*, of the crank. Thus
the link *OQ* will be slowed down and speeded up alternately as the driving crank
exerts first a large moment and then a small moment about the center *O*. Thus the
quick-return motion is a direct consequence of the eccentricity between *C* and *O*.

1.4.5 Ackermann Steering Gear

The Ackermann steering gear is one of the simplest mechanisms invented for the
purposes of exerting directional control over the motor car; it is illustrated in Figure 1.10. This steering gear is based on the four-bar mechanism in which the fixed

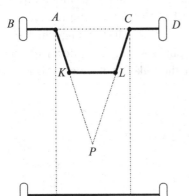

Figure 1.10. The Ackermann steering-gear
mechanism.

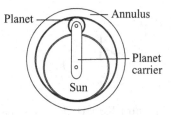

Figure 1.11. Sun and planet gear wheels meshing with an internal gear, the annulus.

link, *AC*, and link *KL* are unequal but the links *AK* and *LC* are equal. In the symmetric state the parallelogram *AKLC* is a trapezium. In this configuration the fixed link, *AC*, and link *KL* are parallel, whereas links *AK* and *LC* are inclined to the vertical at same angle in magnitude to the longitudinal axis.

To steer the car, link *CL* is turned so the angle it makes to the longitudinal axis increases while link *KL* rotates and translates such that the angle that link *AK* makes to the longitudinal axis decreases. As a consequence, link *AB* turns through an angle that is less than the angle of turn of the link *CD*. That is, the left front axle turns through a smaller angle than the right front axle, and consequently the center of rotation is located at point to the right of the car and not at infinity. Thus the car is able to turn smoothly without slipping.

1.4.6 Sun and Planet Epicyclic Gear Train

Figure 1.11 illustrates a typical epicyclic gear train incorporating the Sun and planet-type gears. In the figure, the central gear wheel is the Sun, and the gear wheel meshing with it, which not only rotates about a pin mounted on the carrier link but also revolves about the central gear wheel in a circular orbit, is the planet. The planet also meshes with the internal annulus. When the planet carrier is held stationary and not allowed to rotate, the planet wheel acts as an idler and the gears form a simple gear train with the Sun wheel rotating the annulus. On the other hand, when the Sun wheel is rotated, say, clockwise, the planet wheel is driven about its axis in a counterclockwise direction while rolling inside the annulus and carrying the planet carrier in the clockwise direction. Between these two extreme modes of operation there are an infinite number of other modes in which this epicyclical train could be operated, depending on the speed of the Sun and the planet carrier. Thus the concept may be applied to a host of applications in bicycles, aircraft engines, automobile rear-wheel drives, and a number of machine tools.

1.4.7 Universal Joints

The universal joint is essentially a binary link comprising two revolute joints, with the axes of these joints being two skewed, nonintersecting, and nonparallel lines. A universal joint is used to couple two shafts in different planes. Universal joints have various forms and are used extensively in robot-manipulator-based equipment. An elementary universal joint, sometimes called a Hooke's joint (Figure 1.12) after its

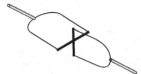

Figure 1.12. Illustration of a Hooke's joint.

inventor Robert Hooke, consists of two U-shaped yokes fastened to the ends of the shafts to be connected (Figure 1.13). Within these yokes is a cross-shaped part that couples the yokes together and allows each yoke to bend, or pivot, in relation to the other. With this arrangement, one shaft can drive the other even when the two shafts are not in alignment.

A variant of the universal joint is a gyroscope-like suspension illustrated in Figure 1.14. This is a two-DOF system in which the twin-gimbaled system takes the form of a Hooke's joint that couples the wheel to the motor driveshaft. The axis of rotation of the wheel need not be aligned to the axis of the driveshaft, which could be flexible.

1.5 Spatial Mechanisms and Manipulators

Several kinematicians have analyzed the geometrical arrangements of a number of in-parallel actuated spatial mechanisms and manipulators recently and established the principles of design of such systems. Thus three links linked together by a Hooke's joint could be supplemented by either two revolute joints, a revolute and a prismatic joint, or by two prismatic joints, thereby synthesizing a spatial manipulator either with three rotational DOFs and an additional optional translation DOF or one with two rotational and two translational DOFs. By the late 1960s spatial manipulators with three rotational DOFs in roll, pitch, and yaw and three translational DOFs in surge, sway, and heave were used in a range of applications. These were based on the Stewart platform, which was developed in 1965.

The six-DOF Stewart platform, illustrated in Figures 1.15(a) and 1.15(b), is also the one parallel kinematic manipulator configuration incorporating multiple closed loops that has been used in a number of recent motion and suspension system designs.

The Stewart platform typically consists of a movable platform connected to a rigid base through six identically jointed and extensible struts. These struts support the platform at three locations by six universal joints. They are also attached and

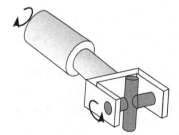

Figure 1.13. U-shaped yoke and cross pin in a typical transmission system.

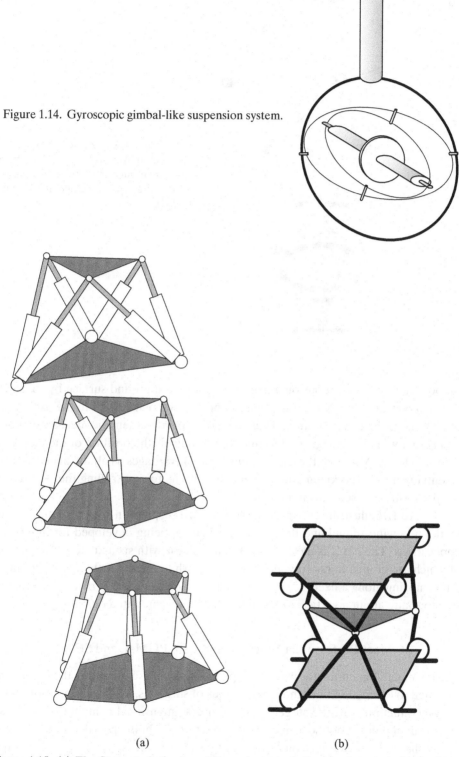

Figure 1.14. Gyroscopic gimbal-like suspension system.

(a)   (b)

Figure 1.15. (a) The Stewart platform and its variants, including the hexapod; (b) a cable-actuated version of the Stewart platform.

The Robot

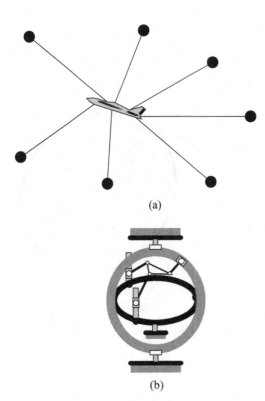

(a)

Figure 1.16. (a) A typical cable-mounted
and actuated wind-tunnel model; (b) a
spatial manipulator with three revolute and three prismatic actuated joints:
Eclipse II.

(b)

supported at three locations on a fixed rigid base on a ground surface by six additional universal joints. No two struts are jointed to the platform or to the fixed rigid
base at more than one location. Thus the struts form a kinematic mechanism consisting of a variable triangulated frame, which controls the position of the platform
with six DOFs. Although there are a number of advantages as all the actuators are
essentially linear, the kinematics and the algorithms for controlling the posture of
the platform are essentially nonlinear.

Figure 1.16 illustrates a typical cable-actuated suspension system of an aircraft
model in a wind tunnel and the spatial mechanism being developed for industrial
applications. The latter is the Eclipse II manipulator with six actuated joints, three
of which are double joints (revolute and slider) while the other three are prismatic
joints moving along a fixed, horizontal circular guide ring. The vertical guide ring is
free to rotate about a vertical axis and the platform is mounted on ball joints.

1.6 Meet Professor da Vinci the Surgeon, PUMA, and SCARA

The da Vinci™ manipulator was developed to assist in the operating theater and is
designed to eliminate the tremor in the hands of surgeons. With the state-of-the-art
da Vinci manipulator, the surgeon uses a three-dimensional computer vision system to manipulate robotic arms. These robotic arms hold special surgical instruments that are inserted into the abdomen through tiny incisions. The robotic arms
can rotate a full 360°, allowing the surgeon to manipulate surgical instruments with

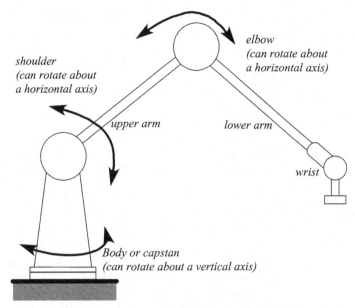

Figure 1.17. Schematic drawing of the Unimation PUMA 560 manipulator.

greater precision and flexibility. Furthermore, the robot's pencil-thin arms and delicate grippers may be inserted through holes as narrow as 8 mm. Two of the hands are used to wield surgical tools that follow finger and wrist movements made by an operator sitting at a nearby console. A third arm carrying a miniature camera is also inserted to track the motions of the surgical tools wielded by the other two.

The da Vinci manipulator has been used, among other tasks, to assist in a successful kidney transplant. Its primary role was in removing the existing kidney from the patient's body, and conventional surgery was then used to implant the new kidney.

Another robot, known as the Penelope Surgical Instrument Server, assisted in the removal of a benign tumor on the forearm of a patient on June 16, 2005. Penelope is now giving doctors at New York-Presbyterian Hospital an extra hand, as the doctors might put it.

Penelope is really a robotic arm equipped with voice-recognition technology that understands commands from doctors and nurses in the operating room. A surgeon can ask for any instrument, a scalpel for example, and Penelope then hands it over.

The Unimation PUMA 560 (PUMA stands for Programmable Universal Machine for Assembly) is a robot manipulator whose joints are all revolute joints and it has six DOFs. It is shown in Figure 1.17 and should be compared with the manipulator illustrated in Figure 1.1. There are three DOFs in the wrist, and a gearing arrangement couples these three DOFs. The nature of the coupling plays an important role in the derivation of the kinematics. Thus the number of joints is not the same as the number of independent actuators. The brief technical specifications of the PUMA 562, a member of the PUMA 560 family, are given in Table 1.9.

Table 1.9. *Brief technical specifications of the Puma 562*

General	Axes	6
	Drives	DC motors
	Control	Numerical
	Positional control	Incremental encoders
	Coordinates	Cartesian
	Configuration	Revolute
	Cables	5 m
Work area	Reach at wrist	878 mm (between joints 1 and 5)
	Work volume	360° working volume in left-arm or right-arm configuration
	Limit joints – 1 to 6	320°, 250°, 270°, 280°, 200°, 532°
Load capacity	Nominal payload	4 kg
	Permitted load at wrist	4 kg at 127 mm from joint 5 and 37.6 mm from joint 6
Performance	Repeatability	762 ± 0.1 mm, 761 ± 0.1 mm
	Maximum speed	1.0 m/s
Universal Controller	Programming and language (VAL II)	Teach pendant; Virtual Device Table (VDT) using VAL II
	VAL II processor	Motorola 68000 32 bit
	Auxiliary processor	Torque processor 68000
	Interface	16 input/output exp 128 Analog I/O Serial I/O
	Serial interface	RS 232 RS 423
	Memory buffer	64 Kb, 512 Kb, 1 Mb, 2 Mb
	Battery buffer	1 year
Environment	Temperature/humidity	5–40°C (EN 60204-1)
	Interface suppression	50% at 40°C, 90% at 20°C
	Power supply	Incorporated 208, 240, or 380 V
	Option	50–60 Hz 0.6 kW – 1 phase clean-room version (class 10)
Weights	Arm	63 kg
	Controller	200 kg

SCARA (which stands for Selectively Compliant Assembly Robot Arm) shown in Figure 1.18 has three parallel revolute joints that allow it to move and orient within a plane, and a fourth prismatic joint moves the end-effector normal to the plane. This is an example of a manipulator that is not fully articulated and is not *anthropomorphic*. Rather, the combination of the revolute and prismatic joints allows this manipulator to operate in a cylindrical workspace. An interesting feature is the fact that only a minimum number of joints are load bearing, depending on the nature of the application.

This brings us to the end of our broad survey of engineering robotic systems. The survey is by no means exhaustive, yet is most relevant to current industrial technology. Robotics plays a key role in current mechatronic systems that are the heart of most automated engineering systems such as unmanned autonomous vehicles or remotely operated manipulators operating in a hazardous environment, manipulators designed to assist surgeons on the operating table by letting them operate with

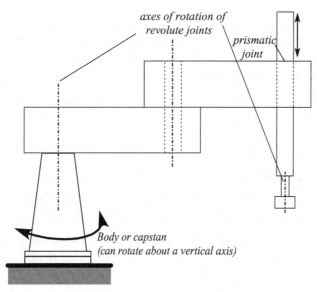

Figure 1.18. Schematic drawing of SCARA.

greater precision and accuracy than ever before, and industrial automated assembly lines for the mass manufacture of a range of engineering products. We are now in position to identify four broad areas that constitute the fascinating field of robotics: mechanisms, control, perception, and computing. In the following chapters we restrict our attention to robot mechanisms and their control.

1.7 Back to the Future

It is indeed quite difficult to look ahead as there is no real magical crystal ball to predict it. Yet there is a prognosis for the future. On one hand, one can predict that the da Vinci-type manipulator would evolve into a robotic octopus. On the other hand, it is not difficult to visualize that the development of multifingered dexterous hands (and legs) and the successful building of walking machines with amphibious capabilities and with machine vision and tactile sensing are set to revolutionize robotics. The engineering approach to robotics is a much more pragmatic one, the emphasis being on the whole of the robot rather than just its computational features.

Several research teams worldwide are involved in the design and development of robot football players based on concepts associated with autonomous systems. They involve a number of areas of robotics including sensor fusion, mobile robotics navigation, nonlinear hybrid feedback control, and coordination. Probably the ultimate test for a walking robot with multifingered dexterous hands is its ability to play cricket; to play the game like an accomplished human player with the ability to run across the wickets at the appropriate speed and with the capacity to grip and spin the ball and participate as a spin or pace bowler, as a batsman, as a wicket keeper, and as a fielder. One need not be surprised if such a complete robot is built within the next 50 years.

EXERCISES

1.1. Study the various kinematic mechanisms that were used in engineering industries in the past 50 years and indicate those that you think are exceptional or novel. Highlight, with the aid of sketches, all the special features of these mechanisms.

1.2. Conduct a survey of industrial robot manipulators currently used and identify at least two different

    (a) anthropomorphic configurations,
    (b) nonanthropomorphic configurations, and
    (c) spatial mechanisms

from those discussed in this chapter. Name the salient features of each of these systems and classify these manipulators based on their distinguishing features.

1.3. Study the various grasping and gripping mechanisms that may currently be in use in various industrial applications and classify these on the basis of their kinematic geometry.

1.4. It is frequently necessary to constrain a point in a mechanism to move along a straight line without the use of a sliding pair. Mechanisms designed for this purpose are known as *straight-line-motion mechanisms*.

    (a) Discuss your thoughts on how an *exact* straight-line-motion mechanism may be designed. Based on your personal research, suggest two alternative mechanisms that are able to generate straight-line motion.
    (b) Often it is not necessary that the motion be an exact straight line. Discuss two or three alternative mechanisms that may be used as *approximate* straight-line-motion mechanisms.
    (c) Sketch three different forms of straight-line-motion mechanisms based on the *four-bar kinematic chain* and the *slider–crank mechanism*.

1.5. You are a member of a team of engineers who have been assigned the task of designing a walking robot, the SPD 408, capable of emulating humans in this respect. Your first task is to establish a kinematically feasible design. Draw up a set of requirements and establish three or four feasible kinematic architectures for a walking robot.

1.6. Isaac Asimov's three laws of robotics are that

    (a) a robot may not injure a human being, or, through inaction, allow a human being to come to harm;
    (b) a robot must obey orders given to it by human beings except where such orders conflict with the first law; and
    (c) a robot must protect its own existence as long as such protection does not conflict with the first or the second law.

Discuss the relevance and adequacy of Asimov's laws, particularly in the context of humans and robots coexisting with each other in the not-too-distant future.

2

Biomimetic Mechanisms

2.1 Introduction

The concept of biomimetic control, i.e., control systems that mimic biological animals in the way they exercise control, rather than just humans, has led to the definition of a new class of biologically inspired robots that exhibit much greater robustness in performance in unstructured environments than the robots that are currently being built. It is believed that there is a duality between engineering and nature that is based on optimum use of energy or an equivalent scarce resource, particularly in exercising control over actions and over interactions with the immediate environment. Biomimesis is generally based on this concept, and it is believed that by mimicking animals that are most capable of performing certain specialized actions, such as the lobster on the seabed, insects and birds in flight, and cockroaches in locomotion, one could build robots that can surpass any other in performance, agility, and dexterity.

A key feature of biomimetic robots is their capacity to adapt to the environment and ability to learn and react fast. However, a biomimetic robot is not just about learning and adaptation but also involves novel mechanisms and manipulator structures capable of meeting the enhanced performance requirements. Thus biomimetic robots are being designed to be substantially more compliant and stable than conventionally controlled robots and will take advantage of new developments in materials, microsystems technology, as well as developments that have led to a deeper understanding of biological behavior.

Roboticists have a lot to learn from animals. Birds have a superior flying machine with multielement "aerofoils" capable of controlling the flow around them quite effortlessly. They use a pair of clap-and-fling or flapping wings to execute the modes such as flapping translation that effectively compensate for the Wagner effect.[1] The Wagner effect manifests itself as a time delay or transport lag in the growth of lift over an impulsively or suddenly started and accelerated aerofoil.

---

[1] Sane, S. P. (2003). Review: The aerodynamics of insect flight, *Exp. Biol.* 206, 4191–4208.

To compensate for this effect, it can be shown that increasing the angle-of-attack aerofoil by a rapid high-frequency flapping action can effectively ensure an almost instantaneous growth in the lift developed. The rapid high-frequency flapping has the added advantage of generating a net mean thrust. Thus a bird can effectively take off by adopting a technique based on rapid flapping of its wings. The clap-and-fling mechanism also has a similar consequence.

Furthermore birds use visual perception based on the concept of optic flow for flight control. One particular control concept that is gaining importance is biomimetic flight control based on the principles of optic flow, which is a visual displacement flow field that can be used to explain changes in an image sequence. The underlying principle that is used to define optic flow is that the intensity level is constant along the visual motion trajectories. Thus there exists a displacement field around the motion trajectories that is similar in structure to a potential flow field. Optic flow can be then defined as the motion of all the surface elements, in a limited view, as perceived by the visual system. As a body moves through the limited view, the objects and surfaces within the visual frame flow around it. The human visual system can determine its current direction of travel from the movement of these surfaces.

The term "optic flow" has been associated with flight and has been used successfully in the design of flight simulators. The initial research on optic flow not only inspired many cognitive psychologists and neurologists[2] with new ways of understanding how animals perceive motion around them but has also led to an increased understanding of how flying animals maintain stability and perceive the world while flying. Migratory birds use the magnetic field as a compass does to fly in the appropriate migratory direction! They seem to know where they want to go by using a magnetic-field map. Light-absorbing molecules present in the retina of a bird's eye help the bird to sense the Earth's magnetic field.

Currently roboticists are exploring ways to use optic flow to provide a robot with visual perception and the associated sensations similar to those of a bird in flight. Although researchers were able to integrate optic-flow-based artificially simulated vision into robotic platforms and to use them to perform simple navigation tasks in real time more than a decade ago, it was only recently that optic-flow sensing was successfully integrated into small autonomously flying aircraft and used to provide it with a basic degree of autonomous flight control. The methodology, which is based on a synergy of techniques that have evolved in computational vision processing and in biologically inspired research into human vision, has led to a better understanding of computer vision that may be implemented. Although computational methods in vision processing have provided a sequence of processes that allows a practical image to be transformed into one of a series of representations, the major focus of computer vision has been the sensation and perception of relevant features in the representations.

---

[2] Tammero, L. and Dickinson, M. H. (2002). The influence of visual landscape on the free flight behavior of the fruit fly *Drosophila melanogaster*, *J. Exp. Biol.* 205, 327–343.

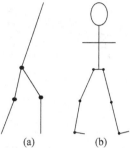

Figure 2.1. (a) Sagittal plane views of biped models of walking, (b) in a ballistic mode.

2.2 Principles of Legged Locomotion

Interest in locomotion in humans and animals has also spurned the development of new biomimetics mechanisms. Primarily this is due to the realization that human joints are not lower kinematic pairs and that they could be modeled as four-bar mechanisms and due to the insight gained into the kinematics of walking. As an example of the former, we observe that typically in the human knee joint between the femur and the tibia, which do not overlap each other but are strapped to each other by two cruciate ligaments that serve as extendable links, so the entire joint functions like a four-bar mechanism. Another example is the foot, which is in fact a serial chain with six links in which the innermost link rolls under the tibia, the lower leg bone, and is also strapped at the other end to the Achilles tendon, so the entire joint functions like a four-bar mechanism. In this case the tendon serves as a spring-loaded link, serving to return the foot to its nominal position when there is force under it.

As for the kinematics of walking, Figure 2.1 illustrates a typical sagittal-plane (normal to the frontal plane) view of a biped model of a walking robot in ballistic motion. The ballistic movements in a dynamic walking mode, while maintaining balance, are only marginally stable and difficult to execute. In the ballistic walking mode each of the two legs are modeled as serial chains with three links and discounting the two rigid-body modes, the two-leg models possess five DOFs.

When the legs are used to support the body in a stationary situation, both legs are locked in a rigid mode, thus losing one DOF each. Of the three remaining DOFs in the supporting mode, one represents the relative motion between the legs whereas the other two represent the motion of the foot at the ankles. When the model is in a walking mode, only one leg assumes the supporting role while the other assumes the role of a swing leg. While the support or stance leg continues to be rigid, the swing leg hinges at the knee. The swing phase then commences with the toe of the swing leg leaving the ground, the body translating forward with the support leg hinged on the ground. The swing phase is completed with the foot returning the ground when the heel of the swing leg is anchored on the ground.

After a short period both legs assume a supporting role and all initiated muscular action is completed. The swing leg then supports the body and the other leg takes the role of the swing leg with gravity acting on it.

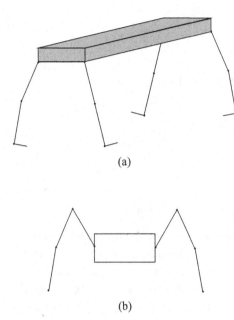

(a)

Figure 2.2. Advanced platform configurations based on biomimetic mechanisms: (a)
a model of a quadruped robot platform
capable of walking, running, and trotting;
(b) a two-dimensional view of a "lobster"
platform.

(b)

The entire dynamical process may be modeled as an energy-conservative process. Only the transfer of support as the swing leg becomes the support leg is modeled as an inelastic collision with just the angular momentum conserved, leading to a
set of constraints. Thus during the walking phase the model has just three DOFs, and
the walking action is performed by this continual process of constraining and releasing a DOF while maintaining balance. In fact, one could consider a loss of balance
to be a loss of effective control, whereas the walking action itself involves coordination. Numerical integration of the equations of motion governing the dynamics
of the model reveals the existence of a stable "limit cycle," representing a periodic,
neutrally stable motion. The cycle is composed of a swing phase and a support phase
with two instantaneous changes in velocity corresponding to the ground collisions.
As the basin of attraction of the stable limit cycle is a narrowband of states surrounding the steady-state trajectory, the goal of control action is to apply torques at
the joints and ankles in order to produce a stable limit cycle gait, with a relatively
larger basin of attraction that is invariant to small slopes. This is a unique application of control engineering, in which the control action is required to increase
the stability margins of what is normally considered to be a dynamical equilibrium
state.

Two examples of typical biomimetic platforms based on the skeletal structure
of mammals are illustrated in Figure 2.2. The first of these concepts has been developed into a 0.7-m-tall robot capable of walking, running, and trotting at a maximum
speed of 5.3 kmph and climbing a slope of 35° while maintaining its balance. An
eight-legged ambulatory vehicle that is based on the lobster has also been built and
is intended for autonomous remote-sensing operations in rivers and on the ocean
bottom. It is capable of robustly adapting to irregular bottom contours, current,
and surge.

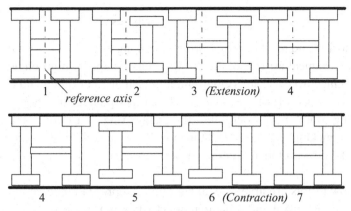

1   reference axis   2   3 *(Extension)*   4

4   5   6 *(Contraction)* 7

Figure 2.3. Operation of an inchworm motor.

Combined with the ballistic action of walking and the energy-exchange mode of running, in which energy is regularly transferred from potential to kinetic energy, these models are capable of moving at high speeds irrespective of the nature of the terrain. Walking robots are better than rolling robots in navigating through uneven terrain, and hopping and running robots are even better than walking robots in this respect.

2.2.1 Inchworm Locomotion

The design and construction of several biomimetic robots have been inspired by the principles of inchworm locomotion. The inchworm-like robots are composed of three primary abilities provided by a corresponding set of subsystems. The ability to stretch provided by a stretching linear actuator that is responsible for creating a step is placed between two blocking elastic bodies that support the main trunk housing the actuator. The robot is able to perform the succession of the six phases of the inchworm locomotion described in Figure 2.3. In the initial position corresponding to phase 1, the robot's main trunk is supported by both the elastic bodies. In the next phase, 2, the frontal body contracts vertically and is moved forward by stretching of the linear actuator in phase 3. In phase 4 both bodies support the main trunk. In the next phase, 5, the rear body contracts vertically and is then moved forward by the linear actuator in the sixth phase, which now retracts to its initial unstretched position. The inchworm robot then returns to its initial configuration in phase 7 and in the process has inched forward by one step. Rapid cyclic execution of these phases is the basis for the locomotion of the inchworm.

A typical inchworm mechanism for locomotion in the intestines consists of two types of actuators: a clamper and an extender. The clamper is used to clamp the device onto the wall of the intestine while the extender generates a forward displacement.

The construction of an inchworm-like robot is feasible because of the availability of smart elastic materials that can be made to contract and expand by the application of an external electric potential.

2.2.2 Walking Machines

Although the first and most important role of walking robots has been to understand several biomechanical aspects of human walking, the applications envisaged in the future is walking into hazardous environments, civil disaster relief, and the scientific exploration of sites on Earth, in space, and on other planets and moons. Although a few walking robots have been built, the engineering of a fully walking machine capable of mimicking all aspects of human walking, capable of interaction with the environment, and with the ability to deal with unexpected disturbance is still a long way off.

As of now robots are able to deal with only preprogrammed activities and are unable to naturally adapt and execute new activities as and when these are required of them. Being able to interact with a community of robots and, more important, with humans requires a degree of intelligence. A notable feature of current walking robots is the fact they offer very high impedance when faced with any external disturbing force. A characteristic feature of humans is their ability offers a low impedance to forces they intend to respond to and a very high impedance to other forces when they so desire. This ability is somewhat akin to the concept of marginal static stability in fighter aircrafts. Fighter aircrafts, which are expected to be very agile and fast in responding to a pilot's command, are for these reasons deliberately designed to be only marginally stable or sometimes even unstable. This lack of a margin of stability greatly enhances the pilot's ability to control the aircraft. A walking robot too must be designed so it can just about maintain its balance.

In the context of individual joints and component mechanisms the requirement reduces to an ability to offer a low impedance to control forces and reject all disturbing forces; this means that the same force when presented as a control force must be offered a relatively low impedance whereas a high impedance is offered when it is construed as a disturbance. It is clear that this means that the joint and component controllers must be endowed with the ability of logical reasoning at a certain level so they can distinguish between a disturbance and a control force.

Biomimicry can a play a significant role in the ability to develop walking robots. Much human capability, particularly in relation to use of limbs, is acquired through learning by imitation. The neural system uses the biologically based central pattern generators and sensory reflexes for learning preferred patterns. The idea of having walking robots learn by imitating a human is a feasible biomimetic concept. The fact that both learning and adaptation can be modeled in a variety of ways facilitates the design and implementation of biomimetic robots capable of learning and adaptation. They are particularly suited for autonomous or semiautonomous tasks involving human and environmental interaction on a much larger scale than is currently considered possible. Thus, based on a variety of imaging techniques, simulations of human walking and other functions involving the use of human limbs can be used in robot implementations.

Walking robots are being developed that attempt to replicate neuroanatomical functional architectures, at different levels of abstraction, and to use them to imitate humanoid robots. The concept of mimicking a neuroanatomical controller for purposes of compensating for the occurrence of tremor in human hands is discussed in the last section of this chapter. In Chapter 12 we discuss the some of the basics of the biomechanics and control of walking locomotion.

2.2.3 Autonomous Footstep Planning

Just as obstacle-avoidance algorithms have been developed for wheeled robots, walking robots have the unique ability to traverse past obstacles by stepping over or upon them. Using appropriate graphical representations and by using a number of heuristics to minimize the number and complexity of the step motions that optimize a cost function, algorithms are being developed to search for a floorplan of a feasible ordered set of footstep locations. It is important to avoid cul-de-sacs and loops as well as meet three-dimensional size and height constraints that the obstacles must satisfy.

One approach is a two-step procedure in which it is assumed that there is no need to step over or upon an obstacle. If a feasible solution is found, the additional maximum cost saving of stepping over or upon an obstacle is computed. If this additional cost saving is within a certain limit or when stepping over or upon an obstacle is a necessity, the feasible solution involving stepping over or upon an obstacle is adopted.

In the next stage the footstep planners are expected to deal with a hazardous terrain where they would be expected to deal with relatively steep climbs and descents. In these environments it may be necessary to use more than two limbs. In realizing these algorithms in practice, a three-dimensional visual image of the environment must first be constructed. Thus this leads to the concept and implementation of autonomous footstep planners. A key requirement for humanoid robot navigation autonomy is the ability to reliably visualize the environment by building representations of obstacles and all other objects and/or the terrain from the data obtained from a variety of sensors. The representation of the environment must be such that it can be used by the footstep planner in generating a feasible ordered set of footstep locations. Moreover these representations must be generated from the data obtained from a diverse set of sensors. Autonomous locomotion with the ability to recognize the surrounding environment and with the ability to traverse it can then be a feasible practical proposition.

2.3 Imitating Animals

Imitation, which is defined as the ability to recognize and reproduce others' actions, is a powerful methodology of learning and developing new skills. Species endowed with this capability are provided with fundamental abilities for learning specialized skills. The ability to imitate reaches its fullest potential in humans. Humans have the capacity to imitate any actions of their own or another's body based on a variety of purposes or goals, such as the goal of reproducing the geometric, performance, precision, or aesthetic aspect of the movement. In terms of biomimetic mechanisms we are not just interested in humans imitating other humans or animals but in developing manipulative robots capable of imitating humans and animals. However, the capacity of humans to recognize biological motions from a limited number of cues and make out the general features of the motion, distinguishing the type of gait or the type of action, as well as its specific features, is a feature that is quite desirable in biomimetic robots and provides the motivation for its study.

Movement recognition among humans has been extensively researched in the context of neurophysiology and visual imaging. The visual recognition of movements and actions is an essential feature in many species. It is fundamental in the learning of complex motor actions by imitation, which in turn forms the basis of communication and recognition at a distance. Neurophysiological modeling for the recognition of biological movements can be based on neural mechanisms that are consistent with existing models of the functioning of the visual cortex. A neural circuit in the inner reaches of the brain composed of "mirror neurons" is the known brain circuitry for consciously analyzing and mimicking other people's actions. It contains special types of basic nerve cells, known as neurons, that become active both when their owner does something and when he or she senses someone else doing the same thing.

Following research in neurobiology during the last decade, mirror neurons had for the first time been directly identified in humans. Previously their existence had been inferred only from primate research. The different actions involving limb motion affect neuron activity in the premotor cortex, which functions in movement control and in the part of the brain responsible for seeing. A human viewing an action in which he or she is an expert in experiences greater activity in the premotor cortex than when they view actions they are not skilled in. By contrast, the nonexpert brains do not experience heightened activity in either case; rather, the nonexpert brains do not exhibit any differential neuron activity regardless of the type of actions viewed.

Visual imaging inputs to the visual cortex are directed to an anterior area with mirror neurons, located in the inferior frontal cortex. The human inferior frontal cortex and the superior parietal lobule are active when the person performs an action and also when the person sees another individual performing an action. These primary motor and premotor cortices are responsible for the execution of bodily actions. They are also used to directly simulate another humans's (or animal's) observed actions. This is largely due to the mirror neurons, which respond to execution and observation of goal-oriented actions.

The brain's neural mechanism allows for a direct matching between the visual description of an action and its execution. The human mirror-neuron networks are stimulated in response to actions that are apparently meaningless. This seems to provide evidence for the theory that there exists a tendency in humans to spontaneously model any and all movements by others. Thus these mechanisms provide the basis and motivation for developing artificial neural models that can be used as computational entities in large connectionist network models of biomimetic robots.

Further, the tendency to imitate has been extensively observed in certain animals other than humans, such as birds that are also endowed with some extraordinary abilities such as flying and being able to propel themselves in the atmosphere.

2.3.1 Principles of Bird Flight

Birds and fish probably provide the best examples of biomimetic mechanisms flight-control engineers can learn from. The bird wing is an exceptional control device used by birds to control all phases of the flight. It may be recalled that, unlike the tail of a human-built aircraft, the bird's tail plays a critical role only in the control of special modes but has no role to play in normal flight. Control over a bird wing is exercised by a complex neuromuscular system that articulates the bones or links and the joints. A bird wing's bone structure is quite similar to that of a human arm, although the free movement around the joint of the wrist is curtailed so that there is only movement in one plane, preventing the wing from bending up or down by the aerodynamic forces exerted in flight. Furthermore, a bird wing's wrist joint is linked to the elbow joint so that the extension of the elbow is transmitted to the wrist, as in a four-bar mechanism.

Like that of humans, the lower part of the bird's forelimb is composed of two bones: the radius and the ulna. The ulna and the radius have their own independent point of articulation (condyle) on the farther (distal) end of the humerus, the long bone in the arm that runs from the shoulder to the elbow. The ulna is farther than the radius from the elbow so that when the elbow is flexed the two bones of the forearm oppose one another. The radius is then pushed into the various carpal bones of the wrist while the ulna is withdrawn. The mechanism linking the wrist makes the hand flex too and the wing is folded. When the wing is unfolded, the opposite movement occurs. In this way the wing is controlled by the flight muscles in a coordinated way.

It must be said that Wilbur Wright himself recognized wing warping when he was able to visualize the top and bottom surfaces of a bicycle's innertube box as wing surfaces. When one end of the box was twisted down, the other end twisted up. In appreciating this motion, Wilbur understood how to control aircraft roll based on the techniques adopted by birds. A bird's wing may be considered as an aerody-namic lifting surface, and it is then easy to recognize that its shape and the section profile will play a critical role in the generation of lift.

In a bird's wing, the lifting surface is covered by feathers that emerge from below a narrow flap of skin called the *propatagium*. This covers the bones of the wing that form the leading edge and is stretched tight along it from the shoulder to the wrist when the wing is unfolded. Being the leading edge of the inner wing, it is an important aerodynamically active component of the wing and supports the main component of the leading-edge suction pressure. It is natural that the wing's leading edge must be sufficiently strong and rigid. Thus the bones of the wing tend to run nearer the leading edge, whereas the trailing edge is made up of a rigid line of feathers. Covert feathers also cover the entire wing and at the trailing edge are found the much larger primary (*remiges*) and secondary feathers (secondary remiges). The

Direction of airflow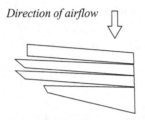

Figure 2.4. Examples of controlled wing-tip fences in aircraft based on the alula.

separated primary feathers at the wing tips have varying degrees of curvature – the leading-edge ones are loaded so that they bend upward; the rest curve downward.

An interesting feature of the primaries that are able to separate and control the gaps in them, in various flight conditions, is that no separated flow regions have ever been observed on the upper surface of the wings, back, or tail. The *alula*, a thumblike feather in certain birds, lies almost at the wrist joint on the leading edge of the wing. Generally it has three or four short vane feathers attached to it. It acts in a similar fashion to the slots on an aircraft, naturally rising into the airstream when the wing approaches stall. The alula is prominent in birds such as *Corvids* (crows) and is almost absent in some soaring seabirds.

High-lift systems have a parallel in the alula; winglets and wing-tip fences (Figure 2.4) are similar in effect to slotted-tip feathers in many soaring birds. *Riblets*, on the other hand, developed to reduce drag in aircraft, are similar to those found in shark skin. Interestingly, although there exists a huge diversity in nature and there are several evolutionary pathways to achieving the same function, nature seems to be able to select the most optimally suitable system for a specific application. By way of example, looking closely at insect and bird flight, we can observe that they fly effortlessly in the low-Reynolds-number region, which is not the case with man-made flight. Our goal is to study nature more fully so as to understand better such solutions that have already been obtained and in so doing become inspired to exercise complete control on biomimetic mechanisms operating in aerodynamic flows.

2.3.2 Mechanisms Based on Bird Flight

One of the objectives of biomimetic control is to surpass and go beyond biomimicry. For this reason it worth considering the various mechanisms adopted by birds in different phases of their flight. Birds have a superior flying machine with multielement "aerofoils" capable of controlling the flow around them quite effortlessly. They use a pair of clap-and-fling or flapping wings to execute the modes such as flapping–translation that effectively reduce the delay in the growth of circulation around an aerofoil by compensating for the Wagner effect.

Birds configure their flying feathers or remiges as a variable-camber aerofoil such that the effective angle of attack is always a constant. It follows that the aerofoil operates so the potential-flow component of the total lift coefficient is almost constant irrespective of the speed of flight. Thus it is possible for a bird to operate at an optimal angle of attack without suffering from the adverse effects of flow separation.

A bird is aware by virtue of its associative memory at what angle of attack it should fly in order to generate a particular magnitude of lift. Thus it operates at a prescribed angle of attack, and any additional lift it may require is generated by controlling the vortex-flow component of the lift.

In one sense birds are different from manmade aircraft in that some birds, such as the sparrowhawk, are able to clap and fling. This is a beating motion about an axis that is pitch forward. Rapid beating results in a quick increase in the angle of attack, which in turn generates lift about an axis normal to the axis of beating. Thus it is able to generate a lift component equal to the weight of the bird and a thrust component that keeps it in aerial equilibrium and hover.

Wing motion of a bird will generally consist of four fundamental motions: (1) flapping motion in a vertical plane, (2) lead–lag motion, which denotes a posterior and anterior motion in the horizontal plane, (3) feathering motion, which denotes a twisting motion of the wing pitch, and (4) spanning motion, which denotes an alternately extending and contracting motion of the wing span.

Another wing motion that is present in larger birds but is technically equivalent to clap and fling is flapping–translation. Flapping–translation involves a combination of the first three modes. A bird's wing is constantly changing its relative forward velocity as it flaps up and down, slowing down at the ends of the downstroke and upstroke, and then accelerating into the next half-stroke. Furthermore, the wing base will always be moving slower than the wing tip, meaning that the wing velocity increases from base to tip. This mode constitutes the flapping–translation mode. In the case of three-dimensional, swept, large-aspect-ratio wings, the flapping–translation mode is invoked to effectively compensate for the Wagner-like effect.

Another mechanism, suitable for low-aspect-ratio wings, is based on stable operation at a maximum constant lift, with the conventional lift being augmented by a leading-edge suction-type vortex-flow-generated lift. Moreover, birds have the ability to align the lift that is due to the vortex flow in any direction so it can be used either to enhance the overall lift or downforce or to act as a propulsive or braking force. Thus birds are able to fly at angles of attack as high as $35°–40°$.

The way a dragonfly controls the generation of vortex lift is probably most representative and generic to all birds. In normal counterstroking flight there is a leading-edge vortex present over the forewing and attached flow on the hindwing. The leading-edge vortex lift, though present in thin-aerofoil theory, does not contribute to the lift as it is directed forward in this case. The vortex lift is formed during rapid increases in angle of attack because of the existence of the rounded leading edge and the aft movement of the leading-edge vortex. In steady flight these rapid increases in angle of attack occur during wing rotation at the start of the downstroke. However, dragonflies can go from attached flow to leading-edge vortex flow at any stage of the wingbeat by rapidly increasing their angle of attack.

Falcons such as the Hobby are able to accelerate rapidly in pursuit of their prey. They have kinked trapezoidal wings with the inner wing nearer the body being swept forward and the outer wings being swept backward. On one hand, they have the ability to soar rapidly to great heights while scanning large areas on the ground for food, and on the other they are able to glide at low speeds.

The capacity of birds to develop high lift is due to their wings being characterized by "emarginations" or gaps in the wing tips that act like wing-tip fences. This facilitates the wing in sustaining a higher magnitude of vortex lift over each and every wing strip in the vicinity of the leading edge and nearer the bird's body. Removal of this leading-edge vortex over a spanwise segment, without stalling, is achieved by spanwise pressure gradients that cause the vortex to travel down the wing to the wing-tip fence, where they are accumulated into a system of tip vortices that are then shed safely. Thus the far end of the wing is used to control and maximize the vortex-lift component in a clever and imaginative way.

At high speeds birds use a complex strategy to clap, fling, and beat their wings in such a way that the gaps between them are able to effectively inhibit the loss of lift. The gaps are also controlled in real time to act as "slats," and most birds are able swing a "canard"-like surface relative to the main wing. These features significantly influence the controllability of the associated vortex flows.

Compensating for the Wagner-like effect eliminates the transport lags without altering the circulatory forces acting on the aerofoil. Compensation for this effect involves using a higher initial value at the start of an impulsive change in the angle of attack to generate a steady lift force. Both the high-frequency clap-and-fling mode and the lower-frequency flapping–translation mode are capable of generating a steady lift force, thus effectively compensating for the transport delays.

The dragonfly, for example, is able to rapidly increase its angle of attack by a process of high-frequency beating, which actually involves a gradual increase in the amplitude accompanied by a relatively rapid increase in the beat frequency. This motion is completely equivalent to a sudden unit-step change in the angle of attack. In fact, what is essential is the angle-of-attack time history that is necessary to generate a step change in the lift distribution, which can be derived from the Wagner problem concerning the growth of lift that is due to a sudden change in the angle of attack.

Birds have an effective and simple way to control the movement of the separation bubble; during the landing approach or in gusty winds, the feathers on the upper-rear surface of bird wings tend to pop up. This mechanism inhibits the movement of the separation bubble.

The separation bubble starts to develop on a wing in the vicinity of the trailing edge, and, following this event, locally reversed flow begins to occur in the separation regime. Under these locally reversed flow conditions, light feathers would pop up, acting like a brake on the upstream travel of the separation bubble toward the leading edge. The effect has been simulated in a wind tunnel by the attachment of self-popping porous movable flaps with notched trailing edges to obtain equal static pressure on both sides of the flap under attached flow conditions. The flaps are attached in the rear part of the aerofoil and could pivot on their leading edges. Unlike active flow-control techniques, this is a passive technique and requires no feedback or feedforward control.

2.3.3 Swimming Like a Fish

Swimming involves the transfer of momentum from the fish to the surrounding water. Momentum transfer, which is always associated with forces, is a direct consequence of Newton's laws of motion. The main momentum forces acting on fish when it is swimming are the drag, lift, thrust, and acceleration reaction forces. Like the drag acting on aircraft, swimming drag consists of the following components:

1. Skin friction drag resulting from the viscous friction between the fish and the boundary layer of water in regions of flow with large velocity gradients. Friction drag depends on the wetted area of contact, the speed of the fish, and the nature of the boundary-layer flow. The use of riblets in aircraft wings to reduce turbulent skin friction arose from the study of large-scale-like surface irregularities known as dermal denticles in modern sharks.

   Riblets are streamwise microgrooves that act as fences to break up spanwise vortices and reduce the surface shear stress and momentum loss by preventing eddies from transporting high-speed fluid close to the surface. Many fish use optimally spaced riblets to reduce drag.
2. Form drag is due to the hydrodynamic pressures that develop when the water is pushed aside by the fish's motion. It is caused by the distortion of flow around bodies and depends on their bulk and shape.
3. Vortex or induced drag is due to energy dissipated in the vortices formed by the fins as they generate lift or thrust. Induced drag depends largely on the shape of these fins and their ability to generate lift. The latter two components are jointly known as pressure drag. Fish capable of cruising fast are endowed with well-streamlined bodies to significantly reduce the form drag while well-shaped fins reduce the vortex drag.

The origin of lift forces is due to the viscosity of the flow that in turn is responsible for the flow's being asymmetric. As the fluid flows past a fin, the velocity pattern may be such that the pressure on one side of the fin is greater than on the other. Lift is then exerted on the fin in a direction perpendicular to it. There is also a net component of force in the direction of the flow that can manifest itself as a drag as in most aircraft or as a thrust as in birds and fishes. A chain of ring vortices shed from behind the tail fin significantly enhances the net thrust generated.

Acceleration reaction is an inertial force, generated by the resistance of the water surrounding a body or an appendage when the velocity of the latter relative to the water is changing. It arises because of the kinetic energy gained by the hydrodynamic flow that is due to the acceleration of the body. Acceleration reaction is often quite negligible in aircraft in steady flight and is more sensitive to size than lift or drag forces, and hence it is a significant factor that influences fish propulsion.

Unlike birds, fish generally do not have to generate both lift and thrust. So they can focus on the issue of thrust generation and drag reduction. Fish are particularly capable in increasing thrust by using a variety of leading-edge devices that

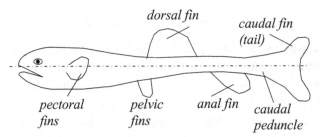

Figure 2.5. Primary thrust effectors in a fish.

generate vortices to enhance the transfer of momentum to the flow. The flipper of the humpback whale is a unique example because of the presence of large, rounded protuberances known as tubercles located on the leading edge of the flipper, an aerofoil-like surface. These act as specialized leading-edge control devices to reduce the drag due to lift ratio on the flipper.

Swimming locomotion again involves two primary types of temporal movements:

1. Periodic swimming, which is characterized by a cyclic movement of thrust effectors to generate the required propulsion. This type of movement is used when a creature must cover a large distance and generally involves constant speed and low accelerations;
2. transient swimming, which almost always involves an impulsive start, maneuvers, and turns. These movements involve very small time constants, of the order of milliseconds, and require rapid generation of thrust and very high accelerations.

Most fish generate thrust by flexing their bodies to generate a backward-moving wave that extends to its caudal or tail fin. The length of the wave could vary from less than half a wavelength to several wavelengths, depending on the anatomical features of the fish. This mechanism of swimming is known as body and/or caudal fin (BCF) propulsion. It is the mechanism used by over 85% of species of fish. An alternative swimming mechanism involves the use of the dorsal and/or the anal fins, which are both referred to as the median fins, and the pectoral and/or pelvic fins, which are referred to as the paired fins. This mechanism of propulsion relies on the ability of the fins to flap and thus generate thrust. As it involves the use of their median and paired fins, it is known as median and/or paired fin (MPF) locomotion. It is a primarily a control mechanism in most fish and is used to direct the thrust in slightly different directions or to enhance stability and balance. Small finlets in the vicinity of the caudal peduncle contribute significantly to the total thrust. The primary thrust effectors in a fish are illustrated in Figure 2.5.

Hydrodynamic drag is a primary mechanism of energy dissipation in determining the performance and thrust generation capabilities of fish. Drag reduction not only involves considerations of fluid dynamics but also of the neural stimulation of muscles, body kinematics, and other anatomical aspects. These mechanisms of generation and control of thrust are analyzed further and discussed in Chapter 12.

2.4 Biomimetic Sensors and Actuators

One of the most significant developments in the emergence of biorobotics as a mature field of study was the synthesis of the biosensor.[3] According to the recent IUPAC[4] definition, a biosensor is "a self-contained integrated device which is capable of providing specific quantitative or semi-quantitative analytical information using a biological recognition element which is in direct spatial contact with a transducer element." Biosensors produce an output (electrical) that is proportional to the concentration of biological analytes. The main components of a biosensor are a specific target analyte (or bioreceptor) that acts as a biological sensing target that needs to be measured and that is in spatial and intimate contact with a recognition element that produces either discrete or continuous measurable signals proportional to the concentration of the analyte. These signals are sensed and measured by a suitable method of transduction.

The recognition elements are the distinguishing elements of a biosensor. They may be catalytic, such as certain enzymes, or affinity types, such as antibodies, receptors, organelles, cells, tissues, or hybrids. The primary recognition element used in most biosensors is an immobilizing layer sandwiched between two semipermeable membrane layers.

Immobilization is defined as a technique used for the physical or chemical fixation of cells, organelles, enzymes, or other proteins (e.g., monoclonal antibodies) onto a solid support, into a solid matrix, or retained by a membrane, in order to increase their stability and make possible their repeated or continued use. The first of the semipermeable membranes allows for the preferential passage of the analyte. The second or other semipermeable membrane permits preferential passage of the by-product of the recognition element or immobilizing layer.

The by-product of the recognition element is sensed by the physical transducer. A typical schematic diagram of a biosensor is illustrated in Figure 2.6.

The approaches to transduction could be electrochemical, thermometric, piezoelectric, magnetic, acoustic, optical, or a combination of these physical principles on which classical transducers are based. Electrochemical transducers (potentiometric, voltammetric, conductimetric types), optical transducers, semiconductor field-effect-transistor- (FET-) based transducers, polymer semiconductor-based transducers, piezoelectric-based quartz crystal microbalance (QCM) transducers, bulk and surface acoustic-wave-based interdigitated transducers, thermal transducers such as thermistors, magnetic transducers, and other mechanical devices have all been developed for biosensing applications.

---

[3] For a review of the recent advances in biosensors the reader is referred to Collings, A. F. and Caruso, F. (1997). Biosensors: Recent advances, *Rep. Prog. Phys.* 60, 1397–1445.

[4] International Union of Pure and Applied Chemistry (IUPAC). (1997). *Compendium of Chemical Terminology*, 2nd ed. IUPAC.

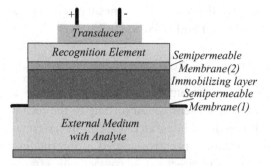

Figure 2.6. A typical schematic diagram of a biosensor.

Electrochemical sensors are the most commonly used as transducers and provide measurements of electrode potential, electrode current, or concentration. They are often obtained by use of a variant of a three-electrode electrochemical cell, shown in Figure 2.7. The electrode potential is dependent on the concentration that can be characterized by the so-called Nernst equation, which is briefly discussed in the following section.

Generally, optical transducers exploit properties such as light absorption, fluorescence/phosphorescence, bioluminescence/chemiluminescence, reflectance, Raman light scattering, and refractive-index variations. Current developments in the sensing of biochemical species by optical-waveguide-based devices include optical immunosensors that transduce antigen–antibody interactions directly into physical signals. Optical transducers use an optical source such as a laser or light-emitting diode (LED) that is directed along a optical fiber that is eventually detected by an optical detector, after it is influenced by the detection event.

The FET harnesses the space-charge layer effect of forward- and reverse-biased *p-n* junctions to make a device in which the current flowing through a wafer of a doped semiconductor is controlled by the voltage applied to a small insert of semiconductor material with reverse doping. In the ion-selective FET (ISFET), the insert or gate is sensitive to the ions or antibodies immobilized on its surface or the surface of a membrane it is connected to.

Fast and accurate measurements of the blood levels of the partial pressures of oxygen (pO_2), carbon dioxide (pCO_2), as well as the concentration of hydrogen ions

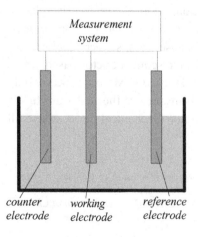

Figure 2.7. Three-electrode electrochemical cell.

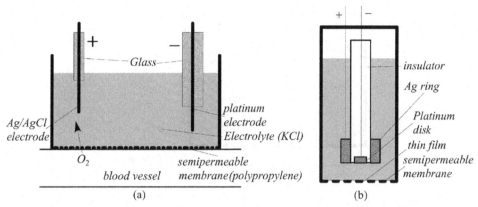

Figure 2.8. Oxygen electrode: (a) principle of operation, (b) practical design.

(pH) are vital in diagnosis. The first successful biosensor was the Clark oxygen-ion-selective electrode developed in 1956 for measurements of oxygen concentrations within the blood in patients prior to undergoing surgery.

The oxygen electrode exploits the reduction of oxygen to water at a platinum cathode (four electrons/oxygen molecule) at a potential of −1.5 V relative to Ag/AgCl. A semipermeable membrane was used to separate the sample of the oxygen-rich analyte from the internal electrolytic blood flow, which contains many other electroactive species that would interfere with the measurement. The ability of the oxygen gas to diffuse through the membrane while preventing the transfer of the solution was the primary factor governing the choice of the polypropylene semipermeable membrane. The membrane is therefore required to have controlled porosity and surface wetting characteristics. The reactions and the output electromotive force (EMF) are governed by the reduction at the platinum cathode:

$$O_2 + 4\,H^+ + 4e^- \leftrightarrow 2\,H_2O \Rightarrow E^{out} = +1.23 \text{ V}, \tag{2.1}$$

and the reaction at the Ag anode:

$$Ag + Cl^- \rightarrow AgCl_{(s)} + e^- \Rightarrow E^{out} = -0.22 \text{ V}. \tag{2.2}$$

The oxygen electrode is shown schematically in Figure 2.8(a), and the practical design of such an electrode is shown in Figure 2.8(b). The output current is proportional to the number of electrons released in the reduction, the sample oxygen partial pressure and properties of the membrane relating to oxygen such as solubility and the diffusion coefficient per unit thickness.

Following the success of the oxygen electrode, Clark[5] and his team went on to develop the glucose enzyme electrode, which led in turn to the development of the glucometer and the glucose pen. A typical glucose enzyme electrode is shown schematically in Figure 2.9. The glucose enzyme electrode consists of a cylinder

---

[5] Hall, E. A. H. (1991). *Biosensors*, Advanced Reference Series, Engineering, Prentice-Hall, Englewood Cliffs, NJ, Chapter 5.

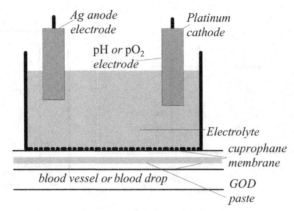

Figure 2.9. The Clark and Lyons glucose electrode. GOD: glucose oxidose.

internally containing a solution of an electrolyte as well as a sensing and a reference electrode. At the base of the cylinder a concentrated solution of the enzyme glucose oxidase (GOD) is held between two semipermeable membranes.

The mechanism of the action of GOD is a complex process in which the enzyme passes through a series of states involving both the oxidation and the reduction of the appropriate substrates. The action of the GOD paste results in a decrease in the concentrations of both glucose and oxygen within the internal electrolyte. It is followed by the formation of gluconic acid and hydrogen peroxide (H_2O_2). The reaction governing the reduction to gluconic acid is

$$\text{glucose} + O_2 + H_2O \xrightarrow[\text{GOD}]{} \text{gluconic acid} + H_2O_2. \tag{2.3}$$

At the anode the breakup of H_2O_2 is governed by

$$\text{anodic}: H_2O_2 \rightarrow O_2 + 2\,H^+ + 2e^-. \tag{2.4}$$

The corresponding changes in the potentials can be measured by an internal sensing (working) electrode. One could measure either the pH change that is due to the production of the acid, the consumption of oxygen molecules, or the production of H_2O_2. Oxygen sensing is prone to interferences from exogenous oxygen. The product H_2O_2 is oxidized at +650 mV versus a Ag/AgCl reference electrode. Thus a potential of +650 mV may be applied and the oxidation of H_2O_2 measured. This current is directly proportional to the concentration of glucose.

Following the development of the oxygen and glucose enzyme electrodes, several recognition elements have been used as receptors, such as other enzymes, antibodies, nucleic acids, and cells. Since that early work relating to the oxygen electrode and on the enzyme-based glucose sensor that was developed by the entrapment of glucose oxidase in a membrane-enclosed sandwich, a number of new methods have evolved for the immobilization of enzymes and proteins. The principal methods of immobilization include physical or chemical adsorption at a solid surface, covalent binding to a surface, entrapment within a membrane or micro-capsule, cross-linking between molecules, sol–gel entrapment, and electropolymer-ization. The actual immobilization method used depends on, among other factors, compatibility with the biomolecule being immobilized, the sensor membrane on which immobilization is to proceed, and the end use of the sensor. By use of these new recognition elements and immobilization techniques, hydrogen and carbon dioxide electrodes have also since been developed.

The equilibrium condition described in the case of the oxygen and glucose enzyme electrodes is not entirely representative of cellular systems because it neglects contributions by other ionic current paths. In the case of mammalian-cell-based biosensors, binding of the analyte to the cell surface receptor triggers a detectable signal. Thus mammalian-cell-based biosensors[6] use a primary transducer (the cell) and a secondary transducer (device that converts cellular/biochemical response into a detectable signal), which may be an electrical or an optical trans-ducer.

2.4.1 Action Potentials

Action potentials are generated in electrically active mammalian cells. The basis for this is the resting potential developed across a plasma membrane of a cell (with its embedded ion). The static electrical state of a cell is one of equilibrium in which the electrochemical forces acting on ions across the cell membrane are completely balanced. Although constant ionic currents flow, the net transfer of ions is zero, resulting in a constant transmembrane potential. This is discussed in this subsection and is followed by a description of the consequences of perturbing the equilibrium state, which results in the generation of an action potential.

If one examines the case of a semipermeable membrane separating two solu-tions (as is the case with a biological cell), such as two salt solutions of different ionic concentrations, a flow of ions will result. In this example, we assume that the membrane is permeable to only one ion (A^+) in the solution. Initially, at the instant the two solutions are brought into contact with the membrane, the transmembrane potential will be zero. However, the A^+ ions will diffuse from the solution with the higher concentration to the one with the lower concentration, causing one side to become biased with respect to the other. This diffusion will continue until the elec-tric field across the membrane is large enough to balance the forces driving diffusion of A^+ across the membrane. When this balance is reached, the system is in equilib-rium with respect to the A^+ ions.

By considering one mole of an ion K with a charge Z_K, one can write an equa-tion giving the equilibrium potential E_K as a function of the ionic concentration ratio and the valence:

$$E_K = E_1 - E_2 = \frac{RT}{Z_K F} \ln \frac{[K]_2}{[K]_1}, \tag{2.5}$$

---

[6] See, for example, Eggins, Brian R. (1997). *Biosensors: An Introduction*, Wiley, New York.

Table 2.1. *Free ionic concentrations and equilibrium potentials*

Ion	Extracellular concentration (mM)	Intracellular concentration (mM)	Equilibrium potential at 37°C (mV)
Na^+	150	13	+61.5
K^+	5.5	150	−88.3
Ca^{2+}	100	0.0001	+184
Cl^-	125	9	−70.3

where $[K]_1$ and $[K]_2$ represent the concentrations of the ion K on each side of the membrane, R is the gas constant (8.315 J/°K mol), and F is Faraday's constant (96 480 C/mol) where K is in degrees Kelvin and C in Coulombs. This equation was first obtained by Nernst over 100 years ago and is therefore known as the Nernst equation. Thus the semipermeable membrane allows flow of certain ions between the intracellular and extracellular fluids, which causes a potential to develop across the membrane as defined by the Nernst equation.

In biological systems, one is concerned with the equilibrium potentials of biologically relevant ions: K^+, Na^+, Ca^{2+}, and Cl^-. Each is associated with its own equilibrium potential calculated based on the assumption that the membrane is permeable only to that ion.

Typical physiologic values for these are given in Table 2.1 for which the standard convention of membrane potentials being measured intracellularly minus extracellularly has been used. Examining the values in Table 2.1, we can see that the limits of the transmembrane potential are set on the positive side by Ca^{2+} and on the negative by K^+. This holds true for most cell types even though the equilibrium potentials are slightly different. In all cases, both K^+ and Cl^- drive the intracellular potential negative with respect to the extracellular, whereas Na^+ and Ca^{2+} attempt to drive it positive. However, because of the presence of multiple ions, a weighted combination of all of these ionic forces must be used to define the actual transmembrane resting potential.

The semipermeable membrane allows certain ions to flow between the intracellular and the extracellular fluids, and consequently a potential is developed across the membrane as defined by the Nernst equation for a single ion. In the case of excitable cells with multiple ions, the net result is a negative transmembrane potential (with respect to an extracellular reference) because the membrane has far more open K^+ channels than Na^+, Cl^-, or Ca^{2+} channels. To obtain a representative estimate of transmembrane potential, the Nernst equation is modified by weighting the concentration of the ions by their permeability constants.

The modified Nernst equation, in which the permeability constants (P_{Na}, P_K, and P_{Cl}) are in proportion to the number of open channels through the membrane, is known as the Goldman–Hodgkin–Katz constant-field equation, which is

$$E_K = \frac{RT}{F} \ln \frac{P_K [K^+]_e + P_{Na} [Na^+]_e + P_{Cl} [Cl^-]_i}{P_K [K^+]_i + + P_{Na} [Na^+]_i + P_{Cl} [Cl^-]_e}. \tag{2.6}$$

The valence term in the Nernst equation is not present, and the negative valence of the Cl^- term is accounted for by inverting the corresponding ratio. Ca^{2+} is neglected in the calculation because there are few normally open Ca channels. The cytoplasm proteins remain within the cell and are not included, although the expulsion of the K^+ ions is controlled by their negative charge. No equilibrium potential is associated with these intercellular anions as they are impermeable. The resting potential is still not accurate as it neglects the effects of other ionic current paths and is determined not only by the leaky K^+ channels, but also by the so-called sodium–potassium pump, which is powered by the hydrolysis of intracellularly present adenosine triphosphate (ATP). This pump actively transports three Na^+ ions out of the cell for every two K^+ ions transported into the cell, thereby lowering the transmembrane potential (inside minus outside).

Any perturbing inputs to the state of equilibrium established across the membrane result in increased (or decreased) ionic flows. These ionic flows counteract the effects of the perturbations. The resting membrane potential is maintained at a constant level within narrow limits. However, when the membrane potential is driven by an external source far beyond its resting state, this equilibrium is disturbed.

In electrically active cells such as neurons and muscle cells, disturbing the equilibrium can result in large, ionically driven excursions from the resting transmembrane potential; this is known as an action potential. In the Goldman–Hodgkin–Katz equation the permeability constants may be interpreted as conductances. It is the changes in these conductances that are responsible for the action potential. Triggering the Na^+ channels by increasing their conductance results in an inward flow of positively charged ions into the cell, rapidly initiating the action potential.

Considering the relative intracellular and extracellular concentrations of the ions in Table 2.1, it follows that triggering the K^+ channels by increasing their conductance results in an outward flow of positively charged potassium from the cell. This causes what is known as hyperpolarization; the transmembrane potential becomes more negative. Increasing the conductance of Cl^- ions will have a similar effect with a large inward flow of negatively charged Cl^- also hyperpolarizing the cell. Although increasing the conductance of both Na^+ and Ca^{2+} channels causes an influx of positively charged ions that tends to depolarize the cell, the latter results in a relatively slower upstroke in the action potential. In both cases the transmembrane potential becomes less negative and may even become positive. The subsequent outward flow of the K^+ ions results in a downstroke in the action potential.

The action potential is the principal electrical signal responsible for the functioning of the brain, the contraction of muscle, beating of the heart, and much of our sensory perception. Different cells generate dissimilar active electrical signals by controlling ionic conductance of membranes. The discovery of voltage-gated ion channels was a major factor in our ability to communicate with neuromotor systems. The voltage-gated or voltage-sensitive ion channels generate and control action potentials. Ion channels are triggered and controlled by proteins with distinctive structures and distributions.

The process of coding an action potential with the information content of an environmental stimulus is known as sensory transduction. It takes place in confined and designated areas of the neuronal membrane known as receptors. It is a two-step process whereby the physical stimulus is first converted to a receptor potential, a small perturbation to the membrane potential. In the second step it is converted to an action potential or modulated onto an existing action potential and transported to an intended recipient. Mechanoreceptors, for example, convert mechanical stimuli into receptor potentials. The receptor area in the terminal axon membrane contains ionic channels that are sensitive to external stimuli such as pressure in the case of mechanoreceptors, which is due to the depolarization of the axon terminal that is due to the flows of both the K^+ and Na^+ ions.

The receptor potential is a direct consequence of the depolarizing effect. It is an analog signal in the sense that its amplitude bears a direct relationship to the magnitude of the stimulus. The receptor potential tends to degrade farther away from the location of the stimulus and therefore cannot be propagated unless it is modulated onto an action potential. Unlike receptor potentials, action potentials are like binary signals that have the same amplitude; they are either present or absent. The intensity of the stimulus cannot be represented as a variation of the amplitude of the action potential. The combined effect of the stimulus and the receptor potential is to generate a number of action potentials at a rate that is dependent on the magnitude of the stimulus. The increasing and decreasing receptor potentials trigger a sequence of action potentials in quick succession. The process is similar to frequency modulation, and it follows that measurements of action potentials may be decoded or encoded to extract or inject appropriate stimuli into the peripheral nervous system.

There are five classes of receptors: chemoreceptors that recognize certain classes of molecules, photoreceptors that are sensitive to light, mechanoreceptors that are sensitive to motion and stress-related stimuli, thermoreceptors that are sensitive to heat, and nocireceptors that are sensitive to stimuli received by tissues that are perceived as pain. Mechanoreceptors, thermoreceptors, and nocireceptors are also collectively known as somatosensory receptors. Proprioceptors are a special class of mechanoreceptors that are present in the muscles and joints that are responsible for the sensation and perception of the position of our limbs. If one is interested in designing a bionic limb, it is essential to interface to the neurons that are providing the efferent signals and provide sensory signals to the afferent neurons, so as to be able to communicate with the central nervous system (CNS).

2.4.2 Measurement and Control of Cellular Action Potentials

Bioelectrical signals, acquired by interfaces such as electrodes, are amplified and filtered by a signal processor for prosthesis control. Aggregates of action potentials are measured and used for purposes of monitoring and diagnosis.

Typical examples of measured aggregates are the electroneurogram (ENG) for propagation of nerve action potential, the electromyogram (EMG) for monitoring electrical activity of the muscle cells, the electrocardiogram (ECG) for monitoring

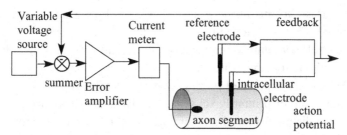

Figure 2.10. Measurement of axon potential by seeking a null voltage in the forward path.

the activity of the cardiac cells, and the electroencephalogram (EEG) for monitoring the brain.

For purposes of measuring action potentials in individual axons, electrodes are used to interface a high number of nerve fibers by using an array of holes with electrodes built around them and implanted between the severed stumps of a peripheral nerve. The method of measurement known as the voltage clamp (Figure 2.10) uses a negative-feedback circuit to establish a null voltage in the forward path.[7] The current in the forward path is zero, and the action potential is proportional to the voltage output.

The intraneural and intracellular electrodes are known as regenerative electrodes. Regenerating axons eventually grow through the holes, making it possible to record the potentials and to stimulate individual axons or small fascicles. These electrodes may be used for acquiring efferent signals. The success of the electrode implantation depends on the axonal regeneration through the perforations or holes. There is the possibility of nerve damage from the mechanical load imposed by the electrode or from constrictive forces within the holes, and this should be avoided.

Cuff electrodes may also be used to stimulate the enclosed nerve leading to the activation of efferent motor or autonomic nerve fibers and to acquire both efferent and afferent signals. They are not intraneural and are bonded to the inside of an insulating tubular sheath that completely encircles the nerve. Electrode contacts exposed at the inner surface of the sheath are connected to insulated leads.

Biocompatibility, which one must consider before interfacing any electrodes, is whether the system performs in its intended manner in the application under consideration. It requires that the implant does not irritate the surrounding structures, provoke an inflammatory response, incite allergic reactions, or cause cancer and does not degrade its functional performance by attacking the body's immune system.

2.4.3 Bionic Limbs: Interfacing Artificial Limbs to Living Cells

The fundamental unit cell in the CNS is the neuron.[8] A characteristic feature of neurons is their extensive dependence on electrical transmission to communicate

[7] For further details the reader is referred to Aidley, D. (1998). *The Physiology of Excitable Cells*, 4th ed., Cambridge University Press, New York.

[8] For a concise introduction to neuroscience the reader is referred to the text by Kingsley, Robert E. (1999). *Concise Text of Neuroscience*, Lippincott Williams and Wilkins, Philadelphia.

coded signals. Basic to this property are the establishment and maintenance of a resting potential. This is achieved through passive and active mechanisms that are present in channels and pumps. Information is integrated by neurons, whereas neural interconnections, together with their specific circuit connections, form circuits and networks that are capable of producing behaviors. Individual neurons communicate with one another primarily via synapses. These are interfaces between cells that have certain distinguishing characteristics. Signals are passed between neurons either electrically, by action potentials being transmitted via "transmission lines," or chemically, by releasing neurotransmitters. A sensory nerve fiber is one that conducts impulses from some sensory ending toward the CNS, and is also referred to as an afferent fiber. Motor fibers, which conduct signals from the CNS to some muscle, are often referred to as efferent fibers.

Our ability to interface with the neural fibers or highways has led to the development of the concept of a neuromotor prosthesis. These are robotic manipulators for rehabilitation that provide people with a high degree of autonomy and mobility. Mechanical prostheses have been devised for many years to complement the physical functions of persons who have, through various accidents or for other reasons, been deprived of their normal measure of physical ability. Following surgical amputation of an arm or leg, prostheses have taken many forms, although in most cases do not provide completely satisfactory solutions. However, several shortcomings of the arm prostheses can now be addressed because of the availability of neuromotor prostheses and the ability to interface a real physical manipulator to human neuromotor signals.

Such prosthetic or bionic devices are intelligent, adaptive devices directly connected to the body's command and control systems capable of both interacting with the human body in a bidirectional manner and replicating a physiological function. The primary shortcomings of mechanical prostheses that can be addressed by these devices are as follows:

1. The prostheses have as many DOFs as the normal arm for which they are intended to substitute. Thus they can perform certain tasks with the same dexterity for the amputee as a normal arm.
2. The controls (i.e., the actions of the amputee that control motions of the prosthesis) for a given prosthesis motion can now be synthesized from the actions of a normal person that cause the corresponding motion of a normal arm. For example, flexion of the "elbow" of a prosthesis could be independent of the movement of the shoulder of the amputee as in a normal person. There is still a need for an amputee to learn an entirely new pattern of activity in order to make the prosthesis useful to him or her, and his or her ultimate performance may often be limited because of the degree of learning that is required is excessive, although the constraints of the control system are not so severe.
3. The sensations that the amputee receives from the prosthesis are not limited to that which he or she can obtain from visual observation of its performance and from any extraneous noises that it makes or mechanical forces that it transmits back through the socket and harness.

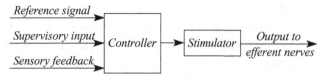

Figure 2.11. Principle of functional electrical stimulation.

We see that there is almost as much sensory information as is present in a real arm. It is not now difficult for an amputee to carry on a conversation or engage in any other activity that reduces his or her conscious concentration on the prosthesis while simultaneously operating his or her prosthesis.

Functional electrical stimulation also provides an alternative to a mechanical hand in cases of injury to the neural connection from the spinal mechanisms and driving the descending neuromotor system. The principle of functional electrical stimulation is illustrated in Figure 2.11. The controller synthesis is based on the supervisory signal, reference trajectories or reference inputs, and sensory feedback. The output to the efferent nerves is a sequence of unidirectional electrical pulses with pulse-width modulation. The currently available functional electrical stimulation systems for grasping may be broadly classified as implanted or surface. Both types of systems provide a basic set of primitives that will improve the user's ability to grasp objects. To understand how the shortcomings of a purely mechanical prosthesis are addressed, consider the CNS or the peripheral nervous system.

The components of a neuron in a nerve fiber are a cell body, dendrites, which are short fiberlike structures responsible for connecting and seeking information from other neurons, and an axon, which is a long cylindrical structure arising from the cell body. The axon is a primary channel for information whereas the dendrites could be considered as gateways for information going into the cell. The axon allows the cell to communicate action potentials over long distances whereas the dendrites are responsible for short-distance interactions with other cells.

Just after a nerve impulse has passed a given point on a nerve axon, there is a period during which that region of the nerve membrane is less sensitive to stimulation than before. The interval is referred to as the refractory period, which determines the bandwidth of the signal. The end of a nerve axon may make contact with either the cell body or dendrites of another neuron. The contacts or junctions between nerves are called synapses, which allow neurons to influence one another. Communication between these junctions is by diffusion of a chemical neurotransmitter. At a pair of nerve endings a synaptic communication takes place, resulting in a postsynaptic transmission based on a chemical messenger or neurotransmitter known as acetylcholine.

A single nerve fiber consists of a cylindrical semipermeable membrane that surrounds the axon of a neuron. The properties of the membrane give rise to a high concentration of potassium ions and a low concentration of sodium ions inside the fiber and a low potassium ion concentration and high sodium ion concentration outside the fiber. This results in a potential difference of about -0.15 V between the inside and the outside of the fiber, which is the resting membrane potential. The nerve is said to be polarized. When the nerve fiber is externally excited, the resulting flow of ionic current causes the potassium ions to be driven out and the sodium ions to enter the fiber. Thus the impedance of the membrane effectively changes as the transmembrane potential is a function of the concentration gradients. Furthermore depolarization takes place as the inside of nerve is less negative than before and a nerve action potential is initiated. As the nerve action potential traverses down the axon, the membrane at the point under consideration is repolarized and the nerve fiber reverts to its resting potential state. The action of the nerve fiber is similar to a transmission line consisting of a series of resistances and capacitances in parallel. The capacitances are the result of the intermittent presence of an insulator known as myelin, which prevents the flow of ions into the extracellular fluid. The transmission fibers or fast fibers are said to be myelinated whereas the slow fibers are nonmyelinated. The speed of signal transmission depends on the capacitances and, together with the resistances, results in a short time constant for a myelinated fiber.

The transmission of the action potentials can be controlled by external excitation. Combined with the fact that the action potentials can also be measured, it is possible to control and measure the nerve impulses being transmitted to and from the brain and from the limbs. The ability to measure and control the action potentials provides a natural way of interfacing an artificial limb to living neuronal cells. By use of concise biomechanical models to correlate neuronal signals and signal rates to the movement kinematics and kinetics, it is possible to synthesize control laws that can control the joint servomotors in an artificial electromechanical limb. As the control system also decodes the efferent neural signals in real time, the artificial limb is directly controllable by the brain.

A typical block diagram illustrating the interfacing of an artificial limb to live nerve endings is shown in Figure 2.12. The system illustrated in Figure 2.12 does not involve direct feedback to the brain via the afferent nerve fibers that provide haptic, tactile, and force feedback. Haptic information is a broad term used to describe both cutaneous (tactile) and kinesthetic (force) information. Both types of information are necessary to form the typical sensations felt with the human limb. The tactile and proprioceptive sensing aspects are discussed in Chapter 11.

As tactile afferents seem to run in separate fascicles from muscle afferents within the nerve trunks, the independent activation of tactile and muscle afferents is considered feasible. It is possible, in principle, for tactile feedback to be routed back via electrodes embedded in afferent tactile nerve fibers, although there are several practical issues, such as the large number of afferents that must be stimulated, that must be addressed before it can be satisfactorily realized. The median nerve is estimated to contain as many as 14 000 tactile afferents connecting the hand. Each fingertip is connected to about 250 afferents, whereas the palm is connected to about 50 afferents. Although it is believed that tactile feedback to provide a prosthesis user with sensations similar to those of a natural hand would require independent microstimulation of hundreds of "on-line" tactile afferents in order to include the various modes and primitives and to reach most of the important contact areas during object manipulation, recent research indicates that feedback to even a few

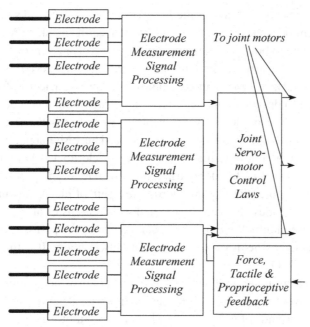

Figure 2.12. Bionic limb control system.

afferents is adequate to give the user a minimal sensation. The stimulation currents are very small and of the order of 30–40 μA. Multiefferent interfaces driven by artificial neural network controllers could provide the answer, and such interfaces with integrated controllers are currently being researched. The sensation received may not only be a function of the coding in and the nerve fiber along which it is received, but also a function of the specific type of nerve ending that produced it. Thus, with multiafferent electrodes inserted to interact with each of specific nerve fiber, the resulting output signals must be adaptively processed with a learning controller to achieve desirable prosthesis control.

Because of the low currents involved, electrical stimulation of small bundles of afferents will be easier to implement in the near future than stimulation of isolated afferents. Multiafferent stimulation has the advantage in that most of the grasping surfaces of a limb such as the hand could be covered with a relatively low number of independent control channels. Currently it is possible to build an artificial mechanical limb and to endow the limb with the essential motor skills and sensory components by defining appropriate reaching and grasping primitives as well as primitives to acquire tactile and proprioceptive information.

2.4.4 Artificial Muscles: Flexible Muscular Motors

An understanding of the bioelectric control of muscles in humans offers the possibility that control of the prosthesis action can be similar to control of the corresponding body action. Such a technique may be applied to the control of a mechanical limb designed to replace a natural human limb that was lost through amputation. In a normal human limb the muscles in the body produce an electrical signal, called the EMG, which can be sensed directly from the surface of the skin. A study of these signals permits one to both generate appropriate control and sensory signals so as to be able to interface an artificial limb to an amputated stump. As an example, a measurement of the surface EMG signals from the biceps and triceps of an amputee's arm can be used to design a controller capable of providing a graded control of an elbow prosthesis. It also allows for the replacement of muscles and tendons on a selective basis by prosthetic muscles and tendons. The EMG signal bears a special relationship to the action potential. When any muscle fiber is stimulated because of a voluntary contraction, the depolarization of its membrane produces an electric field whose action potential appears across any pair of electrodes in the vicinity. However, the electrodes also receive the action potentials of all muscle fibers in a muscle, and consequently the ouput as observed across the pair of electrodes is the aggregate effect of the fields of a collection of muscle fibers. It is this signal that is observed in the EMG.

The use of efferent nerve signals for the control of prosthesis offers many potential advantages over the use of the EMG signal. In a typical nerve trunk in the human arm there are thousands of nerve fibers, which include motor fibers, proprioceptive sensory fibers, and other sensory fibers. The motor fibers drive about 20–50 muscles in the arm.

Intramuscularly inserted electrodes and epimysial electrodes that are attached to the surface of the muscle have been developed to facilitate human muscle activation. These muscle activation electrodes must be placed as close as possible to the motor nerve endings as the excitatory signal satisfies an inverse-square law. Such electrodes are particularly useful for the implementation of functional electrical stimulation of human limbs by prosthetic devices.

The sensory fibers, on the other hand, are responsible for sending a variety of sensory information back to the CNS, not all of them being proprioceptive. Even if one assumes that the sensory afferent fibers were used to send the same signal as each previously did before amputation, there may be some complex coding necessary to properly identify these signals to the CNS. If this can be done, the graded autonomous control of a limb such as the hand or the arm is a distinct possibility.

There are two major classes of muscle in mammals: skeletal and cardiac (smooth) muscles. Skeletal muscles are responsible for all voluntary movements and are of primary interest to us. Contraction of these muscles and tissues is initiated by the action potential. Cardiac muscles are to a greater extent involuntarily controlled; examples are the heart and other visceral bodies. A specific type of nerve transmission is used by an axon terminating on a skeletal muscle fiber, at a specialized structure called the neuromuscular junction. An action potential occurring at this site is known as a neuromuscular transmission. At a neuromuscular junction, the axon subdivides into numerous terminal buttons that reside within depressions formed in the motor end plate. Also at the neuromuscular junction, a synaptic transmission is by the chemical messenger or neurotransmitter. Although the neurotransmitter acetylcholine is responsible for synaptic communications between a nerve ending and a muscle fiber, the neurotransmitter responsible for communications between muscle fibers is noradrenaline.

The presence of both the action potentials that are short period control signals and postsynaptic transmissions that are longer period control signals makes it is possible to interface neuromuscular prosthetic materials to replace damaged muscles in a human. Although much of the early work was related to the use of electropneumatic actuators such as prosthetic artificial muscles, the recent development of muscle fibers from a copolymer of PVDF (poly-vinylidene fluoride trifluoroethylene), a material with electroactive properties related to electrostriction and piezoelectricity, is providing promising prostheses for damaged muscles in humans. This innovative technology is currently being explored, and one can expect artificial muscle prostheses to be available in the not-too-distant future.

2.4.5 Prosthetic Control of Artificial Muscles

There are a small number of transducers distributed within the muscle that are responsible for sensing position but have no role to play in driving the muscle. These are the so-called muscle spindles. The elastic portion of a muscle spindle (called the nuclear bag) at its center can be stretched but cannot actively contract. On the other side of the nuclear bag are two poles that are capable of contraction and are made up of a few small muscle fibers known as intrafusal fibers, as they are internal in the muscle. The nuclear bag is connected to a large fast-conducting sensory nerve fiber. The intrafusal fibers are connected to small, slowly conducting motor nerve fibers known as gamma efferents. Because of the relative ease with which they can be found and worked with, the muscle spindles can be modeled effectively to simulate their functional behavior. The sensory nerve transmits nerve impulses at a rate that is logarithmically proportional to the length of the nuclear bag. The spindles can adapt only slightly after a period of several minutes. The spindles stretch parallel to the primary muscle fibers in the muscle, connecting to the tendons or bones in sympathy with the muscle.

There are two ways in which the nuclear bag can be stretched: (1) by lengthening the muscle that then would lengthen the nuclear bag accordingly; and (2) by excitation of the intrafusal fibers by the gamma efferent nerve fibers, in which case, even if the overall muscle length remains constant, the nuclear bag will be forced to lengthen. Thus the muscle spindle output is relatively steady and contributes to a sense of position.

A number of local circuits emanating from the spinal cord are responsible for regulating and coordinating the action of the muscles connected to it by control signals emerging from efferent neurons and feedback signals to afferent neurons in its roots. These are known as reflexes, and a reflex loop is associated with five subsystems: a sensory receptor, an afferent path, back to the CNS, synaptic connections within the CNS, an efferent path, and a control effector. Typical examples of these reflexes are the muscle stretch reflex, the clasp knife reflex, and the flexion reflex. The first is invoked while a muscle is rapidly stretched, the second when a muscle is stretched during an isometric (constant-length) contraction, and the third when a general flexion of a limb is produced.

The ability to control and regulate the position of an artificial limb grafted to a human depends to very large extent on our understanding of how the position of a natural limb in a human is controlled and regulated. Central to this understanding is the reflex nature of the controller maintaining the position of a limb. One example of such a reflex action or reflex arc is the well-known knee-jerk response to a tap on the tendon just below the kneecap. A commonly known reflex is the knee-jerk reflex. The knee jerk is a deep tendon reflex and is also known as a patellar reflex. When tapped on the patellar tendon over the kneecap, the tendon that connects the quadriceps muscle in the front of the thigh to the lower leg, the quadriceps contracts and involuntarily brings the lower leg forward.

The main function of the reflex arc and the muscle spindle is the ability to hold a joint in some constant position without conscious attention to its position. Thus the reflex arc acts like a regulator and an internal autostabilizing loop, and hopefully can be designed by well-known principles in optimal control theory.

The sensory nerve fibers that are of primary importance in commanding a desired position response and regulating the position are the ones entering the spinal cord by the dorsal root, their cell bodies being located in the dorsal-root ganglion. These are the sensory fibers that are responsible for "innervating" the nuclear bags of the muscle spindles of a given muscle. It is well known among neuroanatomists that these spindle signals enter the spinal cord and synthesize the excitatory synapses with the "alpha motor neurons" for the muscles from which they came.

Muscles receive information from two sets of motor neurons. The extrafusal muscle fibers, external to the muscle, are innervated by the alpha motor neurons. The large myelinated axons, with a diameter in the range of 8–13 μm, associated with them initiate movement through extrafusal muscle contraction. When an external stimulus such as a tap on the tendon is sensed, it momentarily stretches a muscle. The reflex system operates like a feedback controller, and all the elements in the command path and the feedback loop may be explicitly identified. The spindles in that muscle will sense a change in muscle length and return an increased or decreased nerve impulse rate along their respective sensory nerve fibers. This signal will then in turn cause the average impulse rate along the alpha motor nerve fibers of the muscle that was just deformed to change. The net result will be a change in the tension in that muscle, causing the associated joint to move in a direction so as to restore the muscle length by reducing or increasing the spindle discharge rate, and thereby returning the system to its former state.

Unfortunately the proportional control system driven by the outputs of the alpha motor neurons is unable, by itself, to distinguish between set-point changes and feedback signals. The synthesis of the set points is therefore performed independently from the outputs of the gamma motor neurons that innervate only the intrafusal muscle fibers. Set-point commands in the form of the gamma efferents

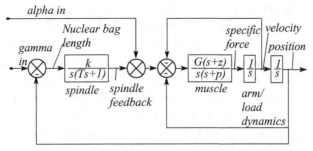

Figure 2.13. Simulink simulation of the proportional position control loop representing the muscle stretch reflex.

tend to induce changes only of the intrafusal muscle length and thus maintain the steady tension on the muscle spindle receptors.

The gamma motor neurons are associated with axons much smaller in diameter than those of the alpha motor neurons, in the range of 3–8 μm, and generate the gamma efferents that innervate only the distal (remote) contractile portion of the intrafusal muscle fibers. The gamma motor neurons can be further grouped into static and dynamic that respectively innervate the static and dynamic muscle fibers.

Muscles generally operate in pairs, and with each muscle is also associated another one in an antagonistic configuration. The alpha motor neurons for the antagonist muscles are forced to change their average discharge rate so as to alter the tension in those muscles, tending once again to assist in returning the joint to its former position.

The simulation of the proportional position control loop representing the muscle stretch reflex is illustrated in Figure 2.13 and is driven by two inputs. The one that is understood to be relevant to the postural setting of the nominal position is the gamma input. This corresponds to a signal applied to the gamma motor fibers from the CNS, and it corresponds to the different set points of desired positions for the joint. This is the postural control loop, with relatively long time delays, that can be used with different input signals to regulate a nominal position at any point.

The proportional control input to the neuromuscular control loop is the alpha input and represents a direct input to the muscle, by which a rapid action can be caused. The alpha subsystem is a fast controller in which the signal is used for rapid voluntary motions, whereas the gamma subsystem may be considered to be its slower counterpart. Probably the most important consequence of our understanding of muscle control is that it is now possible to both simulate and build a muscular prosthesis that can form an important component of a bionic limb.

2.5 Applications in Computer-Aided Surgery and Manufacture

Major applications of the ability to provide haptic feedback are in computer-aided surgery and computer-aided manufacture. In the context of manufacturing, a prototype telemachining system has been developed by use of multiaxis force data and stereo sound information in which the force information from the multiaxis force sensor is used not only to determine the machining and the operation state but is also transmitted back to the operator through a three-DOF joystick to give the operator an artificial feel of the forces involved in manipulating the object. More-over, haptic devices may be used by an operator to train a machine tool, such as a milling machine, to machine a virtual part so that a collision-free tool path can be generated while ensuring accessibility. Further discussion in this chapter is limited to computer-aided surgery applications.

2.5.1 Steady Hands: Active Tremor Compensation

Endoscopy involves a minimally invasive surgical procedure inside the human body. Two examples of such procedures are flexible gastrointestinal endoscopy like gas-troscopy (inside the stomach) and colonoscopy (inside the large intestines). Present-day endoscopes used in minimal-access surgery are rigid, inflexible, and not con-trollable. Furthermore, tremor is a serious problem even in normal humans. It is particularly problematic in people suffering from Parkinson's disease.

Tremor needs to be eliminated, particularly when one is performing surgery around critical regions such as in the presence of tumor near a gland such as the pituitary gland. Laparoscopy is not only a term given to a group of operations that are performed with the aid of a camera placed in the abdomen but also refers to minimally invasive surgery in the abdomen via a small incision. Instruments used in endoscopic surgery via natural incisions have severe limitations such as rigidity and inflexibility and are often not adequately controllable.

A novel approach that can be applied to endoscopy, laproscopy, and for tremor compensation is to use actively controlled micromanipulators. Such micromanipu-lators would use flexible linkage systems that could be completely controlled to any shape or geometry as desired. The control actuation is performed by smart actuators using smart composites with layers of piezopolymers or piezoceramics or embedded shape memory alloy (SMA) wires, in addition to conventional devices. The embed-ded smart actuators are complemented by a matching set of embedded smart sen-sors. Together these systems can be used to achieve the closed-loop servocontrol of the micromanipulator.

Endoscopy is also controlled entirely by vision alone. In terms of laparoscopy, the difficulty is in the coordinated handling of instruments (hand–eye coordination) and the limited workspace that increases the possibility of damage to surrounding organs and vessels either accidentally or through the complexity of the procedures. In typical magnetic-resonance-image- (MRI-) based systems, a solid-state inertial measuring unit embedded at the tip of the endoscope relays the position and ori-entation of the endoscope. The endoscope is then displayed on a diagnostic image relative to the position of the targeted organs.

Thus one of the applications of the vision-based approach is in radiosurgery. Robotic surgical systems are being designed to provide surgeons with full control over instruments that offer precision. Instead of manipulating surgical instruments manually, surgeons move a pair of joysticks that in turn control a pair of robot arms operating miniature instruments that are inserted into the patient. The surgeon's movements of the joysticks transform large motions into miniature movements on the robot arms to greatly improve mechanical precision and safety.

Based on micromanipulation, robot-assisted remote surgical systems have been developed to supplement the ability of surgeons to perform minimally invasive procedures not only by scaling down motions but also by adding additional DOFs to instrument tips. When the user operates these systems by wearing a bionic glove that is appropriately endowed with sensors and microelectronic hardware, the system can identify the user's intentions and generate the relevant control signals to operate the surgical manipulator.

With recently developed surgical systems it is also possible to make measurements of contact forces and provide force feedback so as to control the contact dynamics of the manipulator's end-effector. Yet the application of these systems is limited because of the lack of extensive haptic feedback, including tactile sensations and not just force feedback, to the user.

One of the major issues associated with robotic surgery is the need to be able cancel tremor when fine movements are performed. It is essential to understanding the nature and source of the tremor. In a human the motor cortex in the cerebral hemisphere in the forebrain generates the instructions for fine movement. The basal ganglia and cerebellum are large collections of cells that modify movement on a minute-to-minute basis. The motor cortex sends information to both, and both structures send information right back to the cortex via the thalamus.

The cerebellum coordinates movement by comparing one's intentions with one's actions as evidenced by the responses of the muscles in the limbs. The output of the cerebellum is excitatory, whereas that of the basal ganglia is inhibitory. The antisymmetric nature of the two signals allows a balance to be maintained that culminates in a continuous coordinated movement. Any disturbance in either system will show up as movement disorders. Depending on the type of movement desired, the commands from the motor cortex go through processing units in either the cerebellum (coordinated or repetitive movement instructions) or the diencephalon (complex or fine-movement instructions). The diencephalon is made up of the thalamus and the hypothalamus, which reside just above the brain stem. A simplified model of this control system, shown in block-diagram form in Figure 2.14, in a healthy human illustrates the processing of the movement instructions in the thalamus and the generation of control commands back to the motor cortex. The motor cortex in turn provides the neuromotor actuation signals.

The basal ganglia are a collection of cells deep in the cerebral cortex. They include the caudate and the putamen (CP), nucleus accumbens, globus pallidus, substantia nigra, and the subthalamic nucleus (STN). These structures work in small groups to perform some specialized functions. Thus the fine- or complex-movement instructions are accumulated and sequenced in the CP. The output of the CP is then amplified by the substantia nigra pars compacta (SNpc). The motor cortex generates

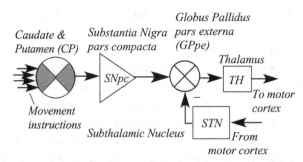

Figure 2.14. Block diagram illustrating the generation of control laws in the motor cortex from fine-movement instructions.

a regulatory feedback signal to the STN. The regulatory feedback signal to the STN triggers an inhibitory control signal that is combined with the excitatory signal from the SNpc by the globus pallidus pars externa (GPpe). The output from the GPpe is transmitted to the thalamus, which then stimulates the motor cortex.

As stated earlier, movement disorders are caused by the lack of balance between the signals from the cerebellum and the basal ganglia. Specifically, the synaptic neurotransmissions from the cerebral hemisphere to the thalamus can have a significantly high noise content, and this manifests itself as a tremor in the muscle. Parkinson's disease, Huntington's disease, and hemiballismus are all disorders that are associated with defects in the neurotransmitters.

This insight into the nature of the tremor suggests that the feedforward control of the tremor by making measurements of early in the signal path and providing an out-of-phase canceling signal to the manipulator close to where it is controlled by the human hand would not only compensate for the tremor in the manipulator but also provide a haptic feedback signal that would eventually compensate for the lack of balance alluded to earlier. The control algorithms may be partially based on known neural properties of this behavior of the basal ganglia but without detailed mimicry of the neurophysiology of the physiological control systems. Thus our understanding of the nature of a tremor can provide for a novel biomimetic approach to its suppression.

2.5.2 Design of Scalable Robotic Surgical Devices

Cable-driven manipulators cannot be really miniaturized, and this is a major disadvantage. However, the use of smart materials for the sensing and actuation of miniature robotic manipulators facilitates the development of a scalable solution to provide motion and force control. SMA wires provide a feasible alternative to cables in the design of manipulators. SMAs are so called because they "remember" their shape while undergoing a solid-to-solid phase transformation during heating. At room temperature, certain nickel–titanium alloys such as Nitinol, are in a low-temperature martensite state. When they are heated beyond a transformation temperature, they undergo the phase transformation to an austenite state. In the martensite phase, the materials exhibit higher flexibility than in the austenite phase and are therefore easily deformed.

When the materials are heated they enter the austenite phase (known as reverse transformation) and regain their original shape as long as permanent plastic deformation does not occur in the martensite phase (around 3%–8% strain for most SMAs). On cooling to martensite again (martensite transformation), the SMA does not change its macroscopic shape unless external loads are applied. In the two-way shape memory effect, both phases remember a unique shape.

An interesting application of the shape memory effect is the atrial septal occlusion device used to seal holes located between the two upper heart chambers upon the septum, which is the surface that splits the upper part of the heart into the right and left atria. The device consists of an assemblage of SMA wires embedded in a waterproof polyurethane film. The shape memory effect is exploited so the device recovers its original shape and seals the holes in the septum. It is a two-part device that is deployed on either side of the septum via a catheter. Another device that is built from an assemblage of SMA wires is the Simon filter. A pulmonary embolism is a sudden blockage in a lung artery. The cause is usually a blood clot in the leg, called a deep vein thrombosis (DVT), that breaks loose and travels upstream to the lung. The Simon filter is used to act as a filter to blood clots, prevent their upstream movement, and thus prevent pulmonary embolism.

Moreover, with the availability of microelectromechanical sensors and actuators such as single-chip accelerometers and motors, the design and construction of miniature surgical instruments is a feasible proposition. A number of miniaturized tools using mechanisms such as grippers, tongs, scissors, and others are currently being developed based on the active control of SMA wires to assist in surgery. A major difficulty is in the precise control of these smart wires that is due to the inherent hysteresis during the phase transformation and the fact that the controllers must be able to precisely predict the inverse dynamics in these situations. However, recent advances in the control of systems that exhibit hysteresis facilitates the design of reasonably accurate inverse model controllers.

The development of systems for endoscopic cardiovascular surgery evolved with the development of instrumented catheters in the 1960s. Catheters are capable of introducing diagnostic sensors at the tip of a long wire or tube able to reach the most remote regions of the cardiovascular system as well as being capable of being introduced via a patient's veins. It is then possible to perform a number of measurements *in vivo* by using miniaturized pressure and flow sensors and ultrasound probes. The measured data are transmitted back via optical fibers to a measuring system that is in a remote location. The optical fiber is embedded within the catheter and is capable of transmitting the measured signal over a long and flexible catheter.

Catheters are now also used in conjunction with deployable stents to repair aneurysms, which are localized, blood-filled bulgings of a blood vessel caused by the local weakening of the vessel wall. Self-expanding stents, named after the dentist C. P. Stent who invented them, are mainly used to maintain the size of a blood vessel. Catheters are also used with deployable superelastic balloons for angioplasty, which is the mechanical widening of a narrowed or obstructed blood vessel. In these interventional procedures, the medically qualified operator or surgeon activates the balloon or releases the stent while viewing the on-line image provided by x rays or fluoroscopy. Real-time MRI-guided coronary catheterization is a potential alternative to conventional x-ray imaging. It is made possible by a coaxially inserted active guide wire containing a loopless antenna mounted on a nitinol ring serving as an internal radio-frequency (RF) receiver coil for enhanced direct guide wire visualization. The advancement of the catheter is totally performed by the surgeon, who has to insert the device and visually guide the catheter to the correct destination before deploying the device.

Catheters may be modeled as multilink robotic manipulators, and currently there are a number of dynamic simulators that are commercially available. These can be used as dynamical training systems to train surgeons in the use of active catheter-based systems.

2.5.3 Robotic Needle Placement and Two-Hand Suturing

The processes of implementing robotic needle placement or two-hand suturing, which are typical examples of micromanipulation procedures, are based on basic suturing processes. The basic principles of suturing involve not just the choice of the needles, sutures, and suturing materials but certain generic as well as some specific procedures. Among the generic procedures are the techniques of needle holding, needle driving, and knot placement. Among the specific suturing techniques are simple or running sutures, vertical, half-buried vertical, and modified vertical mattress sutures, horizontal mattress sutures, buried sutures, pulley sutures, variations of running sutures, and suture-removal techniques.

The development of automatic suturing evolved from developments related to the sewing machine. The first sewing machines used a needle with an eye at the point. The needle was pushed through the fabric and created a loop on the other side; a shuttle on a track then slipped the second thread through the loop, creating what is called the lockstitch. With the inventions of the up-and-down-motion mechanism, and the rotary hook shuttle, automatic fabric sewing machines were built. These are basic in the implementation robotic suturing mechanisms.

For the suturing techniques to be performed by a robot surgeon, they are analyzed and broken up into basic primitives. Thus one can identify five basic primitives: select, involving the selection of a suitable entry and a suitable exit point; align, involving grasping, navigation to a certain location, and alignment of the needle tip in a certain direction; bite, when entry and exit "bites" are made as the needle passes from one tissue to the other; loop, to create a suturing loop; and knot, to tie a knot. Given the appropriate manipulator and/or gripper, a constrained controller is designed to implement each of these primitives. With the primitives implemented, robotic suturing may be achieved by the process of scheduling and sequencing the controlled primitives so as to execute the desired tasks.

EXERCISES

2.1. State and discuss six basic features of biosensors. Discuss the types of most common biological agents (e.g., chromorphores, fluorescence dyes) and the ways in which they can be interfaced with a variety of transducers to create a biosensor for biomedical applications.

2.2. Study the characteristics of a range of biosensors that are currently available and identify and classify the various immobilization techniques in each of these sensor types.

2.3. Research the principles of fiber-optic-based biosensing techniques and identify five optical biosensors and systems (e.g., fluorescence spectroscopy, microscopy) for biomedical applications.

2.4. Explain briefly the governing principles and applications of

- (a) molecular beacons,
- (b) fluorescent semiconductor quantum dots,
- (c) bioMEMS,
- (d) confocal and multiphoton microscopy,
- (e) quantum computing.

2.5. Study the muscle contraction process and draw an equivalent circuit model for it by combining the electrical and mechanical components and representing them with suitable electrical analogs.

2.6. Discuss the basic elements of a typical implementation of a bionic arm and identify the primary computational tasks involved in the implementation of such prostheses.

3

Homogeneous Transformations and Screw Motions

3.1 General Rigid Motions in Two Dimensions

Pure rotations of rigid bodies can be described by the application of rotational transformations to the reference frames. In the Appendix, the various transformations representing pure rotations of rigid bodies are considered in detail. However, for a two- or three-dimensional system of reference axes, general transformations must simultaneously involve both rotation and translation of the reference axes.

To begin, we consider simultaneous rotation and translation of two-dimensional reference axes. In two dimensions, considering simultaneous rotation and translation of the reference axes, we may express the components of a position vector as

$$\begin{bmatrix} R'_{1x} \\ R'_{1y} \end{bmatrix} = \begin{bmatrix} R_{1x} \\ R_{1y} \end{bmatrix} + \begin{bmatrix} x_d \\ y_d \end{bmatrix}, \tag{3.1}$$

where

$$\begin{bmatrix} R_{1x} \\ R_{1y} \end{bmatrix} = \begin{bmatrix} \cos\varphi & \sin\varphi \\ -\sin\varphi & \cos\varphi \end{bmatrix} \begin{bmatrix} R_{0x} \\ R_{0y} \end{bmatrix} \tag{3.2}$$

are the components following a simple rotation, and the vector

$$\mathbf{d} = [x_d \quad y_d]^T \tag{3.3}$$

represents the translational displacement of the origin of the reference axes. In general, the preceding transformation represents a simultaneous rotation and translation that may be considered to be an *affine* transformation due to a general rigid motion of the reference axes.

Introducing a similar change in notation as was defined in the Appendix, we have

$$\begin{bmatrix} x'_1 \\ y'_1 \end{bmatrix} = \mathbf{R} \begin{bmatrix} x_0 \\ y_0 \end{bmatrix} + \mathbf{d}, \tag{3.4}$$

62

where

$$\mathbf{R} = \begin{bmatrix} \cos\varphi & \sin\varphi \\ -\sin\varphi & \cos\varphi \end{bmatrix} \quad \text{and} \quad \mathbf{d} = \begin{bmatrix} x_d \\ y_d \end{bmatrix}.$$

Thus the transformation, $\mathbf{T}$, is uniquely defined by the matrix–vector pair, $(\mathbf{R}, \mathbf{d})$. It is the direct sum of a pure rotation and a displacement. This is denoted as

$$(\mathbf{R}, \mathbf{d}) = (\mathbf{R}, \mathbf{0}) \oplus (\mathbf{I}, \mathbf{d}). \tag{3.5}$$

The identity transformation, the one that leaves the original vector unchanged, corresponds to the pair $(\mathbf{I}, \mathbf{0})$.

The inverse relationship is

$$\begin{bmatrix} x_0 \\ y_0 \end{bmatrix} = \mathbf{R}^T \begin{bmatrix} x_1' \\ y_1' \end{bmatrix} - \mathbf{R}^T \mathbf{d}, \tag{3.6}$$

which may be represented by the matrix–vector pair $(\mathbf{R}^T, -\mathbf{R}^T\mathbf{d})$.

This fact is stated as

$$(\mathbf{R}^T, -\mathbf{R}^T\mathbf{d}) \otimes (\mathbf{R}, \mathbf{d}) = (\mathbf{R}, \mathbf{d}) \otimes (\mathbf{R}^T, -\mathbf{R}^T\mathbf{d}) = (\mathbf{I}, \mathbf{0}). \tag{3.7}$$

However, it is quite obvious that it is not possible to add or compose the transformations by the application of the rules of matrix–vector algebras.

To overcome this shortcoming, we define a matrix representation of the matrix–vector pair $(\mathbf{R}, \mathbf{d})$. Together they may be used to determine the position and orientation of a reference frame. Consider the matrix transformation

$$\mathbf{T} = \begin{bmatrix} \mathbf{R} & \mathbf{d} \\ \mathbf{0} & 1 \end{bmatrix}, \tag{3.8}$$

and let the position vector be initially represented by $[x_0 \quad y_0 \quad 1]^T$.

Applying the transformation results in

$$\begin{bmatrix} \mathbf{R} & \mathbf{d} \\ \mathbf{0} & 1 \end{bmatrix} \begin{bmatrix} x_0 \\ y_0 \\ 1 \end{bmatrix} = \begin{bmatrix} \mathbf{R} \begin{bmatrix} x_0 \\ y_0 \end{bmatrix} + \mathbf{d} \\ 1 \end{bmatrix} = \begin{bmatrix} x_1 \\ y_1 \\ 1 \end{bmatrix}; \tag{3.9}$$

i.e.,

$$\begin{bmatrix} x_1 \\ y_1 \\ 1 \end{bmatrix} = \begin{bmatrix} \mathbf{R} & \mathbf{d} \\ \mathbf{0} & 1 \end{bmatrix} \begin{bmatrix} x_0 \\ y_0 \\ 1 \end{bmatrix} = \mathbf{T} \begin{bmatrix} x_0 \\ y_0 \\ 1 \end{bmatrix}. \tag{3.10}$$

The transformation of the components is achieved by a simple matrix transformation. Interestingly, considering the inverse relation, we have

$$\begin{bmatrix} \mathbf{R} & \mathbf{d} \\ \mathbf{0} & 1 \end{bmatrix} \begin{bmatrix} \mathbf{R}^T & -\mathbf{R}^T\mathbf{d} \\ \mathbf{0} & 1 \end{bmatrix} = \begin{bmatrix} \mathbf{R}^T & -\mathbf{R}^T\mathbf{d} \\ \mathbf{0} & 1 \end{bmatrix} \begin{bmatrix} \mathbf{R} & \mathbf{d} \\ \mathbf{0} & 1 \end{bmatrix} = \begin{bmatrix} \mathbf{I} & \mathbf{0} \\ \mathbf{0} & 1 \end{bmatrix}; \tag{3.11}$$

i.e., the inverse transformation is the inverse of the matrix transformation.

3.1.1 Instantaneous Centers of Rotation

Before any of the features of the matrix transformation introduced in the previous section are enunciated, it is instructive to consider a rotation of the reference axes about another axis that does pass through the origin. To consider such a rotation, we may translate the reference axis to a point on the axis of rotation, rotate the axes, and translate back by the same translation vector. The net result of this operation is

$$\mathbf{T} = \begin{bmatrix} \mathbf{I} & \mathbf{p} \\ \mathbf{0} & 1 \end{bmatrix} \begin{bmatrix} \mathbf{R} & \mathbf{0} \\ \mathbf{0} & 1 \end{bmatrix} \begin{bmatrix} \mathbf{I} & -\mathbf{p} \\ \mathbf{0} & 1 \end{bmatrix} = \begin{bmatrix} \mathbf{R} & \mathbf{p} - \mathbf{Rp} \\ \mathbf{0} & 1 \end{bmatrix}. \tag{3.12}$$

If this composite transformation is equivalent to a general simultaneous rotation and translation, it follows that

$$\begin{bmatrix} \mathbf{R} & \mathbf{d} \\ \mathbf{0} & 1 \end{bmatrix} = \begin{bmatrix} \mathbf{R} & \mathbf{p} - \mathbf{Rp} \\ \mathbf{0} & 1 \end{bmatrix}; \tag{3.13}$$

i.e.,

$$\mathbf{d} = \mathbf{p} - \mathbf{Rp} = [\mathbf{I} - \mathbf{R}]\mathbf{p}. \tag{3.14}$$

Hence it follows that

$$\mathbf{p} = [\mathbf{I} - \mathbf{R}]^{-1}\mathbf{d}. \tag{3.15}$$

When the vector $\mathbf{p}$ satisfies the preceding equation and the general motion transformation is applied to a point defined by it, the observed result is

$$\begin{bmatrix} \mathbf{R} & \mathbf{d} \\ \mathbf{0} & 1 \end{bmatrix} \begin{bmatrix} \mathbf{p} \\ 1 \end{bmatrix} = \begin{bmatrix} \mathbf{p} \\ 1 \end{bmatrix}, \tag{3.16}$$

and hence the point defined by the vector $\mathbf{p}$ is invariant. Thus the point $\mathbf{p}$ defines the *center of rotation*. In the case of a pure rotational transformation the center is at the origin whereas in the case of pure translation it is at an infinite distance from it.

Some important applications of the center of rotation of a rigid body follow from the *Arnold–Kennedy three-centers-in-line theorem*. When two rigid bodies, b_1 and b_2, are moving in the same plane relative to a third rigid body, b_3, then the instantaneous centers of b_2 relative to b_3, of b_1 relative to b_3, and of b_2 relative to b_1 all lie in a straight line. The theorem is particularly useful in finding all the instantaneous centers of rotation of rigid links in a planar mechanism relative to each other. With the positions of these centers known, it is also possible to determine all the velocities of the links.

3.2 Rigid Body Motions in Three Dimensions: Definition of *Pose*

The position and orientation of a body define the *pose* of the body, i.e., the pose of a body is completely determined by its position and orientation. In general this involves six independent parameters or DOFs. However, three of these pertain to the orientation of the body and any rotational transformation in three dimensions may be expressed in terms of these orientation parameters. Thus, in three dimensions, as in the case of two dimensions, the position coordinates of a general vector may be expressed as

$$\begin{bmatrix} x_1' \\ y_1' \\ z_1' \end{bmatrix} = \mathbf{R} \begin{bmatrix} x_0 \\ y_0 \\ z_0 \end{bmatrix} + \mathbf{d}, \tag{3.17}$$

where $\mathbf{R}$ is a 3×3 rotation matrix of the type discussed in Chapter 2 and $\mathbf{d}$ is the displacement vector by which the origin of the reference axes is displaced. As in the two-dimensional case, the transformation may be represented by the single matrix

$$\mathbf{T} = \begin{bmatrix} \mathbf{R} & \mathbf{d} \\ \mathbf{0} & 1 \end{bmatrix}. \tag{3.18}$$

When the transformation is applied to a position vector,

$$\begin{bmatrix} \mathbf{R} & \mathbf{d} \\ \mathbf{0} & 1 \end{bmatrix} \begin{bmatrix} x_0 \\ y_0 \\ z_0 \\ 1 \end{bmatrix} = \begin{bmatrix} \mathbf{R} \begin{bmatrix} x_0 \\ y_0 \\ z_0 \end{bmatrix} + \mathbf{d} \\ \mathbf{1} \end{bmatrix} = \begin{bmatrix} x_1 \\ y_1 \\ z_1 \\ 1 \end{bmatrix}; \tag{3.19}$$

i.e.,

$$\begin{bmatrix} x_1 \\ y_1 \\ z_1 \\ 1 \end{bmatrix} = \begin{bmatrix} \mathbf{R} & \mathbf{d} \\ \mathbf{0} & 1 \end{bmatrix} \begin{bmatrix} x_0 \\ y_0 \\ z_0 \\ 1 \end{bmatrix}. \tag{3.20}$$

As in the two-dimensional case, the inverse transformation is given by

$$\mathbf{T}^{-1} = \begin{bmatrix} \mathbf{R}^T & -\mathbf{R}^T\mathbf{d} \\ \mathbf{0} & 1 \end{bmatrix} \tag{3.21}$$

and satisfies

$$\begin{bmatrix} \mathbf{R} & \mathbf{d} \\ \mathbf{0} & 1 \end{bmatrix} \begin{bmatrix} \mathbf{R}^T & -\mathbf{R}^T\mathbf{d} \\ \mathbf{0} & 1 \end{bmatrix} = \begin{bmatrix} \mathbf{R}^T & -\mathbf{R}^T\mathbf{d} \\ \mathbf{0} & 1 \end{bmatrix} \begin{bmatrix} \mathbf{R} & \mathbf{d} \\ \mathbf{0} & 1 \end{bmatrix} = \begin{bmatrix} \mathbf{I} & \mathbf{0} \\ \mathbf{0} & 1 \end{bmatrix}. \tag{3.22}$$

3.2.1 Homogeneous Coordinates: Transformations of Position and Orientation

The four-component vector representation of the position coordinates is known as the *homogeneous coordinate vector* and the vector components are known as the homogeneous coordinates. The matrix $\mathbf{T}$ is known as the *homogeneous matrix transformation*. When the transformation operation is applied successively to a homogeneous coordinate vector $\mathbf{h}$ with two transformation matrices, first by $\mathbf{T}_1$ and then by $\mathbf{T}_2$, the composite effect of the two transformations may be realized by the application of a single transformation,

$$\mathbf{h}' = \mathbf{T}_2 \times \mathbf{T}_1 \times \mathbf{h} = \mathbf{T} \times \mathbf{h}, \ \ \mathbf{T} = \mathbf{T}_1 \times \mathbf{T}_2. \tag{3.23}$$

However, the composite transformation $\mathbf{T}$ is not equal to the matrix product:

$$\mathbf{T} \neq \mathbf{T}_1 \times \mathbf{T}_2 \neq \mathbf{T}_2 \times \mathbf{T}_1. \tag{3.24}$$

The direct sum of two homogeneous transformations is not a homogeneous transformation. The composition of two homogeneous transformations is

$$\mathbf{T} = \begin{bmatrix} \mathbf{R}_2 & \mathbf{d}_2 \\ \mathbf{0} & 1 \end{bmatrix} \times \begin{bmatrix} \mathbf{R}_1 & \mathbf{d}_1 \\ \mathbf{0} & 1 \end{bmatrix} = \begin{bmatrix} \mathbf{R}_1\mathbf{R}_2 & \mathbf{R}_2\mathbf{d}_1 + \mathbf{d}_1 \\ \mathbf{0} & 1 \end{bmatrix}. \tag{3.25}$$

3.3 General Motions of Rigid Frames in Three Dimensions: Frames with *Pose*

When a homogeneous transformation is applied to a reference frame, the pose of the frame changes from one to the other. Because some of the reference axes before and after transformation may be located arbitrarily, the pose of a frame may be parameterized with a minimal set of parameters. By choosing an initial set of reference axes and aligning the final set of reference axes relative to the initial set in a certain way, Denavit and Hartenberg were able parameterize a homogeneous transformation in terms of just four parameters. This parameterization is discussed in the next subsection.

3.3.1 The Denavit–Hartenberg Decomposition

The *Denavit–Hartenberg* (DH) decomposition has become the standard way of representing reference frames and modeling their motions. The method begins with a systematic approach to assigning and labeling an orthonormal (X, Y, Z) right-handed coordinate system to each point representing the origin of a reference frame. It is then possible to relate one frame to the next and ultimately to assemble a complete representation of the motion of the final frame relative to the first. The frame assignment method is very application dependent and will be discussed in a later chapter. The homogeneous transformation is composed of four successive transformations that will make these frames coincident with each other. The transformations, in reverse order, are

1. a rotation θ_{n+1} about the Z_n axis to bring X_n parallel with X_{n+1},
2. a translation d_{n+1} along the Z_n axis to make the x axes collinear,
3. a translation a_{n+1} along the X_n axis to make the z axes coincide, and,
4. a rotation α_{n+1} about the X_n axis to bring Z_n parallel with Z_{n+1}.

The rotations and displacements are illustrated in Figure 3.1. The four parameters a_i, α_i, d_i, and θ_i in the transformation are generally given the names *link length*, *link twist*, *link offset*, and *joint angle*, respectively. The transformations may be represented as

$$\mathbf{T}_{\mathrm{DH}} = \mathbf{T}_{n,n+1} = \mathrm{rot}\,(Z_n, \theta_{n+1})\,\mathrm{trans}\,(Z_n, d_{n+1})\,\mathrm{trans}\,(X_n, a_{n+1})\,\mathrm{rot}\,(X_n, \alpha_{n+1})\,.$$

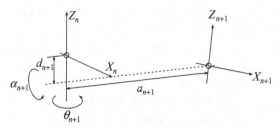

Figure 3.1. The DH decomposition.

Thus it follows that

$$
\mathbf{T}_{\text{DH}} = \mathbf{T}_{n,n+1} = \begin{bmatrix} \cos\theta_{n+1} & -\sin\theta_{n+1} & 0 & 0 \\ \sin\theta_{n+1} & \cos\theta_{n+1} & 0 & 0 \\ 0 & 0 & 1 & 0 \\ 0 & 0 & 0 & 1 \end{bmatrix} \begin{bmatrix} 1 & 0 & 0 & 0 \\ 0 & 1 & 0 & 0 \\ 0 & 0 & 1 & d_{n+1} \\ 0 & 0 & 0 & 1 \end{bmatrix}
$$

$$
\times \begin{bmatrix} 1 & 0 & 0 & a_{n+1} \\ 0 & 1 & 0 & 0 \\ 0 & 0 & 1 & 0 \\ 0 & 0 & 0 & 1 \end{bmatrix} \begin{bmatrix} 1 & 0 & 0 & 0 \\ 0 & \cos\alpha_{n+1} & -\sin\alpha_{n+1} & 0 \\ 0 & \sin\alpha_{n+1} & \cos\alpha_{n+1} & 0 \\ 0 & 0 & 0 & 1 \end{bmatrix}. \tag{3.26}
$$

When the transformations are applied, the last of the matrices is multiplied with the initial homogeneous position vector, followed by the third, the second, and finally the first. When several links are present, the composite homogeneous transformation is a product defined by

$$
\mathbf{T}_{\text{DH}} = \prod \mathbf{T}_{n,n+1}. \tag{3.27}
$$

3.3.2 Instantaneous Axis of Screw Motion

There is yet another physically meaningful interpretation of the four-parameter representation of a homogeneous transformation. We have already observed in the previous chapter that any rotational transformation may be expressed as a single rotation, the Euler angle φ about some axis, the Euler axis $\mathbf{e}_\varphi$. This representation accounts for three parameters, and any rotation matrix may be expressed as

$$
\mathbf{R} = \mathbf{R}(\varphi, \mathbf{e}_\varphi). \tag{3.28}
$$

Consider a general homogeneous transformation with the rotation matrix expressed in terms of the Euler angle and the Euler axis:

$$
\mathbf{T} = \begin{bmatrix} \mathbf{R}(\varphi, \mathbf{e}_\varphi) & \mathbf{d} \\ \mathbf{0} & 1 \end{bmatrix}, \tag{3.29}
$$

and let the vector $\mathbf{d}$ be selected such that it is proportional to $\mathbf{e}_\varphi$. Specifically, we let

$$
\mathbf{d} = \frac{\varphi}{2\pi} \mathbf{e}_\varphi p; \tag{3.30}
$$

the homogeneous transformation represents a rotation and uniform displacement along the same axis passing through the origin of the reference frame. Thus we have *Chasles' Theorem*, which states that the most general displacement of a rigid body is equivalent to a translation of an arbitrary reference point on the body and a simultaneous rotation about an axis through that reference point. Together the two motions constitute a single screw motion. In fact, the homogeneous transformation represents a transformation of a point to such a screw-type motion, in which the parameter p is the pitch of the screw, the distance traveled along the axis for every complete rotation about the axis. When the pitch is zero, the screw motion is a pure rotation.

To show that every general rigid motion corresponds to an instantaneous screw motion, the general homogeneous transformation must be shown to be equivalent to that corresponding to screw motion. To do this, consider a screw motion of the reference axes about another axis that does pass through the origin. To consider such a motion, we may translate the reference axis to a point on the axis of rotation, perform a screw motion on the axes, and translate back by the same translation vector. The net result of this operation is

$$\mathbf{T} = \begin{bmatrix} \mathbf{I} & \mathbf{p} \\ \mathbf{0} & 1 \end{bmatrix} \begin{bmatrix} \mathbf{R} & \mathbf{d}_s \\ \mathbf{0} & 1 \end{bmatrix} \begin{bmatrix} \mathbf{I} & -\mathbf{p} \\ \mathbf{0} & 1 \end{bmatrix} = \begin{bmatrix} \mathbf{R} & \mathbf{p} - \mathbf{R}\mathbf{p} + \mathbf{d}_s \\ \mathbf{0} & 1 \end{bmatrix}, \tag{3.31}$$

where

$$\mathbf{d}_s = \frac{\varphi}{2\pi}\mathbf{e}_\varphi p. \tag{3.32}$$

If this composite transformation is equivalent to a general simultaneous rotation and translation, it follows that

$$\begin{bmatrix} \mathbf{R} & \mathbf{d} \\ \mathbf{0} & 1 \end{bmatrix} = \begin{bmatrix} \mathbf{R} & \mathbf{p} - \mathbf{R}\mathbf{p} + \frac{\varphi}{2\pi}\mathbf{e}_\varphi p \\ \mathbf{0} & 1 \end{bmatrix}; \tag{3.33a}$$

i.e.,

$$\mathbf{d} = \mathbf{p} - \mathbf{R}\mathbf{p} + \frac{\varphi}{2\pi}\mathbf{e}_\varphi p = [\mathbf{I} - \mathbf{R}]\mathbf{p} + \frac{\varphi}{2\pi}\mathbf{e}_\varphi p. \tag{3.33b}$$

If we now impose the requirement that the vector $\mathbf{p}$ be normal to $\mathbf{e}_\varphi$ and obtain the dot product of the preceding equation with $\mathbf{e}_\varphi$, we have

$$\mathbf{d} \cdot \mathbf{e}_\varphi = [\mathbf{I} - \mathbf{R}]\mathbf{p} \cdot \mathbf{e}_\varphi + \frac{\varphi}{2\pi}\mathbf{e}_\varphi \cdot \mathbf{e}_\varphi p = \frac{\varphi}{2\pi}p. \tag{3.34}$$

Hence,

$$p = \frac{2\pi\,(\mathbf{d} \cdot \mathbf{e}_\varphi)}{\varphi} \tag{3.35}$$

and

$$\mathbf{d} = [\mathbf{I} - \mathbf{R}]\mathbf{p} + (\mathbf{d} \cdot \mathbf{e}_\varphi)\,\mathbf{e}_\varphi. \tag{3.36}$$

This is a set of homogeneous equations for the vector **p** that is given as the solution of

$$[\mathbf{I} - \mathbf{R}]\,\mathbf{p} = \mathbf{d} - (\mathbf{d} \cdot \mathbf{e}_\varphi)\,\mathbf{e}_\varphi. \tag{3.37}$$

Thus it is always possible to find the vector **p** and the pitch of the screw motion that will render a homogeneous transformation equivalent to a screw motion about some axis in space. Any general rigid body motion may therefore be represented as a spatial screw displacement. The vector **p**, which is assumed to be normal to $\mathbf{e}_\varphi$, defines the axis of the screw motion.

In a time-dependent situation it can be seen that this axis will continuously change with time. However, at every instant there is an equivalent axis of rotation that renders a general homogeneous transformation equivalent to a screw motion about that axis. It follows from the preceding analysis that, when the pitch of the screw is fixed and the axis of the screw motion is known *a priori* or is fixed, just four parameters are adequate to parameterize the transformation. This is the basis for the DH decomposition. It is also apparent from the preceding analysis that the general three-dimensional representation of the axis of the screw motion is critical to the parameterization of the general transformation of a set of reference axes.

Several alternative representations, notably the *dual quaternion* and *Plücker coordinates* and their associated algebras, have been put forward with a view to obtaining an ideal representation of a general rigid body spatial screw displacement. A discussion of these alternative representations is well beyond the scope of this chapter, and the reader is referred to the more advanced texts referenced in the bibliography.

3.3.3 A Screw from a Twist

Given a directional unit vector of a rotational axis, $\mathbf{w} = [w_1 \quad w_2 \quad w_3]^T$, with an associated rotation of magnitude φ and a general positional vector $\mathbf{u} = [u_1 \quad u_2 \quad u_3]^T$, the *twist coordinates* are defined by the vector $[\mathbf{u}^T \quad \mathbf{w}^T]^T$. A *twist* is defined as a rotation about an axis and a translation along the same axis by an amount equal to the product of the pitch and the angle of rotation. The twist may also be expressed as a 4×4 matrix:

$$\tilde{\mathbf{s}} = \begin{bmatrix} \hat{\mathbf{w}} & \mathbf{u} \\ \mathbf{0} & 0 \end{bmatrix}, \tag{3.38}$$

where

$$\hat{\mathbf{w}} = \begin{bmatrix} 0 & -w_z & w_y \\ w_z & 0 & -w_x \\ -w_y & w_x & 0 \end{bmatrix}.$$

An interesting feature of the twist that we shall state without proof is that

$$\exp(\tilde{\mathbf{s}}\varphi) = \begin{bmatrix} \mathbf{E} & \mathbf{A} \\ \mathbf{0} & 1 \end{bmatrix}, \tag{3.39}$$

where

$$\mathbf{E} = \exp(\hat{\mathbf{w}}\varphi) = \mathbf{I} + \hat{\mathbf{w}}\sin(\varphi) + \hat{\mathbf{w}}^2[1 - \cos(\varphi)],$$

$$\mathbf{A} = \varphi\mathbf{I} + \hat{\mathbf{w}}[1 - \cos(\varphi)] + \hat{\mathbf{w}}^2[\varphi - \sin(\varphi)].$$

Thus $\exp(\tilde{\mathbf{s}}\varphi)$ may be interpreted as a screw motion. This is an important result, and as a consequence it follows that every homogeneous transformation may be expressed as the exponential of the twist. Because a general homogeneous transformation is decomposed into a product of component homogeneous transformations, a general homogeneous transformation may be expressed as a *product of exponentials*. This leads to an alternative and useful representation of the time-dependent homogeneous transformations.

EXERCISES

3.1. A homogeneous transform is given as

$$\mathbf{T} = \begin{bmatrix} \dfrac{\sqrt{3}}{2} & -\dfrac{1}{2} & 0 & 11 \\ \dfrac{1}{2} & \dfrac{\sqrt{3}}{2} & 0 & -1 \\ 0 & 0 & 1 & 9 \\ 0 & 0 & 0 & 1 \end{bmatrix}.$$

A vector given by $[1 \quad 3 \quad 2]^T$ is transformed by this transformation. Find the transformed vector.

3.2. Find the result of a homogeneous transformation acting on a vector $[a \quad b \quad c]^T$ and defined by

    (a) a rotation of the frame by 90° about the z axis, followed by

    (b) a rotation of the frame by 90° about the x axis, followed by

    (c) a translation of the frame by a displacement vector given by $[d \quad e \quad f]^T$. Hence show that when $[d \quad e \quad f]^T = [b \quad c \quad -a]$ the result is the null vector.

3.3. Find the origin and coordinate directions of a frame resulting from a rotation of 90° about the z axis, followed by a displacement of $[1 \quad 7 \quad 3]^T$. Hence find the position, in the original frame, of the vector $[3 \quad 8 \quad 1]^T$, defined in the resulting frame.

3.4. Obtain a homogeneous transformation acting on a vector and defined by

    (a) a rotation of the frame by 90° about the z axis, followed by

    (b) a rotation of the frame by −90° about the x axis, followed by

    (c) a translation of the frame by a displacement vector given by $[6 \quad 8 \quad 7]^T$.

Compare the result with another homogeneous transformation defined by

(a) a translation of the frame by a displacement vector given by $[6 \quad 8 \quad 7]^T$, followed by

(b) a rotation of the frame by $90°$ about the z axis, followed by

(c) a rotation of the frame by $-90°$ about the x axis.

3.5. Which of the following matrices represents a homogeneous transformation?

$$\begin{bmatrix} 0 & 0 & -1 & 0 \\ 1 & 0 & 0 & 1 \\ 0 & -1 & 0 & 3 \\ -1 & 0 & 0 & 1 \end{bmatrix}, \begin{bmatrix} 0 & 0 & -1 & 0 \\ 1 & 0 & 0 & 1 \\ 0 & -1 & 0 & 3 \\ 0 & 0 & 0 & 1 \end{bmatrix}, \begin{bmatrix} 0 & 0 & -1 & 0 \\ 2 & 0 & 0 & 1 \\ 0 & -1 & 0 & 3 \\ 0 & 0 & 0 & 1 \end{bmatrix}, \begin{bmatrix} 0 & 0 & -1 & 0 \\ 1 & 0 & 0 & 1 \\ 0 & -1 & 0 & 3 \\ 0 & 0 & 0 & 0 \end{bmatrix}.$$

3.6. Find the inverse of the homogeneous transformation

$$\begin{bmatrix} 0 & 1 & 0 & 2 \\ 0 & 0 & -1 & 1 \\ -1 & 0 & 0 & 3 \\ 0 & 0 & 0 & 1 \end{bmatrix}.$$

3.7. A part is located relative to the base frame with its location defined by the homogeneous transform,

$$\mathbf{T}_{\text{PB}} = \begin{bmatrix} 0 & 1 & 0 & 5 \\ 0 & 0 & -1 & 2 \\ -1 & 0 & 0 & 1 \\ 0 & 0 & 0 & 1 \end{bmatrix}.$$

A robot end-effector is parked at a point defined by the homogeneous transform

$$\mathbf{T}_{\text{EB}} = \begin{bmatrix} 1 & 0 & 0 & 1 \\ 0 & 1 & 0 & 11 \\ 0 & 0 & 1 & 7 \\ 0 & 0 & 0 & 1 \end{bmatrix}.$$

It is desired to move the end-effector and colocate and align it with the part. Determine the transformation that must be applied to the end-effector.

3.8. The parameters for a two-link serial manipulator are given in Table 3.1.

Assume that the end of link 1 is coincident with the start of link 2. Employ the DH decomposition and determine a homogeneous transformation relating the position at the end of link 2 to that at the start of link 1.

Table 3.1. *DH parameters for two-link serial manipulator*

Link No.	a_i	α_i	d_i	θ_i
1	L	0	0	Θ_1
2	L	0	0	Θ_2

Table 3.2. *DH parameters for a three-link*
spherical wrist

Link No.	a_i	α_i	d_i	θ_i
4	0	$-90°$	0	θ_4
5	0	$90°$	0	θ_5
6	0	0	d_6	θ_6

3.9. The parameters for a three-link spherical wrist are given in Table 3.2, where it is assumed that the links are serially connected. Employ the DH decomposition to determine a homogeneous transformation relating the position at the end wrist to that at the start of link 4.

3.10. The Euler angle is obtained from the matrix $\mathbf{T}$ that transforms a vector in the reference inertial coordinate system to representations in the body-fixed coordinate system by the following relation:

$$\cos(\varphi) = [\text{trace}(\mathbf{T}) - 1]/2,$$

and the Euler axis is defined by

$$e_1 = (T_{23} - T_{32})/(2s),\ e_2 = (T_{31} - T_{13})/(2s)\ \text{and}\ e_3 = (T_{12} - T_{21})/(2s),$$

where $s = \sin\varphi$.

A rigid body transformation transforms the three points $(0, 0, 0)$, $(1, 0, 0)$, and $(0, 1, 0)$ to the points $(1, -1, 1)$, $(1, 0, 1)$, and $(0, -1, 1)$, respectively. Determine the 4×4 homogeneous transformation corresponding to these initial and transformed points and the axis and pitch of the transformation.

3.11. Find the axis of the composite rotation of a rigid body, rotated by $90°$ about the x axis followed by a $90°$ rotation about the y axis.

State the most general three-dimensional homogeneous transformation representing a screw motion with the same rotation angle and axis.

3.12. Find the axis and pitch of the following homogeneous transformation:

$$\mathbf{T} = \frac{1}{12\sqrt{2}} \begin{bmatrix} 6\sqrt{2} + 3\sqrt{6} & 6\sqrt{2} - 3\sqrt{6} & 6 & -2 \\ 6\sqrt{2} - 3\sqrt{6} & 6\sqrt{2} + 3\sqrt{6} & -6 & 10 \\ -6 & 6 & 6\sqrt{6} & 12\sqrt{2} - 6\sqrt{6} \\ 0 & 0 & 0 & 12\sqrt{2} \end{bmatrix}.$$

3.13. Given that $\mathbf{P}^{-1}$ is a quadratic matrix polynomial in $\hat{\mathbf{w}}$, with,

$$\hat{\mathbf{w}} = \begin{bmatrix} 0 & -w_z & w_y \\ w_z & 0 & -w_x \\ -w_y & w_x & 0 \end{bmatrix},\ \mathbf{P}^{-1} = \mathbf{I} - \frac{1}{2}\hat{\mathbf{w}}\varphi + \hat{\mathbf{w}}^2 \left(1 - \frac{[1 + \cos(\varphi)]\varphi}{2\sin(\varphi)}\right)\varphi^2,$$

show that

$$\log \begin{bmatrix} \mathbf{R} & \mathbf{d} \\ \mathbf{0} & 1 \end{bmatrix} = \begin{bmatrix} \log \mathbf{R} & \mathbf{P}^{-1}\mathbf{d} \\ \mathbf{0} & 0 \end{bmatrix},$$

where

$$\log \mathbf{R} = \frac{\varphi}{2\sin\varphi} \left(\mathbf{R} - \mathbf{R}^T \right), \text{ trace } (\mathbf{R}) = 1 + 2\cos\varphi \text{ and } \left| \log \mathbf{R} \right|^2 = \varphi^2.$$

4

Direct Kinematics of Serial Robot Manipulators

4.1 Definition of Direct or Forward Kinematics

Kinematics is the study of the motion attributes of a collection of bodies without any reference to the forces or moments that may have caused it. In the context of robotics in general and robot manipulators in particular, it is the geometry of the mechanism that plays a pivotal role in the determination of the kinematics and the associated kinematic relations. Of primary importance are the relations expressing the position and orientation coordinates of the end-effector in terms of the DOFs at each joint. The objective is to establish parameters characterizing each link in a general coordinate system for satisfying the kinematic requirements of a manipulator. To find the location of the end-effector, only the DOFs at the joints, the joint variables, are given, and the problem is to express the position and orientation of the end-effector, the pose of the end-effector, in terms of the given variables. This is the direct or forward kinematics problem. For serial manipulators the direct kinematics is a unique relation, i.e., the location of the end-effector may be uniquely represented in terms of the joint variables.

Although the joint variables may be defined in an arbitrary fashion, a method of frame assignment proposed by Denavit and Hartenberg has emerged as a standard. The advantage of this systematic labeling of joint variables is that the associated homogeneous transformations may be decomposed into a product form that was discussed in the last chapter.

4.2 The Denavit–Hartenberg Convention

Robot manipulators are mechanisms involving a sequence of links connected together at the joints. To analyze the motion of each link in such manipulators, successive reference frames are attached, starting with a frame F_o, to the first link that is chosen to be a fixed link and followed by the frames F_i, $i = 1, 2, 3, \ldots, n$, attached to the consecutively connected links, all the way to the robot end-effector. The transformations relating consecutive frames may then be expressed as homogeneous transformations. To completely realize the advantages of the DH decomposition of

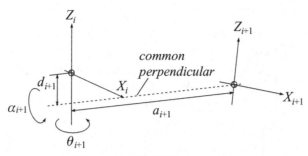

Figure 4.1. The DH convention.

each of these transformations discussed in the previous chapter, the assignment of the frames and the joint variables at each of the joints must be done in a systematic manner. When this is done each of the transformations may be represented in terms of the four DH parameters. However, there are certain conventions that must be followed in this process.

Given two consecutive reference frames, $F_i = \{X_i, Y_i, Z_i\}$ and $F_{i+1} = \{X_{i+1}, Y_{i+1}, Z_{i+1}\}$, with each rigidly attached to a link in a mechanism, frame F_{i+1} will be uniquely determined from frame F_i in terms of the parameters the four parameters $d_{i+1}, a_{i+1}, \alpha_{i+1}$, and θ_{i+1}, illustrated in Figure 4.1.

The four parameters a_i, α_i, d_i, and θ_i in the transformation are given the names *link length*, *link twist*, *link offset*, and *joint angle*, respectively.

There is one main rule that must be adhered to in the process of frame assignment:

> The Z axis of a frame is always aligned with a joint axis unless the link does not have one nullary link.

The four parameters may then be defined in a systematic manner as indicated in what follows:

1. The parameter α_{i+1} is the angle about X_{i+1} that axis Z_{i+1} makes with axis Z_i.
2. The common perpendicular to axes Z_i and Z_{i+1} is then established. The parameter a_{i+1} is the length of this common perpendicular.
3. The parameter d_{i+1} is the algebraic distance along axis Z_i to the point where the common perpendicular drawn from Z_{i+1}, at its origin, intersects the axis Z_i.
4. The parameter θ_{i+1} is the angle about Z_i that the common perpendicular makes with axis X_i.

When these rules are adhered to and the parameters are defined as previously stated, the composite homogeneous transformation is

$$\mathbf{T}_{\mathrm{DH}} = \mathbf{T}_{n,n+1} \cdots \mathbf{T}_{i,i+1} \mathbf{T}_{i-1,i} \mathbf{T}_{i-2,i-1} \cdots \mathbf{T}_{1,2} = \prod_{i=1}^{n} \mathbf{T}_{i,i+1}, \qquad (4.1)$$

where

$$\mathbf{T}_{i,i+1} = \begin{bmatrix} \cos\theta_{i+1} & -\sin\theta_{i+1} & 0 & 0 \\ \sin\theta_{i+1} & \cos\theta_{i+1} & 0 & 0 \\ 0 & 0 & 1 & 0 \\ 0 & 0 & 0 & 1 \end{bmatrix} \begin{bmatrix} 1 & 0 & 0 & 0 \\ 0 & 1 & 0 & 0 \\ 0 & 0 & 1 & d_{i+1} \\ 0 & 0 & 0 & 1 \end{bmatrix}$$

$$\times \begin{bmatrix} 1 & 0 & 0 & a_{i+1} \\ 0 & 1 & 0 & 0 \\ 0 & 0 & 1 & 0 \\ 0 & 0 & 0 & 1 \end{bmatrix} \begin{bmatrix} 1 & 0 & 0 & 0 \\ 0 & \cos\alpha_{i+1} & -\sin\alpha_{i+1} & 0 \\ 0 & \sin\alpha_{i+1} & \cos\alpha_{i+1} & 0 \\ 0 & 0 & 0 & 1 \end{bmatrix}. \qquad (4.2)$$

The structure of the composite homogeneous transformation is such that the matrix product may be computed by use of custom VLSI- (very large system integrated-) based special-purpose semiconductor array processors such as systolic array processors. This feature has led to the adoption of the DH convention as a standard.

The introduction of the DH parameters greatly simplifies the geometric design problem of n-DOF spatial manipulators ($n < 6$). Studying the geometric design problem of spatial manipulators, one may also discover important properties of spatial mechanisms in general and overconstrained spatial mechanisms in particular. DH parameters and 4×4 homogeneous transformation matrices may be used to obtain the relevant design equations. Additional design equations are obtained by use of the inverse kinematics problem of general six-DOF serial link manipulators. Then the joint variables are eliminated so that the remaining equations have only unknowns related to the geometric design problem.

The application of the DH convention to the direct kinematic analysis of certain generic classes of manipulators is discussed in the following sections.

4.3 Planar Anthropomorphic Manipulators

Planar anthropomorphic manipulators are a class of biomimetic manipulators based on the human arm and involve only revolute and spherical or ball-and-socket joints.

A typical configuration is the planar open-loop chain with only revolute joints. A two-link planar arm of this type is shown in Figure 4.2. The joint axes Z_i are normal to the plane of the paper. The reference frames, with the first fixed to the base and the other two fixed to the links, are chosen according to the DH convention.

The link parameters are summarized in Table 4.1.

The homogeneous transformations are

$$\mathbf{T}_{i,i+1} = \begin{bmatrix} \cos\theta_{i+1} & -\sin\theta_{i+1} & 0 & 0 \\ \sin\theta_{i+1} & \cos\theta_{i+1} & 0 & 0 \\ 0 & 0 & 1 & 0 \\ 0 & 0 & 0 & 1 \end{bmatrix} \begin{bmatrix} 1 & 0 & 0 & a_{i+1} \\ 0 & 1 & 0 & 0 \\ 0 & 0 & 1 & 0 \\ 0 & 0 & 0 & 1 \end{bmatrix}, \qquad (4.3)$$

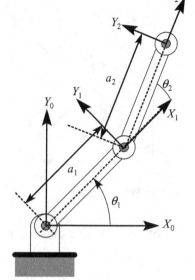

Figure 4.2. Two-link planar anthropomorphic manipulator; the Z axes are all aligned normal to the plane of the paper.

that is,

$$\mathbf{T}_{i,i+1} = \begin{bmatrix} \cos\theta_{i+1} & -\sin\theta_{i+1} & 0 & a_{i+1}\cos\theta_{i+1} \\ \sin\theta_{i+1} & \cos\theta_{i+1} & 0 & a_{i+1}\sin\theta_{i+1} \\ 0 & 0 & 1 & 0 \\ 0 & 0 & 0 & 1 \end{bmatrix}, \tag{4.4}$$

where $i = 0, 1$. Hence it follows that the end-effector coordinates may be expressed in terms of the base coordinates as

$$\begin{bmatrix} x_0 \\ y_0 \\ z_0 \\ 1 \end{bmatrix} = \mathbf{T}_{0,1}\mathbf{T}_{1,2} \begin{bmatrix} x_2 \\ y_2 \\ z_2 \\ 1 \end{bmatrix} = \begin{bmatrix} c_{12} & -s_{12} & 0 & a_2c_{12} + a_1\cos\theta_1 \\ s_{12} & c_{12} & 0 & a_2s_{12} + a_1\sin\theta_1 \\ 0 & 0 & 1 & 0 \\ 0 & 0 & 0 & 1 \end{bmatrix} \begin{bmatrix} x_2 \\ y_2 \\ z_2 \\ 1 \end{bmatrix}, \tag{4.5}$$

where $c_{12} = \cos(\theta_1 + \theta_2)$ and $s_{12} = \sin(\theta_1 + \theta_2)$.

The rotational part of the homogeneous transformation that may be used to determine the orientation of the end-effector is

$$\mathbf{R} = \begin{bmatrix} c_{12} & -s_{12} & 0 \\ s_{12} & c_{12} & 0 \\ 0 & 0 & 1 \end{bmatrix}. \tag{4.6}$$

Table 4.1. *Link parameters for the two-link planar manipulators*

Link no.	a_i	α_i	d_i	θ_i
1	a_1	0	0	θ_1
2	a_2	0	0	θ_2

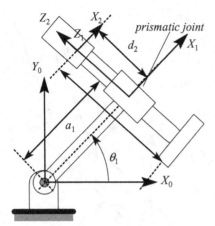

Figure 4.3. Two-link planar nonanthropomorphic manipulator; the Z axes are all aligned normal to the plane of the paper.

The origin of the end-effector in the fixed base frame is

$$
\begin{bmatrix} x_{02} \\ y_{02} \\ z_{02} \\ 1 \end{bmatrix} = \begin{bmatrix} c_{12} & -s_{12} & 0 & a_2 c_{12} + a_1 \cos\theta_1 \\ s_{12} & c_{12} & 0 & a_2 s_{12} + a_1 \sin\theta_1 \\ 0 & 0 & 1 & 0 \\ 0 & 0 & 0 & 1 \end{bmatrix} \begin{bmatrix} 0 \\ 0 \\ 0 \\ 1 \end{bmatrix} = \begin{bmatrix} a_2 c_{12} + a_1 \cos\theta_1 \\ a_2 s_{12} + a_1 \sin\theta_1 \\ 0 \\ 1 \end{bmatrix}.
$$

$$(4.7)$$

4.4 Planar Nonanthropomorphic Manipulators

A typical planar arm, a two-link kinematic chain that is representative of a nonanthropomorphic manipulator, is illustrated in Figure 4.3. The prismatic joint is such that the second link moves in a direction normal to the axis of the first link.

The reference frames are chosen according to the DH convention as in the anthropomorphic example considered in the preceding section.

The link parameters are summarized in Table 4.2.

The first homogeneous transformation is

$$
\mathbf{T}_{0,1} = \begin{bmatrix} \cos\theta_1 & -\sin\theta_1 & 0 & 0 \\ \sin\theta_1 & \cos\theta_1 & 0 & 0 \\ 0 & 0 & 1 & 0 \\ 0 & 0 & 0 & 1 \end{bmatrix} \begin{bmatrix} 1 & 0 & 0 & a_1 \\ 0 & 1 & 0 & 0 \\ 0 & 0 & 1 & 0 \\ 0 & 0 & 0 & 1 \end{bmatrix} \begin{bmatrix} 1 & 0 & 0 & 0 \\ 0 & \cos\alpha_1 & -\sin\alpha_1 & 0 \\ 0 & \sin\alpha_1 & \cos\alpha_1 & 0 \\ 0 & 0 & 0 & 1 \end{bmatrix}.
$$

$$(4.8)$$

Table 4.2. *Link parameters for the two-link planar manipulators*

Link no.	a_i	α_i	d_i	θ_i
1	a_1	$-90°$	0	θ_1
2	0	0	d_2	0

The transformation simplifies to

$$\mathbf{T}_{0,1} = \begin{bmatrix} \cos\theta_1 & -\sin\theta_1 & 0 & a_1\cos\theta_1 \\ \sin\theta_1 & \cos\theta_1 & 0 & a_1\sin\theta_1 \\ 0 & 0 & 1 & 0 \\ 0 & 0 & 0 & 1 \end{bmatrix} \begin{bmatrix} 1 & 0 & 0 & 0 \\ 0 & 0 & 1 & 0 \\ 0 & -1 & 0 & 0 \\ 0 & 0 & 0 & 1 \end{bmatrix} \qquad (4.9)$$

and is given by

$$\mathbf{T}_{0,1} = \begin{bmatrix} \cos\theta_1 & 0 & -\sin\theta_1 & a_1\cos\theta_1 \\ \sin\theta_1 & 0 & \cos\theta_1 & a_1\sin\theta_1 \\ 0 & -1 & 0 & 0 \\ 0 & 0 & 0 & 1 \end{bmatrix}. \qquad (4.10)$$

The second homogeneous transformation is

$$\mathbf{T}_{1,2} = \begin{bmatrix} 1 & 0 & 0 & 0 \\ 0 & 1 & 0 & 0 \\ 0 & 0 & 1 & d_2 \\ 0 & 0 & 0 & 1 \end{bmatrix}. \qquad (4.11)$$

Hence it follows that the end-effector coordinates may be expressed in terms of the base coordinates as

$$\begin{bmatrix} x_0 \\ y_0 \\ z_0 \\ 1 \end{bmatrix} = \mathbf{T}_{0,1}\mathbf{T}_{1,2} \begin{bmatrix} x_2 \\ y_2 \\ z_2 \\ 1 \end{bmatrix} = \begin{bmatrix} \cos\theta_1 & 0 & -\sin\theta_1 & a_1\cos\theta_1 - d_2\sin\theta_1 \\ \sin\theta_1 & 0 & \cos\theta_1 & a_1\sin\theta_1 + d_2\cos\theta_1 \\ 0 & -1 & 0 & 0 \\ 0 & 0 & 0 & 1 \end{bmatrix} \begin{bmatrix} x_2 \\ y_2 \\ z_2 \\ 1 \end{bmatrix}. \qquad (4.12)$$

The rotational part of the homogeneous transformation that may be used to determine the orientation of the end-effector is

$$\mathbf{R} = \begin{bmatrix} \cos\theta_1 & 0 & -\sin\theta_1 \\ \sin\theta_1 & 0 & \cos\theta_1 \\ 0 & -1 & 0 \end{bmatrix}. \qquad (4.13)$$

The origin of the end-effector in the fixed base frame is

$$\begin{bmatrix} x_{02} \\ y_{02} \\ z_{02} \\ 1 \end{bmatrix} = \begin{bmatrix} \cos\theta_1 & 0 & -\sin\theta_1 & a_1\cos\theta_1 - d_2\sin\theta_1 \\ \sin\theta_1 & 0 & \cos\theta_1 & a_1\sin\theta_1 + d_2\cos\theta_1 \\ 0 & -1 & 0 & 0 \\ 0 & 0 & 0 & 1 \end{bmatrix} \begin{bmatrix} 0 \\ 0 \\ 0 \\ 1 \end{bmatrix} = \begin{bmatrix} f_1(\theta_1, d_2) \\ f_2(\theta_1, d_2) \\ 0 \\ 1 \end{bmatrix}, \qquad (4.14)$$

where

$$f_1(\theta_1, d_2) = a_1\cos\theta_1 - d_2\sin\theta_1, \ f_2(\theta_1, d_2) = a_1\sin\theta_1 + d_2\cos\theta_1.$$

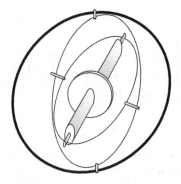

Figure 4.4. The yaw–pitch–roll-sequence-based triple-gimbaled support.

4.5 Kinematics of Wrists

Wrists are essentially nonplanar kinematic chains. A typical example is the spherical wrist. There are classes of robot components associated with wrists that are of zero link length and zero link offset. In these cases the use of the DH decomposition is probably superfluous and the rotational kinematics or the rotational component of the homogeneous transformation may be directly established either in terms of Euler angles or equivalently in terms of quaternions (see Appendix).

Considering the Euler angle representations of rotational kinematics, there are two basic classes of Euler angle sequences. Examples of each of these were presented in the previous chapter. With each of these classes are associated wrist components known as the *yaw–pitch–roll* and the $3R$ wrists. The rotations in the two cases are respectively given by

$$\mathbf{R}_{\text{YPR}} = \begin{bmatrix} \cos\psi & -\sin\psi & 0 \\ \sin\psi & \cos\psi & 0 \\ 0 & 0 & 1 \end{bmatrix} \begin{bmatrix} \cos\theta & 0 & \sin\theta \\ 0 & 1 & 0 \\ -\sin\theta & 0 & \cos\theta \end{bmatrix} \begin{bmatrix} 1 & 0 & 0 \\ 0 & \cos\phi & -\sin\phi \\ 0 & \sin\phi & \cos\phi \end{bmatrix}$$

$$(4.15)$$

and

$$\mathbf{R}_{3R} = \begin{bmatrix} \cos\gamma & -\sin\gamma & 0 \\ \sin\gamma & \cos\gamma & 0 \\ 0 & 0 & 1 \end{bmatrix} \begin{bmatrix} \cos\beta & 0 & \sin\beta \\ 0 & 1 & 0 \\ -\sin\beta & 0 & \cos\beta \end{bmatrix} \begin{bmatrix} \cos\alpha & -\sin\alpha & 0 \\ \sin\alpha & \cos\alpha & 0 \\ 0 & 0 & 1 \end{bmatrix}.$$

$$(4.16)$$

The first has a structure very similar to that of a platform mounted on a triple-gimbaled support, as illustrated in Figure 4.4. The outer gimbals and the inner wheel are rigidly attached to the input and output links.

A typical $3R$ wrist based on the second rotational sequence is shown in Figure 4.5. It has three co-located rotational joints, and hence it is referred to as the $3R$ wrist. It is a typical example of a *spherical wrist*. The composite homogeneous transformation for the wrists is given by

$$\mathbf{T} = \begin{bmatrix} \mathbf{R}(\varphi, \mathbf{e}_\varphi) & \mathbf{d} \\ \mathbf{0} & 1 \end{bmatrix},$$

$$(4.17)$$

where $\mathbf{R}(\varphi, \mathbf{e}_\varphi)$ is equal to either $\mathbf{R}_{\text{YPR}}$ or $\mathbf{R}_{3R}$ and the vector $\mathbf{d} = \mathbf{0}$.

Figure 4.5. Example of a 3*R* wrist.

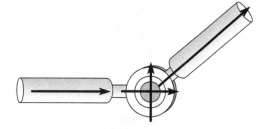

4.6 Direct Kinematics of Two Industrial Manipulators

The Unimation PUMA 560 (Figure 1.19) is a manipulator with six rotational joints and six corresponding DOFs. Hence it is also referred to as an *R-R-R*-3*R* mechanism and is shown in Figure 4.6. (The difference between the 3*R* joints and the *R-R-R* joints is that the *R-R-R* joints are not colocated whereas the 3*R* joints are.) Although the direct kinematics of the PUMA 560 may be obtained by adopting the DH convention, we will demonstrate an alternative technique and consider it to be an assemblage of two kinematic chains.

The first chain consists of the first three links. The second is the spherical wrist. Associated with first kinematic chain are five coordinate frames, which are also shown in Figure 4.6.

The first is the base or fixed frame, whereas the next three are fixed to the links in accordance with the DH convention. The final frame is fixed to a dummy link that is rigidly attached to the third link but is oriented in such a way as to facilitate the interfacing of the 3*R* wrist.

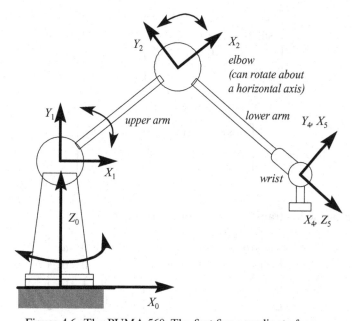

Figure 4.6. The PUMA 560: The first five coordinate frames.

Table 4.3. *Link parameters for the first three links of the PUMA 560 manipulator*

Link no.	a_i	α_i	d_i	θ_i
1	0	90°	h	θ_1
2	a_1	0	0	θ_2
3	a_2	0	0	θ_3
Dummy	0	−90°	0	−90°

The first kinematic chain is considered, and the link parameters are given in the Table 4.3.

The homogeneous transformation for the first link is

$$\mathbf{T}_{0,1} = \begin{bmatrix} \cos\theta_1 & -\sin\theta_1 & 0 & 0 \\ \sin\theta_1 & \cos\theta_1 & 0 & 0 \\ 0 & 0 & 1 & 0 \\ 0 & 0 & 0 & 1 \end{bmatrix} \begin{bmatrix} 1 & 0 & 0 & 0 \\ 0 & 1 & 0 & 0 \\ 0 & 0 & 1 & h \\ 0 & 0 & 0 & 1 \end{bmatrix} \begin{bmatrix} 1 & 0 & 0 & 0 \\ 0 & 0 & -1 & 0 \\ 0 & 1 & 0 & 0 \\ 0 & 0 & 0 & 1 \end{bmatrix},$$

that is,

$$\mathbf{T}_{0,1} = \begin{bmatrix} \cos\theta_1 & -\sin\theta_1 & 0 & 0 \\ \sin\theta_1 & \cos\theta_1 & 0 & 0 \\ 0 & 0 & 1 & h \\ 0 & 0 & 0 & 1 \end{bmatrix} \begin{bmatrix} 1 & 0 & 0 & 0 \\ 0 & 0 & -1 & 0 \\ 0 & 1 & 0 & 0 \\ 0 & 0 & 0 & 1 \end{bmatrix} = \begin{bmatrix} \cos\theta_1 & 0 & \sin\theta_1 & 0 \\ \sin\theta_1 & 0 & -\cos\theta_1 & 0 \\ 0 & 1 & 0 & h \\ 0 & 0 & 0 & 1 \end{bmatrix}.$$

$$(4.18)$$

The homogeneous transformation for the second link is

$$\mathbf{T}_{1,2} = \begin{bmatrix} \cos\theta_2 & -\sin\theta_2 & 0 & 0 \\ \sin\theta_2 & \cos\theta_2 & 0 & 0 \\ 0 & 0 & 1 & 0 \\ 0 & 0 & 0 & 1 \end{bmatrix} \begin{bmatrix} 1 & 0 & 0 & a_1 \\ 0 & 1 & 0 & 0 \\ 0 & 0 & 1 & 0 \\ 0 & 0 & 0 & 1 \end{bmatrix},$$

that is,

$$\mathbf{T}_{1,2} = \begin{bmatrix} \cos\theta_2 & -\sin\theta_2 & 0 & a_1\cos\theta_2 \\ \sin\theta_2 & \cos\theta_2 & 0 & a_1\sin\theta_2 \\ 0 & 0 & 1 & 0 \\ 0 & 0 & 0 & 1 \end{bmatrix}.$$

$$(4.19)$$

The homogeneous transformation for the third link is

$$\mathbf{T}_{2,3} = \begin{bmatrix} \cos\theta_3 & -\sin\theta_3 & 0 & 0 \\ \sin\theta_3 & \cos\theta_3 & 0 & 0 \\ 0 & 0 & 1 & 0 \\ 0 & 0 & 0 & 1 \end{bmatrix} \begin{bmatrix} 1 & 0 & 0 & a_1 \\ 0 & 1 & 0 & 0 \\ 0 & 0 & 1 & 0 \\ 0 & 0 & 0 & 1 \end{bmatrix},$$

that is,

$$
\mathbf{T}_{2,3} = \begin{bmatrix} \cos\theta_3 & -\sin\theta_3 & 0 & a_2\cos\theta_3 \\ \sin\theta_3 & \cos\theta_3 & 0 & a_2\sin\theta_3 \\ 0 & 0 & 1 & 0 \\ 0 & 0 & 0 & 1 \end{bmatrix}. \tag{4.20}
$$

Finally, the homogeneous transformation for the dummy link is

$$
\mathbf{T}_{3,4} = \begin{bmatrix} 0 & -1 & 0 & 0 \\ 1 & 0 & 0 & 0 \\ 0 & 0 & 1 & 0 \\ 0 & 0 & 0 & 1 \end{bmatrix} \begin{bmatrix} 1 & 0 & 0 & 0 \\ 0 & 0 & 1 & 0 \\ 0 & -1 & 0 & 0 \\ 0 & 0 & 0 & 1 \end{bmatrix},
$$

that is,

$$
\mathbf{T}_{3,4} = \begin{bmatrix} 0 & 0 & -1 & 0 \\ 1 & 0 & 0 & 0 \\ 0 & -1 & 0 & 0 \\ 0 & 0 & 0 & 1 \end{bmatrix}. \tag{4.21}
$$

For this component kinematic chain, interface coordinates at the end of link 3 may be expressed in terms of the base coordinates as

$$
\begin{bmatrix} x_0 \\ y_0 \\ z_0 \\ 1 \end{bmatrix} = \mathbf{T}_{0,1}\mathbf{T}_{1,2}\mathbf{T}_{2,3}\mathbf{T}_{3,4} \begin{bmatrix} x_4 \\ y_4 \\ z_4 \\ 1 \end{bmatrix}. \tag{4.22}
$$

For the spherical $3R$ wrist, the wrist coordinates transform as

$$
\begin{bmatrix} x_4 \\ y_4 \\ z_4 \\ 1 \end{bmatrix} = \mathbf{T}_w \begin{bmatrix} x_w \\ y_w \\ z_w \\ 1 \end{bmatrix} = \begin{bmatrix} \mathbf{R}_{3R} & \mathbf{0} \\ \mathbf{0} & 1 \end{bmatrix} \begin{bmatrix} x_w \\ y_w \\ z_w \\ 1 \end{bmatrix}, \tag{4.23}
$$

where $\mathbf{R}_{3R}$ is

$$
\mathbf{R}_{3R} = \begin{bmatrix} \cos\gamma & -\sin\gamma & 0 \\ \sin\gamma & \cos\gamma & 0 \\ 0 & 0 & 1 \end{bmatrix} \begin{bmatrix} \cos\beta & 0 & \sin\beta \\ 0 & 1 & 0 \\ -\sin\beta & 0 & \cos\beta \end{bmatrix} \begin{bmatrix} \cos\alpha & -\sin\alpha & 0 \\ \sin\alpha & \cos\alpha & 0 \\ 0 & 0 & 1 \end{bmatrix}.
$$

Hence it follows that

$$
\begin{bmatrix} x_0 \\ y_0 \\ z_0 \\ 1 \end{bmatrix} = \mathbf{T}_{0,1}\mathbf{T}_{1,2}\mathbf{T}_{2,3}\mathbf{T}_{3,4} \begin{bmatrix} \mathbf{R}_{3R} & \mathbf{0} \\ \mathbf{0} & 1 \end{bmatrix} \begin{bmatrix} x_w \\ y_w \\ z_w \\ 1 \end{bmatrix}, \tag{4.24}
$$

which defines the complete forward kinematics of the PUMA 560. The type of decomposition of the manipulator adopted in this case is particularly advantageous

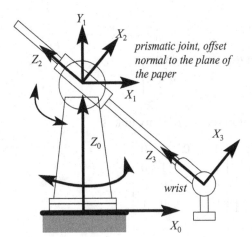

Figure 4.7. The Stanford R-R-P-$3R$ manipulator.

for solving the inverse kinematic problem, which will be studied in the next chapter and is concerned with determining values for the joint variables that achieve a desired position and orientation for the end-effector of the robot. The decomposition is adopted when the inverse kinematics problem decouples into two independent subproblems, one for the estimation of joint variables that influence the position and the other for the estimation of the orientation of the end-effector.

Unlike the PUMA 560, the wrist of the Stanford manipulator is not a spherical wrist, although it is may be approximated as one. The Stanford manipulator is illustrated in Figure 4.7. Thus the DH procedure is much more appropriate to this manipulator, which may be described as an R-R-P-$3R$ manipulator.

The link parameters for the Stanford manipulator are given in Table 4.4.

The homogeneous transformation for the first link is

$$\mathbf{T}_{0,1} = \begin{bmatrix} \cos\theta_1 & -\sin\theta_1 & 0 & 0 \\ \sin\theta_1 & \cos\theta_1 & 0 & 0 \\ 0 & 0 & 1 & 0 \\ 0 & 0 & 0 & 1 \end{bmatrix} \begin{bmatrix} 1 & 0 & 0 & 0 \\ 0 & 1 & 0 & 0 \\ 0 & 0 & 1 & h \\ 0 & 0 & 0 & 1 \end{bmatrix} \begin{bmatrix} 1 & 0 & 0 & 0 \\ 0 & 0 & -1 & 0 \\ 0 & 1 & 0 & 0 \\ 0 & 0 & 0 & 1 \end{bmatrix},$$

Table 4.4. *Link parameters for the links of the Stanford manipulator*

Link no.	a_i	α_i	d_i	θ_i
1	0	90°	h	θ_1
2	0	−90°	d_2	θ_2
3	0	0	d_3	0
4	0	90°	0	θ_4
5	0	−90°	0	θ_5
6	0	0°	d_6	θ_6

that is,

$$\mathbf{T}_{0,1} = \begin{bmatrix} \cos\theta_1 & 0 & \sin\theta_1 & 0 \\ \sin\theta_1 & 0 & -\cos\theta_1 & 0 \\ 0 & 1 & 0 & h \\ 0 & 0 & 0 & 1 \end{bmatrix}. \tag{4.25}$$

The homogeneous transformation for the second link is

$$\mathbf{T}_{1,2} = \begin{bmatrix} \cos\theta_2 & -\sin\theta_2 & 0 & 0 \\ \sin\theta_2 & \cos\theta_2 & 0 & 0 \\ 0 & 0 & 1 & 0 \\ 0 & 0 & 0 & 1 \end{bmatrix} \begin{bmatrix} 1 & 0 & 0 & 0 \\ 0 & 1 & 0 & 0 \\ 0 & 0 & 1 & d_2 \\ 0 & 0 & 0 & 1 \end{bmatrix} \begin{bmatrix} 1 & 0 & 0 & 0 \\ 0 & 0 & 1 & 0 \\ 0 & -1 & 0 & 0 \\ 0 & 0 & 0 & 1 \end{bmatrix},$$

that is,

$$\mathbf{T}_{1,2} = \begin{bmatrix} \cos\theta_2 & 0 & -\sin\theta_2 & 0 \\ \sin\theta_2 & 0 & \cos\theta_2 & 0 \\ 0 & -1 & 0 & d_2 \\ 0 & 0 & 0 & 1 \end{bmatrix}. \tag{4.26}$$

For link 3 the homogeneous transformation is

$$\mathbf{T}_{2,3} = \begin{bmatrix} 1 & 0 & 0 & 0 \\ 0 & 1 & 0 & 0 \\ 0 & 0 & 1 & d_3 \\ 0 & 0 & 0 & 1 \end{bmatrix}. \tag{4.27}$$

For link 4 the homogeneous transformation is

$$\mathbf{T}_{3,4} = \begin{bmatrix} \cos\theta_4 & -\sin\theta_4 & 0 & 0 \\ \sin\theta_4 & \cos\theta_4 & 0 & 0 \\ 0 & 0 & 1 & 0 \\ 0 & 0 & 0 & 1 \end{bmatrix} \begin{bmatrix} 1 & 0 & 0 & 0 \\ 0 & 0 & -1 & 0 \\ 0 & 1 & 0 & 0 \\ 0 & 0 & 0 & 1 \end{bmatrix} = \begin{bmatrix} \cos\theta_4 & 0 & \sin\theta_4 & 0 \\ \sin\theta_4 & 0 & -\cos\theta_4 & 0 \\ 0 & 1 & 0 & 0 \\ 0 & 0 & 0 & 1 \end{bmatrix}. \tag{4.28}$$

For link 5 the homogeneous transformation is,

$$\mathbf{T}_{4,5} = \begin{bmatrix} \cos\theta_5 & -\sin\theta_5 & 0 & 0 \\ \sin\theta_5 & \cos\theta_5 & 0 & 0 \\ 0 & 0 & 1 & 0 \\ 0 & 0 & 0 & 1 \end{bmatrix} \begin{bmatrix} 1 & 0 & 0 & 0 \\ 0 & 0 & 1 & 0 \\ 0 & -1 & 0 & 0 \\ 0 & 0 & 0 & 1 \end{bmatrix} = \begin{bmatrix} \cos\theta_5 & 0 & -\sin\theta_5 & 0 \\ \sin\theta_5 & 0 & \cos\theta_5 & 0 \\ 0 & -1 & 0 & 0 \\ 0 & 0 & 0 & 1 \end{bmatrix}. \tag{4.29}$$

Finally, for link 6 the homogeneous transformation is

$$\mathbf{T}_{5,6} = \begin{bmatrix} \cos\theta_6 & -\sin\theta_6 & 0 & 0 \\ \sin\theta_6 & \cos\theta_6 & 0 & 0 \\ 0 & 0 & 1 & 0 \\ 0 & 0 & 0 & 1 \end{bmatrix} \begin{bmatrix} 1 & 0 & 0 & 0 \\ 0 & 1 & 0 & 0 \\ 0 & 0 & 1 & d_6 \\ 0 & 0 & 0 & 1 \end{bmatrix}, \tag{4.30}$$

that is,

$$
\mathbf{T}_{5,6} = \begin{bmatrix} \cos\theta_6 & -\sin\theta_6 & 0 & 0 \\ \sin\theta_6 & \cos\theta_6 & 0 & 0 \\ 0 & 0 & 1 & d_6 \\ 0 & 0 & 0 & 1 \end{bmatrix}.
\tag{4.31}
$$

Hence, for the Stanford manipulator, the end-effector coordinates may be expressed in terms of the base coordinates as

$$
\begin{bmatrix} x_0 \\ y_0 \\ z_0 \\ 1 \end{bmatrix} = \mathbf{T}_{0,1}\mathbf{T}_{1,2}\mathbf{T}_{2,3}\mathbf{T}_{3,4}\mathbf{T}_{4,5}\mathbf{T}_{5,6} \begin{bmatrix} x_6 \\ y_6 \\ z_6 \\ 1 \end{bmatrix}.
\tag{4.32}
$$

It must be said that the DH convention does not lead to a unique parameterization, as in the case of the Stanford manipulator.

EXERCISES

4.1. Consider the three-link planar manipulator shown in Figure 4.8. Derive the forward kinematic equations by using the DH convention.

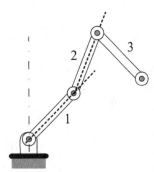

Figure 4.8. Three-link planar arm of Exercise 4.1.

4.2. Consider the two-link Cartesian manipulator shown in Figure 4.9. Derive the forward kinematic equations using the DH-convention.

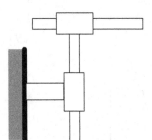

Figure 4.9. Two-link Cartesian manipulator of Exercise 4.2.

4.3. Consider the two-link manipulator of Figure 4.10, which has one revolute joint and one prismatic joint. Derive the forward kinematic equations by using the DH convention.

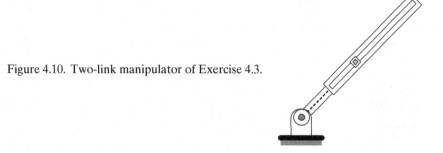

Figure 4.10. Two-link manipulator of Exercise 4.3.

4.4. Consider the three-link planar manipulator shown in Figure 4.11. Derive the forward kinematic equations by using the DH convention.

Figure 4.11. Three-link planar arm of Exercise 4.4.

4.5. Consider the slider–crank mechanism in Figure 4.12. Derive the forward kinematic equations by using the DH convention.

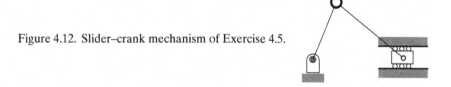

Figure 4.12. Slider–crank mechanism of Exercise 4.5.

4.6. Consider the planar manipulator shown in Figure 4.13. Assume the wrist is a $3R$ wrist (3–2–3 Euler angle sequence). Derive the forward kinematic equations by using the DH convention.

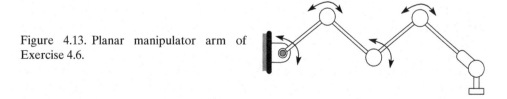

Figure 4.13. Planar manipulator arm of Exercise 4.6.

4.7. Consider the SCARA manipulator discussed in Chapter 1 (Figure 1.18). Derive the forward kinematic equations by using the DH convention.

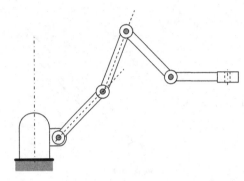

Figure 4.14. Nonplanar manipulator arm of Exercise 4.8.

4.8. Consider the four-link manipulator illustrated in Figure 4.14, which is attached to capstan capable of rotating about the vertical axis. The wrist is attached to the free end where the joint axis is perpendicular to the preceding four revolute joints. The manipulator has six DOFs. Derive the forward kinematic equations.

5

Manipulators with Multiple Postures
and Compositions

5.1 Inverse Kinematics of Robot Manipulators

In Chapter 4 we showed how to determine the end-effector position and orientation in terms of the joint variables in a frame fixed in space. This was the forward kinematics problem, which is to determine the position and orientation of the end-effector, given the values for the joint variables of the robot. The joint variables are the angles between the links in the case of revolute or rotational joints, and the extension of the links themselves or link segments in the case of prismatic or sliding joints. This chapter is concerned with the inverse problem of finding the joint variables in terms of the position and orientation of the end-effector. The inverse problem is concerned with determining values for the joint variables that achieve a desired position and orientation or a desired pose for the end-effector of the manipulator. This is the problem of inverse kinematics of position and orientation, and it is, in general, more difficult than the forward kinematics problem.

In this chapter, we begin by formulating the general inverse kinematics problem. In the last chapter it was shown that the end-effector coordinates may be expressed in terms of the base coordinates in terms of a homogeneous transformation matrix given by

$$\mathbf{T}_{\text{DH}} = \mathbf{T}_{n,n+1} \cdots \mathbf{T}_{i,i+1} \mathbf{T}_{i-1,i} \mathbf{T}_{i-2,i-1} \cdots \mathbf{T}_{1,2} = \prod_{i=1}^{n} \mathbf{T}_{i,i+1}. \tag{5.1}$$

In general, it is often desired that the position and orientation of the end-effector be defined according to certain requirements, and this corresponds to a certain homogeneous transformation applied to the current end-effector position vector in the end-effector fixed frame. Such a homogeneous transformation may be assumed to be given by

$$\mathbf{T}_{\text{ED}} = \begin{bmatrix} \mathbf{R}_{\text{ED}}\left(\varphi, \mathbf{e}_{\varphi}\right) & \mathbf{d}_{\text{ED}} \\ \mathbf{0} & 1 \end{bmatrix}. \tag{5.2}$$

The inverse kinematic problem is to determine the joint variables in $\mathbf{T}_{DH}$ such that

$$\mathbf{T}_{DH} = \mathbf{T}_{ED} = \begin{bmatrix} \mathbf{R}_{ED}(\varphi, \mathbf{e}_\varphi) & \mathbf{d}_{ED} \\ \mathbf{0} & 1 \end{bmatrix}. \tag{5.3}$$

This is equivalent to the matrix–vector pair of equations,

$$\mathbf{R}_{DH} = \mathbf{R}_{ED}, \quad \mathbf{d}_{DH} = \mathbf{d}_{ED}. \tag{5.4}$$

In general, the first of these, the inverse orientation equation, is equivalent to nine scalar equations whereas the second, the inverse position vector equation, is equivalent to three. However, there are only six independent equations for solving for the six unknowns.

It has been shown that for certain kinematic configurations the inverse orientation and inverse position equation decouple and may be solved sequentially. These kinematic configurations involve one of the following:

1. any three joints are translational,
2. any three rotational joint axes cointersecting at a common point,
3. any two translational joints normal to a rotational joint,
4. a translational joint normal to two parallel joints,
5. any three rotational joints in parallel.

Two types of decoupling may be identified. In the first, the more common type, the inverse position vector equation is solved first to determine the position-related joint variables. These are then used to solve the inverse orientation equations. Typical examples in this category are manipulators with spherical wrists.

The second type of decoupling permits the solution of the inverse orientation equation without any knowledge of the position-related joint variables. The solution is then used to solve the inverse position equations.

To show this formally, the homogeneous transformation representing the direct kinematics is assumed to be the product of two components. The first represents the kinematic chain leading to the wrist whereas the second represents the kinematic chain from the wrist to the end-effector. Thus,

$$\mathbf{T}_{DH} = \begin{bmatrix} \mathbf{R}_L & \mathbf{d}_L \\ \mathbf{0} & 1 \end{bmatrix} \times \begin{bmatrix} \mathbf{R}_W & \mathbf{d}_W \\ \mathbf{0} & 1 \end{bmatrix}. \tag{5.5}$$

Thus the direct kinematic equations for the end-effector position in the base coordinates are

$$\begin{bmatrix} \mathbf{R}_L & \mathbf{d}_L \\ \mathbf{0} & 1 \end{bmatrix} \times \begin{bmatrix} \mathbf{R}_W & \mathbf{d}_W \\ \mathbf{0} & 1 \end{bmatrix} \begin{bmatrix} \mathbf{0} \\ 1 \end{bmatrix} = \begin{bmatrix} \mathbf{d}_{ED} \\ 1 \end{bmatrix}. \tag{5.6}$$

Hence, for the position,

$$\mathbf{R}_L \mathbf{d}_W + \mathbf{d}_L = \mathbf{d}_{ED}. \tag{5.7}$$

For the orientation,

$$\mathbf{R}_L \mathbf{R}_W = \mathbf{R}_{ED}. \tag{5.8}$$

In the case of a spherical wrist it can be shown that

$$\mathbf{R}_L \mathbf{d}_W = \mathbf{R}_L \mathbf{R}_W \mathbf{N}, \tag{5.9}$$

and the position equation may be expressed as

$$\mathbf{d}_{ED} = \mathbf{d}_L + \mathbf{R}_{ED} \mathbf{N}. \tag{5.10}$$

It depends on only the three joint variables not associated with the wrist and may be solved first, and the orientation equation, which depends on the three remaining joint variables associated with the wrist, is solved next.

On the other hand, when

$$\mathbf{R}_L = \mathbf{U}_L \tag{5.11}$$

is a constant matrix with a determinant equal to unity, the joint variables associated with the wrist may be solved first by

$$\mathbf{R}_W = \mathbf{U}_L^{-1} \mathbf{R}_{ED}, \tag{5.12}$$

followed by

$$\mathbf{U}_L \mathbf{d}_W + \mathbf{d}_L = \mathbf{d}_{ED}, \tag{5.13}$$

to obtain the remaining joint variables.

Of the five kinematic configurations associated with decoupling of the inverse kinematic problem, closed-form solutions are possible only in the case of the first two. In solving the inverse kinematics problem one is interested, in the first instance, in finding a closed-form solution of the equations rather than a numerical solution. Closed-form solutions are almost always preferable as they may be evaluated in real time with sufficient speed and accuracy. Further, the inverse kinematic equations in general have multiple solutions. When closed-form solutions are available, it is possible to choose a particular solution from among several with relative ease.

5.1.1 The Nature of Inverse Kinematics: Postures

In most cases the solution of the inverse kinematic problem is based on an intuitive approach involving the kinematic geometry of the mechanism. Generally, the problem of solving an inverse kinematic problem reduces to the algebraic or geometric problem of solving an oblique triangle (Figure 5.1).

Associated with oblique triangles are a number of advanced trigonometric formulas that are summarized in Table 5.1 and are useful in the solution of the inverse kinematic problem. Of these, the first two are the most commonly adopted formulas, and the solution for the joint angles is normally expressed as an inverse sine, inverse cosine, or inverse tangent. However, because solutions of inverse trigonometric functions are nonunique, the general solutions to the inverse kinematic problems are nonunique.

A typical example illustrates the procedures involved. Consider the two-link manipulator shown in Figure 4.2. For this simple planar anthropomorphic and

Table 5.1. *Summary of nonstandard trigonometric formulas used in the closed-form solution of the inverse kinematics problems (other formulas follow by cyclic letter changes, continued)*

No.	Name	Formulas
1	Projection formulas	$a = b \cos C + c \cos B$
2	Law of sines	$\dfrac{a}{\sin A} = \dfrac{b}{\sin B} = \dfrac{c}{\sin C}$
3	Law of cosines	$a^2 = b^2 + c^2 - 2bc \cos A$
4	Mollweide's formulas	$\dfrac{a+b}{c} = \dfrac{\cos\left(\frac{A-B}{2}\right)}{\sin\left(\frac{C}{2}\right)}$
		$\dfrac{a-b}{c} = \dfrac{\sin\left(\frac{A-B}{2}\right)}{\cos\left(\frac{C}{2}\right)}$
5	Law of tangents	$\dfrac{a-b}{a+b} = \dfrac{\tan\left(\frac{A-B}{2}\right)}{\tan\left(\frac{A+B}{2}\right)}$
6	Inscribed circle radius	$r_i = \sqrt{\dfrac{(s-a)(s-b)(s-c)}{s}}$
		$s = \dfrac{a+b+c}{2}$
7	Half-angle formulas	$\tan\left(\dfrac{A}{2}\right) = \dfrac{r_i}{s-a}$
8	Circumscribed circle radius	$R_c = \dfrac{a}{2\sin A} = \dfrac{b}{2\sin B} = \dfrac{c}{2\sin C}$
9	Area formulas (area $= S$)	$S = \dfrac{bc \sin A}{2}$
		$S = \dfrac{a^2}{2}\dfrac{\sin B \sin C}{\sin A}$
		$S = s \times r_i$
10	Derived formulas	$S = 2R_c^2 \sin A \sin B \sin C$
		$S = \dfrac{abc}{4R_c}$
		$S = r_i R_c (\sin A + \sin B + \sin C)$

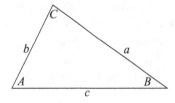

Figure 5.1. A typical oblique triangle with sides a, b, and c, including angles A, B, and C and area S.

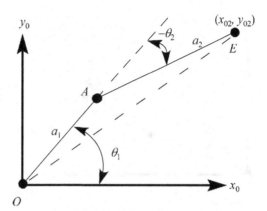

Figure 5.2. Geometry of the two-link planar manipulator.

biomimetic manipulator, the end-effector coordinates were expressed in terms of the joint angles and from Equation (4.7) we have

$$\begin{bmatrix} x_{02} \\ y_{02} \\ z_{02} \end{bmatrix} = \begin{bmatrix} a_2 c_{12} + a_1 \cos \theta_1 \\ a_2 s_{12} + a_1 \sin \theta_1 \\ 0 \end{bmatrix}, \tag{5.14}$$

where $c_{12} = \cos(\theta_1 + \theta_2)$ and $s_{12} = \sin(\theta_1 + \theta_2)$.

Although one can adopt a purely algebraic approach to solve for $\cos \theta_1$ and $\sin \theta_1$, an algebrogeometric approach is much more appealing. The geometry of the two-link manipulator is illustrated in Figure 5.2.

Applying the law of cosines to the triangle OAE in Figure 5.2, we have

$$x_{02}^2 + y_{02}^2 = a_1^2 + a_2^2 - 2a_1 a_2 \cos(180° + \theta_2). \tag{5.15}$$

Solving for $\cos(\theta_2)$, we have

$$\cos(\theta_2) = \frac{x_{02}^2 + y_{02}^2 - a_1^2 - a_2^2}{2a_1 a_2}, \tag{5.16a}$$

and $\sin(\theta_2)$ is

$$\sin(\theta_2) = \pm\sqrt{1 - \cos(\theta_2)^2}. \tag{5.16b}$$

The solution for $\sin(\theta_2)$ is not unique and there are two possible solutions for θ_2. Further,

$$\begin{bmatrix} x_{02} \\ y_{02} \end{bmatrix} = \begin{bmatrix} a_2 c_{12} + a_1 \cos \theta_1 \\ a_2 s_{12} + a_1 \sin \theta_1 \end{bmatrix}, \tag{5.17}$$

where $c_{12} = \cos(\theta_1 + \theta_2)$ and $s_{12} = \sin(\theta_1 + \theta_2)$.

Thus, if we let $\cos(\theta_2) = \sigma$ and $\sin(\theta_2) = \rho$,

$$\begin{bmatrix} x_{02} \\ y_{02} \end{bmatrix} = \begin{bmatrix} a_1 + a_2\sigma & -a_2\rho \\ a_2\rho & a_1 + a_2\sigma \end{bmatrix} \begin{bmatrix} \cos \theta_1 \\ \sin \theta_1 \end{bmatrix}, \tag{5.18}$$

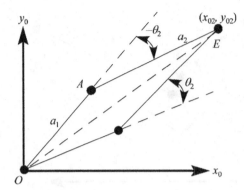

Figure 5.3. Inverse position solutions of the two-link planar manipulator depicting the two possible solutions for θ_2.

it follows that

$$\begin{bmatrix} \cos\theta_1 \\ \sin\theta_1 \end{bmatrix} = \begin{bmatrix} a_1 + a_2\sigma & -a_2\rho \\ a_2\rho & a_1 + a_2\sigma \end{bmatrix}^{-1} \begin{bmatrix} x_{02} \\ y_{02} \end{bmatrix}. \qquad (5.19)$$

Corresponding to each of the solutions for θ_2 there is a unique solution for θ_1. For this planar manipulator there are two possible solutions for the inverse position corresponding to two kinematic configurations that are both illustrated in Figure 5.3. These configurations are known as the *postures* of the manipulator; the first is an "elbow-up" configuration whereas the second is an "elbow-down" configuration. Yet, when $\theta_2 = 0$, the postures are unique and the *workspace* of the manipulator is then obtained as a circle of radius equal to the sum of the link lengths.

Before concluding this section, we also briefly address the topic of redundant manipulators. A redundant manipulator has more joint variables than the number of DOFs necessary to specify the end-effector's pose in relation to a particular task. Thus a redundant manipulator is one that has more internal DOFs than are required for performing a specified task. For example, a three-link planar arm is redundant for the task of positioning in a plane. Although in such cases there is no unique solution for the inverse kinematics problem, the availability of multiple postures permits one to eliminate those solutions that may possess certain undesirable features. One achieves this kind of elimination by imposing one of several workspace constraints involving either the position or both the position and the velocity.

Considering the nonanthropomorphic manipulator illustrated in Figure 4.2, the end-effector coordinates were expressed in terms of the joint angles in Equation (4.14). From Equation (4.14),

$$\begin{bmatrix} x_{02} \\ y_{02} \\ z_{02} \end{bmatrix} = \begin{bmatrix} a_1\cos\theta_1 - d_2\sin\theta_1 \\ a_1\sin\theta_1 + d_2\cos\theta_1 \\ 0 \end{bmatrix}, \qquad (5.20)$$

which may be expressed as

$$
\begin{bmatrix} x_{02} \\ y_{02} \\ z_{02} \end{bmatrix} = \sqrt{\left(a_1^2 + d_2^2\right)} \begin{bmatrix} \dfrac{a_1}{\sqrt{\left(a_1^2 + d_2^2\right)}} \cos\theta_1 - \dfrac{d_2}{\sqrt{\left(a_1^2 + d_2^2\right)}} \sin\theta_1 \\[2ex] \dfrac{a_1}{\sqrt{\left(a_1^2 + d_2^2\right)}} \sin\theta_1 + \dfrac{d_2}{\sqrt{\left(a_1^2 + d_2^2\right)}} \cos\theta_1 \\[2ex] 0 \end{bmatrix} . \qquad (5.21)
$$

If we let

$$
\cos\alpha = \frac{a_1}{\sqrt{\left(a_1^2 + d_2^2\right)}}, \quad \sin\alpha = \frac{d_2}{\sqrt{\left(a_1^2 + d_2^2\right)}}, \qquad (5.22)
$$

it follows that

$$
x_{02}^2 + y_{02}^2 = \left(a_1^2 + d_2^2\right). \qquad (5.23)
$$

The solution for the joint variables d_2 and θ_1 may be expressed as

$$
d_2 = \pm\sqrt{x_{02}^2 + y_{02}^2 - a_1^2}, \qquad (5.24)
$$

and

$$
\begin{bmatrix} \cos\left(\alpha + \theta_1\right) \\ \sin\left(\alpha + \theta_1\right) \\ 0 \end{bmatrix} = \frac{1}{\sqrt{\left(x_{02}^2 + y_{02}^2\right)}} \begin{bmatrix} x_{02} \\ y_{02} \\ z_{02} \end{bmatrix} . \qquad (5.25)
$$

In this case also there are two sets of solutions indicating the existence of multiple postures.

5.1.2 Some Practical Examples

For the first example we consider the PUMA 560 (Figure 1.19). The component kinematic chain leading to the spherical wrist, the interface coordinates at the end of link 3, may be expressed in terms of the base coordinates as

$$
\begin{bmatrix} x_0 \\ y_0 \\ z_0 \\ 1 \end{bmatrix} = \mathbf{T}_{0,1}\mathbf{T}_{1,4} \begin{bmatrix} x_4 \\ y_4 \\ z_4 \\ 1 \end{bmatrix} , \qquad (5.26)
$$

where

$$
\mathbf{T}_{0,1} = \begin{bmatrix} \cos\theta_1 & 0 & \sin\theta_1 & 0 \\ \sin\theta_1 & 0 & -\cos\theta_1 & 0 \\ 0 & 1 & 0 & h \\ 0 & 0 & 0 & 1 \end{bmatrix} , \qquad (5.27)
$$

$$\mathbf{T}_{1,4} = \begin{bmatrix} -s_{23} & 0 & -c_{23} & a_2 c_{23} + a_1 c_2 \\ c_{23} & 0 & -s_{23} & a_2 s_{23} + a_1 s_2 \\ 0 & -1 & 0 & 0 \\ 0 & 0 & 0 & 1 \end{bmatrix}, \tag{5.28}$$

and $c_i = \cos \theta_i$, $s_i = \sin \theta_i$, $c_{ij} = \cos(\theta_i + \theta_j)$, and $s_{ij} = \sin(\theta_i + \theta_j)$.

Hence the origin of the spherical wrist in the base coordinates is

$$\begin{bmatrix} x_{03} \\ y_{03} \\ z_{03} \\ 1 \end{bmatrix} = \begin{bmatrix} \cos \theta_1 & 0 & \sin \theta_1 & 0 \\ \sin \theta_1 & 0 & -\cos \theta_1 & 0 \\ 0 & 1 & 0 & h \\ 0 & 0 & 0 & 1 \end{bmatrix} \begin{bmatrix} -s_{23} & 0 & -c_{23} & a_2 c_{23} + a_1 c_2 \\ c_{23} & 0 & -s_{23} & a_2 s_{23} + a_1 s_2 \\ 0 & -1 & 0 & 0 \\ 0 & 0 & 0 & 1 \end{bmatrix} \begin{bmatrix} 0 \\ 0 \\ 0 \\ 1 \end{bmatrix}. \tag{5.29}$$

Hence,

$$\begin{bmatrix} x_{03} \\ y_{03} \\ z_{03} \\ 1 \end{bmatrix} = \begin{bmatrix} (a_2 c_{23} + a_1 c_2) c_1 \\ (a_2 c_{23} + a_1 c_2) s_1 \\ a_2 s_{23} + a_1 s_2 + h \\ 1 \end{bmatrix}. \tag{5.30}$$

They may be expressed as

$$\begin{bmatrix} \dfrac{x_{03}}{\sqrt{x_{03}^2 + y_{03}^2}} \\ \dfrac{y_{03}}{\sqrt{x_{03}^2 + y_{03}^2}} \\ \sqrt{x_{03}^2 + y_{03}^2} \\ z_{03} \end{bmatrix} = \begin{bmatrix} c_1 \\ s_1 \\ a_2 c_{23} + a_1 c_2 \\ a_2 s_{23} + a_1 s_2 + h \end{bmatrix}. \tag{5.31}$$

These are the inverse position equations that must be solved for the first three joint variables, θ_1, θ_2, and θ_3. The last two of these equations are similar to the equations for a planar two-link manipulator, which one can solve quite easily by using the law of cosines. The inverse orientation equations for the last three joint variables are

$$\mathbf{R}_{3R} = \begin{bmatrix} -s_{23} & 0 & -c_{23} \\ c_{23} & 0 & -s_{23} \\ 0 & -1 & 0 \end{bmatrix}^{-1} \begin{bmatrix} \cos \theta_1 & 0 & \sin \theta_1 \\ \sin \theta_1 & 0 & -\cos \theta_1 \\ 0 & 1 & 0 \end{bmatrix}^{-1} \mathbf{R}_{ED}, \tag{5.32}$$

where $\mathbf{R}_{3R}$ is

$$\mathbf{R}_{3R} = \begin{bmatrix} \cos \gamma & -\sin \gamma & 0 \\ \sin \gamma & \cos \gamma & 0 \\ 0 & 0 & 1 \end{bmatrix} \begin{bmatrix} \cos \beta & 0 & \sin \beta \\ 0 & 1 & 0 \\ -\sin \beta & 0 & \cos \beta \end{bmatrix} \begin{bmatrix} \cos \alpha & -\sin \alpha & 0 \\ \sin \alpha & \cos \alpha & 0 \\ 0 & 0 & 1 \end{bmatrix}. \tag{5.33}$$

The inverse orientation equations are solved by the methods introduced in Chapter 2, i.e., by expressing both sides in terms of the equivalent quaternion representations.

Another interesting and generic example is the Stanford manipulator introduced in Chapter 4 and illustrated in Figure 4.7. For the Stanford manipulator, the end-effector coordinates may be expressed in terms of the base coordinates as

$$
\begin{bmatrix} x_0 \\ y_0 \\ z_0 \\ 1 \end{bmatrix} = \mathbf{T}_{0,1}\mathbf{T}_{1,2}\mathbf{T}_{2,3}\mathbf{T}_{3,4}\mathbf{T}_{4,5}\mathbf{T}_{5,6} \begin{bmatrix} x_6 \\ y_6 \\ z_6 \\ 1 \end{bmatrix} = \mathbf{T}_{0,3}\mathbf{T}_{3,6} \begin{bmatrix} x_6 \\ y_6 \\ z_6 \\ 1 \end{bmatrix},
\tag{5.34}
$$

where

$$
\mathbf{T}_{0,3} = \begin{bmatrix} c_1 c_2 & -s_1 & -c_1 s_2 & d_2 s_1 - d_3 s_2 c_1 \\ s_1 c_2 & c_1 & s_1 s_2 & -d_2 c_1 + d_3 s_1 s_2 \\ s_2 & 0 & c_2 & d_3 c_2 + h \\ 0 & 0 & 0 & 1 \end{bmatrix},
\tag{5.35}
$$

$$
\mathbf{T}_{3,6} = \begin{bmatrix} c_4 & 0 & s_4 & 0 \\ s_4 & 0 & -c_4 & 0 \\ 0 & 1 & 0 & 0 \\ 0 & 0 & 0 & 1 \end{bmatrix} \begin{bmatrix} c_5 & 0 & -s_5 & 0 \\ s_5 & 0 & c_5 & 0 \\ 0 & -1 & 0 & 0 \\ 0 & 0 & 0 & 1 \end{bmatrix} \begin{bmatrix} c_6 & -s_6 & 0 & 0 \\ s_6 & c_6 & 0 & 0 \\ 0 & 0 & 1 & d_6 \\ 0 & 0 & 0 & 1 \end{bmatrix},
\tag{5.36}
$$

and $c_i = \cos\theta_i$ and $s_i = \sin\theta_i$.

It follows that the origin of the end-effector in the base coordinates is

$$
\begin{bmatrix} x_{06} \\ y_{06} \\ z_{06} \\ 1 \end{bmatrix} = \begin{bmatrix} c_1 c_2 & -s_1 & -c_1 s_2 & d_2 s_1 - d_3 s_2 c_1 \\ s_1 c_2 & c_1 & s_1 s_2 & -d_2 c_1 + d_3 s_1 s_2 \\ s_2 & 0 & c_2 & d_3 c_2 + h \\ 0 & 0 & 0 & 1 \end{bmatrix} \begin{bmatrix} -d_6 c_4 s_5 \\ -d_6 s_4 s_5 \\ d_6 c_5 \\ 1 \end{bmatrix},
\tag{5.37}
$$

where the right-hand vector

$$
\begin{bmatrix} -d_6 c_4 s_5 \\ -d_6 s_4 s_5 \\ d_6 c_5 \\ 1 \end{bmatrix} = \mathbf{T}_{3,4} \begin{bmatrix} \mathbf{0} \\ 1 \end{bmatrix} = \begin{bmatrix} c_4 & 0 & s_4 & 0 \\ s_4 & 0 & -c_4 & 0 \\ 0 & 1 & 0 & 0 \\ 0 & 0 & 0 & 1 \end{bmatrix} \begin{bmatrix} c_5 & 0 & -s_5 & 0 \\ s_5 & 0 & c_5 & 0 \\ 0 & -1 & 0 & 0 \\ 0 & 0 & 0 & 1 \end{bmatrix}
$$

$$
\times \begin{bmatrix} c_6 & -s_6 & 0 & 0 \\ s_6 & c_6 & 0 & 0 \\ 0 & 0 & 1 & d_6 \\ 0 & 0 & 0 & 1 \end{bmatrix} \begin{bmatrix} 0 \\ 0 \\ 0 \\ 1 \end{bmatrix}.
$$

Hence,

$$
\begin{bmatrix} x_{06} \\ y_{06} \\ z_{06} \end{bmatrix} = d_6 \begin{bmatrix} c_1 c_2 & -s_1 & -c_1 s_2 \\ s_1 c_2 & c_1 & s_1 s_2 \\ s_2 & 0 & c_2 \end{bmatrix} \begin{bmatrix} -c_4 s_5 \\ -s_4 s_5 \\ c_5 \end{bmatrix} + \begin{bmatrix} d_2 s_1 - d_3 s_2 c_1 \\ -d_2 c_1 + d_3 s_1 s_2 \\ h + d_3 c_2 \end{bmatrix}.
\tag{5.38}
$$

These are the inverse position equations that must be solved for the first three joint variables, θ_1, θ_2, and d_3. However, they cannot be solved unless the orientation equations are solved first as the right-hand side of these equations is a function of c_4, s_4, c_5, and s_5. The inverse orientation equations for the last three joint variables are again

$$\begin{bmatrix} c_1 c_2 & -s_1 & -c_1 s_2 \\ s_1 c_2 & c_1 & s_1 s_2 \\ s_2 & 0 & c_2 \end{bmatrix} \mathbf{R}_{3R\text{-DH}} = \mathbf{R}_{\text{ED}}, \tag{5.39}$$

where

$$\mathbf{R}_{3R\text{-DH}} = \begin{bmatrix} c_4 & 0 & s_4 \\ s_4 & 0 & -c_4 \\ 0 & 1 & 0 \end{bmatrix} \begin{bmatrix} c_5 & 0 & -s_5 \\ s_5 & 0 & c_5 \\ 0 & -1 & 0 \end{bmatrix} \begin{bmatrix} c_6 & -s_6 & 0 \\ s_6 & c_6 & 0 \\ 0 & 0 & 1 \end{bmatrix},$$

which reduces to

$$\mathbf{R}_{3R\text{-DH}} = \begin{bmatrix} c_4 c_5 & -s_4 & -c_4 s_5 \\ s_4 c_5 & c_4 & -s_4 s_5 \\ s_5 & 0 & c_5 \end{bmatrix} \begin{bmatrix} c_6 & 0 & s_6 \\ s_6 & 0 & -c_6 \\ 0 & 1 & 0 \end{bmatrix}$$

and $\mathbf{R}_{\text{ED}}$ is the desired orientation transformation matrix.

Thus,

$$\begin{bmatrix} c_1 c_2 & -s_1 & -c_1 s_2 \\ s_1 c_2 & c_1 & s_1 s_2 \\ s_2 & 0 & c_2 \end{bmatrix} \begin{bmatrix} c_4 c_5 & -s_4 & -c_4 s_5 \\ s_4 c_5 & c_4 & -s_4 s_5 \\ s_5 & 0 & c_5 \end{bmatrix} = \mathbf{R}_{\text{ED}} \begin{bmatrix} c_6 & s_6 & 0 \\ 0 & 0 & 1 \\ s_6 & -c_6 & 0 \end{bmatrix}, \tag{5.40}$$

and it follows that

$$\begin{bmatrix} c_1 c_2 & -s_1 & -c_1 s_2 \\ s_1 c_2 & c_1 & s_1 s_2 \\ s_2 & 0 & c_2 \end{bmatrix} \begin{bmatrix} -c_4 s_5 \\ -s_4 s_5 \\ c_5 \end{bmatrix} = \mathbf{R}_{\text{ED}} \begin{bmatrix} 0 \\ 1 \\ 0 \end{bmatrix}. \tag{5.41}$$

In this case, the preceding subset of the inverse orientation equations are used to eliminate c_4, s_4, c_5, and s_5 in the inverse position equations. Thus the inverse position equations are

$$\begin{bmatrix} x_{06} \\ y_{06} \\ z_{06} \end{bmatrix} = d_6 \, \mathbf{R}_{\text{ED}} \begin{bmatrix} 0 \\ 1 \\ 0 \end{bmatrix} + \begin{bmatrix} d_2 s_1 - d_3 s_2 c_1 \\ -d_2 c_1 + d_3 s_1 s_2 \\ h + d_3 c_2 \end{bmatrix}. \tag{5.42}$$

The inverse position may be solved for the first three joint variables, θ_1, θ_2, and d_3. Subsequently the inverse orientation equations are solved for the wrist joint variables.

These examples demonstrate the importance of kinematic decoupling in the determination of the joint variables by solving the inverse kinematics equations. They also demonstrate a number of unique features of manipulator systems. The analyses presented in this and the preceding chapter involve a number of generic steps. Starting from a description of an open-loop kinematic chain in terms of its link components, kinematic pairs, link–pair connections, and driving constraints, the mechanisms are decomposed into single components. For each component a unique reference frame is set up, and the transformation relating the displacement vector of each to another connected component is established. Based on the connectivity criteria and the associated invariant geometric properties, a mathematical model for determining the positions of the mechanism joints and links is established. It is the basis for the DH convention. The model naturally leads to the direct kinematics, and it involves writing and solving a simultaneous set of nonlinear congruence relations in terms of body positions, orientations, and driving constraints. The inverse kinematics involves the solution or inversion of these equations. However, the general inversion of these equations is almost never attempted.

The two industrial manipulator examples considered also suggest that, even at the design stage, by specific choice of the open-loop architecture, the synthesis and hence the analysis problems are simplified. Several open-loop kinematic mechanisms of any complexity are obtained by sequential addition of certain open-loop determined chains. In an open loop-kinematic chain, the position and orientation of any kinematic element are independent of the positions and orientations of all successively added elements. It is this feature that is fully exploited in the analysis of the direct kinematics and the solution of the inverse kinematic problems. Thus, whereas the establishment of direct kinematics requires a complete decomposition of the entire mechanism into constituent elements, inverse kinematics is derived by successive partial assembly of some of the elements of the mechanism followed by compilation of the subassemblies. Thus the process of deriving direct and inverse kinematics is not purely mathematical but relies on the physical decomposition and assembly of the kinematic chains.

5.2 Parallel Manipulators: Compositions

The problem of positioning and orienting a rigid body in spatial motion can be accomplished by means of open- and closed-loop kinematic chains, referred to as serial and parallel manipulators, respectively. Serial manipulators are open chains with all joints or kinematic pairs actively controlled, whereas parallel manipulators consist of one or more closed loops with only some pairs actuated. The direct kinematic analysis of parallel mechanisms aims to estimate the position and orientation of an output link or platform with a given set of actuator displacements. This is a difficult problem in general as several nonlinear equations are involved and the solutions are not unique.

In a parallel manipulator of the type shown in Figure 5.4, the calculation of inverse kinematics is relatively simpler than in serial manipulators. However, inverting the inverse kinematics to reconstruct the forward or direct kinematics even in this simple mechanism involves the solution of coupled nonlinear equations with nonunique solutions and is a relatively lengthy process. To illustrate these features, we consider a simple planar parallel manipulator, shown in Figure 5.4.

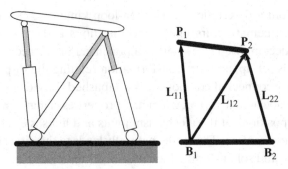

Figure 5.4. Two-dimensional parallel manipulator and its representation in a vector diagram.

Considering the vector diagram representing the planar manipulator, the lengths of the linearly actuated legs may be expressed as

$$|\mathbf{B}_i - \mathbf{P}_j| = |\mathbf{L}_{ij}| = L_{ij}. \tag{5.43}$$

Hence it follows that

$$(\mathbf{B}_i - \mathbf{P}_j) \cdot (\mathbf{B}_i - \mathbf{P}_j) = |\mathbf{L}_{ij}|^2. \tag{5.44}$$

Expanding the preceding vector equation and writing it in terms of the scalar components of each of the vectors,

$$\mathbf{B}_i = b_{ix}\mathbf{i} + b_{iy}\mathbf{j}, \mathbf{P}_i = p_{ix}\mathbf{i} + p_{iy}\mathbf{j}, \tag{5.45}$$

we have

$$(b_{ix} - p_{jx})^2 + (b_{iy} - p_{jy})^2 = |\mathbf{L}_{ij}|^2. \tag{5.46}$$

Further, we assume that the platform length is

$$(p_{1x} - p_{2x})^2 + (p_{1y} - p_{2y})^2 = p^2. \tag{5.47}$$

To simplify the equations the origin is assumed to be at $\mathbf{B}_1$. Hence,

$$b_{ix}^2 + b_{iy}^2 + p_{jx}^2 + p_{jy}^2 - 2b_{ix}p_{jx} - 2b_{iy}p_{jy} = L_{ij}^2. \tag{5.48}$$

Assuming $i = 1$ and $j = 1$,

$$p_{1x}^2 + p_{1y}^2 = L_{11}^2. \tag{5.49a}$$

Assuming $i = 1, 2$ and $j = 2$,

$$p_{2x}^2 + p_{2y}^2 - 2b_{1x}p_{2x} - 2b_{1y}p_{2y} = L_{12}^2, \tag{5.49b}$$

$$b_{2x}^2 + b_{2y}^2 + p_{2x}^2 + p_{2y}^2 - 2b_{2x}p_{2x} - 2b_{2y}p_{2y} = L_{22}^2. \tag{5.49c}$$

The equations for L_{11}, L_{12}, and L_{22} are the primary inverse kinematic relations.

Subtracting the second of the preceding pair of equations from the first and rearranging, we obtain

$$(b_{1x} - b_{2x}) p_{2x} + (b_{1y} - b_{2y}) p_{2y} = \frac{1}{2}(L_{22}^2 - L_{12}^2 - b_{2x}^2 - b_{2y}^2).$$

Assuming the base to be level, and $b_{2x} = b$,

$$p_{2x} = -\frac{1}{2b}(L_{22}^2 - L_{12}^2 - b_{2x}^2) \tag{5.50a}$$

and

$$p_{2y}^2 = L_{22}^2 - (p_{2x} - b)^2. \tag{5.50b}$$

Assuming $i = 1$ and $j = 1, 2$, and subtracting the second from the first and rearranging the terms, we have

$$p_{1x} p_{2x} + p_{1y} p_{2y} = \frac{1}{2}(L_{11}^2 - L_{12}^2 - p^2).$$

The four equations to solve for the position of the platform points are

$$\begin{aligned}
p_{2x} &= -\frac{1}{2b}(L_{22}^2 - L_{12}^2 - b_{2x}^2), \\
p_{2y}^2 &= L_{22}^2 - (p_{2x} - b)^2, \\
p_{1x}^2 + p_{1y}^2 &= L_{11}^2, \\
p_{1x} p_{2x} + p_{1y} p_{2y} &= \frac{1}{2}(L_{11}^2 - L_{12}^2 - p^2).
\end{aligned} \tag{5.51}$$

Because of the quadratic nature of these equations, there are two solutions for p_{2y} and a further pair of solutions for p_{1x} and p_{1y} corresponding to each. These correspond to four different platform configurations.

Such multiple configurations corresponding to the same inverse kinematics are generally referred to as *compositions*. The compositions of the planar parallel manipulator in Figure 5.4 are shown in Figure 5.5. It is quite obvious that the situation is not unexpected, considering that these configurations may be generated by the application of certain symmetry transformations to the original configurations.

5.2.1 Parallel Spatial Manipulators: The Stewart Platform

Parallel spatial manipulators have the capacity to handle higher loads with greater accuracy, higher speeds, and lighter weight; however, a major drawback is that the workspace of parallel spatial manipulators is severely restricted compared with that of equivalent serial manipulators. Parallel spatial manipulators are used not only in expensive flight simulators but also as machining tools and are used in high-accuracy, high-repeatability, high-precision robotic surgery. The main advantages

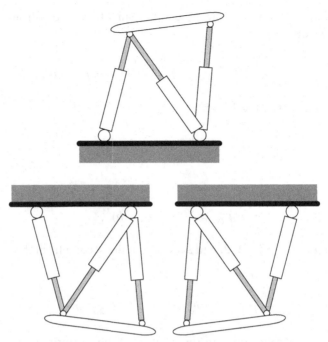

Figure 5.5. Compositions of the manipulator in Figure 5.4.

of parallel manipulators are the high nominal load/weight ratio and the high positioning accuracy that is due to their inherent rigidity.

The Stewart platform is a typical example of a parallel spatial manipulator. It is
used extensively in the design of flight simulator motion systems. Compositions and
inversions of the Stewart platform have also been used to develop a range of measurement and instrumentation application platforms. In the general Stewart platform, the base and the platform can be of any shape as long as the entire structure
remains stable. A structure with regular hexagons for both the base and the platform
is an example of an unstable configuration.

A six-DOF parallel spatial manipulator generally consists of a moving platform
connected by six links or legs to a fixed base. The legs are connected to the base
and platform by either ball joints and/or universal (Hooke) joints. Most commonly,
the joints are fixed on both the platform and the base. They are linked to each other
respectively by kinematic links of variable length. Changing the link lengths controls
the position and orientation of the platform. Thus, by changing the lengths of the
legs, the pose of the platform can be changed with respect to the base, thus causing
the relative motion of the platform.

The inverse kinematics problem of the Stewart platform deals with calculating
the leg lengths when the pose is given. In effect, it is a transformation relating the
global pose to local actuator lengths. The inverse kinematics of parallel manipulators is almost straightforward. The equations for the lengths of the variable links,
which are the joint variables, can very easily be expressed in terms of the position

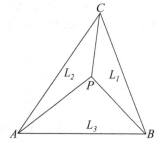

Figure 5.6. Definition of triangular coordinates.

coordinates of the points where the joints are fixed to the platform and to the base. However, in their most general form, the equations for the lengths of the variable links are quadratic and are in the form

$$l_{PB}^2 = (x_P - x_B)^2 + (y_P - y_B)^2 + (z_P - z_B)^2, \tag{5.52}$$

where P is a point on the platform and B is a point on the base where the joints are located and the coordinates are in terms of a reference frame fixed in the base. Although these could, in principle, be solved for the platform coordinates, x_P, y_P, and z_P, or the direct kinematics by use of spherical trigonometric relations, they are quite difficult to solve in practice. Apart from the use of advanced kinematic representations such as *dual quaternions* and *Plücker coordinates* referred to earlier in Subsection 3.3.2, there are a number of other practical approaches for simplifying the solution of the direct kinematics problem.

Generally these approaches define the platform position and orientation, without the use of any angles, in terms of the position of three fixed points on the platform forming a triangle. Given the distances of any point in space from each of the vertices of the triangle, the coordinates of the point may be found exactly and uniquely. The three fixed points on the platform are chosen such that the points on the platform where the joints are located are within this triangle. The coordinates of the points on the platform where the joints are located may be conveniently expressed as linear relations in terms of the so-called *triangular coordinates* or *area coordinates*.

Referring to Figure 5.6 we see that these are defined by the three coordinates, L_1, L_2, and L_3 of point P:

$$L_1(P) = \frac{\text{area } PBC}{\text{area } ABC}, \quad L_2(P) = \frac{\text{area } PCA}{\text{area } ABC}, \quad L_3(P) = \frac{\text{area } PAB}{\text{area } ABC}. \tag{5.53}$$

Because there are only two coordinates for point P, there is an implicit relation satisfied by the three coordinates L_1, L_2, and L_3 of point P, given by

$$L_1 + L_2 + L_3 = 1. \tag{5.54}$$

Using (5.54) as a constraint equation, one may eliminate L_3 or any of the other two triangular coordinates, in principle.

The Cartesian coordinates of point P in the plane of the platform, in a coordinate system fixed in the base, are given by the linear interpolation relations as

$$x_P = x_A L_1 + x_B L_2 + x_C L_3,$$

$$y_P = y_A L_1 + y_B L_2 + y_C L_3, \qquad (5.55)$$

$$z_P = z_A L_1 + z_B L_2 + z_C L_3.$$

The inverse relations may be found by inverting these equations. Thus,

$$\begin{bmatrix} L_1 \\ L_2 \\ L_3 \end{bmatrix} = \begin{bmatrix} x_A & x_B & x_C \\ y_A & y_B & y_C \\ z_A & z_B & z_C \end{bmatrix}^{-1} \begin{bmatrix} x_P \\ y_P \\ z_P \end{bmatrix}. \qquad (5.56)$$

The point P must satisfy the constraint that it lie in the same plane as points A, B, and C. This may be expressed in determinant form as

$$\begin{vmatrix} x_P & x_A & x_B & x_C \\ y_P & y_A & y_B & y_C \\ z_P & z_A & z_B & z_C \\ 1 & 1 & 1 & 1 \end{vmatrix} = 0. \qquad (5.57)$$

This constraint reduces to the implicit relation that the three area coordinates must satisfy. For an initial choice of point P, selected relative to points A, B, and C, the three area coordinates L_1, L_2, and L_3 of point P may be computed. They depend on the geometry of only the initial configuration and the relative location of P with reference to the three points A, B, and C. They are not affected by the subsequent translation and rotation of the plane containing the four points, P, A, B, and C.

For a parallel spatial manipulator, it is therefore possible to express the coordinates of the six points, P_i, $i = 1, 2, \ldots, 6$, on the platform where the joints are located, as linear interpolation relations in terms of the coordinates of the three fixed reference points, R_i, $i = 1, 2, 3$, on the platform forming a triangle, as illustrated in Figure 5.7. Moreover, the six points, P_i, $i = 1, 2, \ldots, 6$, also lie on a circle. The coordinates of the vertices of the triangle also generate three constraint equations because the sides of this triangle are constants. Thus, when appended to the inverse kinematic relations for the lengths of the variable links, there are in general nine quadratic equations that must solved for the nine coordinates defining the platform's position and orientation in terms of the coordinates on the three points on the platform.

In principle, nine quadratic equations for the nine coordinates are also solvable by the Lyapunov approach discussed in Subsection 7.3.2. However, there is some flexibility in the choice of the fixed points, which could be located arbitrarily to facilitate a recursive minimum-time solution. Specific choices for the location of the three points, while sacrificing the generality of the method, facilitate a minimum-time solution of the direct kinematics problem. Once these points are determined, the relative position and orientation of the platform may be estimated from the

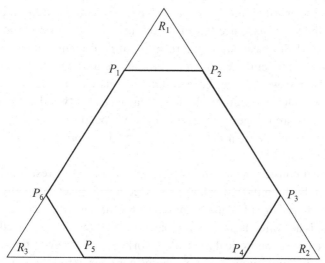

Figure 5.7. Diagram showing a typical example of the Stewart platform's joint locations relative to three fixed points in the platform frame.

initial and time-dependent estimates of the three coordinates of points. Thus the direct kinematics of the platform in terms of the translation and rotation coordinates of the center of mass of the platform may be estimated, in principle.

5.3 Workspace of a Manipulator

The workspace is defined as the space enclosing the entire set of points representing the maximum extent or reach of the robot end-effector in all three spatial directions. The workspace of a manipulator is also defined as the total volume swept out by the end-effector as the manipulator executes all possible motions. The workspace describes the working volume of the manipulator. It defines what positions the manipulator can and cannot reach in space, with the former being included within the workspace boundary.

The workspace is often broken down into a reachable workspace and a dexterous workspace. The reachable workspace is the entire set of points reachable by the manipulator, whereas the dexterous workspace consists of those points that the manipulator can reach with an arbitrary orientation of the end-effector. The reachable workspace describes the volume in space within which the manipulator end-effector's *tool center point* (TCP) can reach. The dexterous workspace is a subset of the reachable workspace based on both the position and orientation reachability of the end-effector. That is, at each point in the dexterous workspace, the end-effector can be arbitrarily oriented. Therefore the dexterous workspace is a subset of the reachable workspace.

The size and shape of the workspace depend on the coordinate geometry of the manipulator arm and also on the number of DOFs. Some workspaces are flat and confined almost entirely to one horizontal plane, whereas others are cylindrical;

still others are spherical. Some workspaces have very complicated shapes. The size and shape of the workspace are therefore the primary design criteria, particularly in the selection of the basic architecture of a parallel manipulator. The choice of the manipulator architecture and manipulator arm must be such that its workspace matches the desired workspace, both in terms of its extent and shape, as well as provide adequate DOFs within it. The load-bearing capacity is then used to size the entire manipulator and, together with the desired speed of response, an appropriate drive system, including an actuator, is selected to complete the design of the manipulator mechanism.

For practical considerations the workspace is limited or restricted by limiting devices that establish limits that will not be exceeded in the event of any foreseeable failure of the manipulator. The maximum distance that the robot can travel after the limiting device is activated is used as the basis for defining the restricted workspace. It is a subspace of the theoretical reachable workspace. However, for all theoretical purposes the space requirement of a mechanism is considered the *reachable* or *total workspace*, which specifies the space that is reachable by the TCP of the end-effector of a mechanical manipulator.

Similar to approaches in the analysis of inverse position kinematics, the investigation of workspaces is based on invoking special properties of mechanical manipulators. Frequently, additional assumptions are made about the DOFs or about the type of joints. The geometric shape of the parts is neglected in almost all cases. The reachable workspace of the TCP is the union of all reachable workspaces when the mechanism adopts any of the allowable postures. Further, as the coordinates of the TCP are determined by an inverse nonlinear transformation, the reachable workspace may also be defined as the range space of all points traversed in the vicinity of singular points; that is, points where some of the joint variables are a maximum that correspond to points where the corresponding joint velocities are zero.

Kinematic singularities of robotic manipulators correspond to regions in the manipulator's workspace where execution of a prescribed spatial motion may lead to extreme values of the joint variables, their velocities, accelerations, and higher time derivatives. Typically a kinematic configuration is the combined set of positions and orientations of all physical links in the manipulator at any particular instant in time. A singular configuration is one in which the complete instantaneous motion of the mechanism (all joint velocities) cannot be determined. In particular, for revolute-jointed serial robots, the outer workspace boundary usually corresponds to a singular configuration, making it quite difficult to perform tasks there. This boundary singularity reduces the usable workspace of the manipulator and is particularly evident to users of telerobotic manipulators.

The determination of the outer workspace boundary must involve considerations of the joint variable velocities and hence can be considered only in conjunction with inverse velocity kinematics, which is dealt with in a later chapter. There exist a number of heuristic methods of determining feasible workspace boundaries that may be analyzed to determine whether the points interior to the boundary are reachable. Although it is well nigh impossible to describe the whole range of heuristic methods that have been developed for this purpose, the basis of these heuristic methods is briefly presented. To determine the workspace of a typical manipulator such as a parallel manipulator, a computer program that uses a heuristic Monte Carlo method to calculate the volume of the workspace and condition number at different positions of the moving platform may be used. A box is defined and 10^6 or more points are randomly chosen inside the box. At each of these points the lengths of the legs are calculated and verified if they are inside the minimum and maximum imposed limits. For all the points inside the workspace of the parallel platform, the joint angles are verified to be between predefined limits. If this condition is satisfied the condition number of the Jacobian is determined. The condition number is defined as the ratio between the biggest and smallest eigenvalues of the manipulator's Jacobian matrix. The workspace volume is then estimated.

To determine the workspace boundary of the manipulator one may begin with a known point, the first critical point, on the workspace boundary. This could be a point of maximum reach with all constraints on the joints satisfied and one or more of the joints at their limits. Several feasible workspace boundary segments are then generated. One does this by varying a minimum number of joints at one time and tracing the path of the end-effector, while ensuring that the generated segment is not within the partially known workspace boundary. After a sufficiently large number of candidate boundary segments have been generated, additional critical points on the workspace boundary are identified. Any candidate critical points within the workspace boundary are discarded. One then generates new families of feasible boundary segments by tracing the envelopes of several critical points and by using the process of identifying further critical points. This process of envelope generation is repeated until the entire workspace boundary is effectively determined.

EXERCISES

5.1. In the Stanford manipulator considered earlier in this chapter, the kinematics of the orientation of the spherical wrist is expressed in terms of the three DH angle variables and the rotation matrix is expressed as

$$\mathbf{R}_{3R-DH} = \begin{bmatrix} c_4 & 0 & s_4 \\ s_4 & 0 & -c_4 \\ 0 & 1 & 0 \end{bmatrix} \begin{bmatrix} c_5 & 0 & -s_5 \\ s_5 & 0 & c_5 \\ 0 & -1 & 0 \end{bmatrix} \begin{bmatrix} c_6 & -s_6 & 0 \\ s_6 & c_6 & 0 \\ 0 & 0 & 1 \end{bmatrix}$$

where $c_i = \cos \theta_i$ and $s_i = \sin \theta_i$.

Determine equivalence relations relating the joint angles θ_i to the components of a quaternion representing the orientation of the wrist. Also obtain formulas for the joint angles in terms of the quaternion components.

5.2. Reconsider Exercises 4.1–4.6 and derive the equivalent inverse kinematic equations in each case.

5.3. Consider the cable-stayed square planar platform, $ABCD$, of side a illustrated in Figure 5.8. The cable lengths may be altered to position the platform at any location within the square of side L and with any desired rotation in the plane.

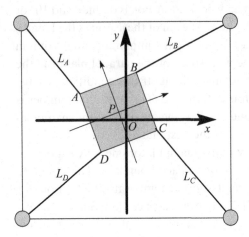

Figure 5.8. Cable-stayed planar platform.

Derive the direct and inverse kinematic equations. Comment on the compositions, if any, of the platform.

5.4. (a) Consider the triangular planar platform of side a illustrated in Figure 5.9. The link lengths may be altered to position the platform at any location within the outer triangle of side L because $L \gg a$ and with any desired rotation of the moving platform in the plane.

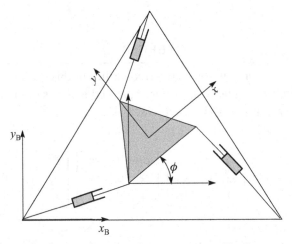

Figure 5.9. Triangular planar platform supported by linear extendable links.

Derive the inverse kinematic equations. Comment on the compositions, if any, of the platform.

(b) Consider the triangular planar platform of side a in Figure 5.10. The link lengths are fixed and of length b in each of the three double pendulums. All joints are planar revolute joints.

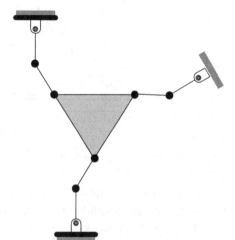

Figure 5.10. Triangular planar platform supported by linear fixed-length links.

Derive the inverse kinematic equations. Comment on the compositions, if any, of the platform.

5.5. Consider the SCARA manipulator discussed in Chapter 1 (Figure 1.18). The forward kinematics was the subject of Exercise 4.7. Derive the corresponding inverse kinematic equations.

5.6. A model is suspended in a wind tunnel by three variable-length links, each of which is anchored at an arbitrary but fixed location in space. The links are all attached to the model at almost the same point, P. The coordinates of point P are x_P, y_P, and z_P in a fixed reference frame. The coordinates of the three anchor points are x_i, y_i, and z_i, $i = 1, 2, 3$. The link lengths are given by ρ_i, $i = 1, 2, 3$ and satisfy the relations

$$\rho_i^2 = (x_P - x_i)^2 + (y_P - y_i)^2 + (z_P - z_i)^2.$$

The distances of the anchor points are given by L_i, $i = 1, 2, 3$. Show that the coordinates x_P, y_P, and z_P are given by

$$\begin{bmatrix} x_P \\ y_P \\ z_P \end{bmatrix} = \frac{1}{2} \begin{bmatrix} (x_2 - x_1) & (y_2 - y_1) & (z_2 - z_1) \\ (x_3 - x_1) & (y_3 - y_1) & (z_3 - z_1) \\ (x_3 - x_2) & (y_3 - y_2) & (z_3 - z_2) \end{bmatrix}^{-1} \begin{bmatrix} \rho_1^2 - \rho_2^2 - L_1^2 + L_2^2 \\ \rho_1^2 - \rho_3^2 - L_1^2 + L_3^2 \\ \rho_2^2 - \rho_3^2 - L_2^2 + L_3^2 \end{bmatrix}.$$

5.7. Consider the mechanisms illustrated in Figures 4.8 and 4.9. Derive expressions for the boundary of the reachable workspace in each case.

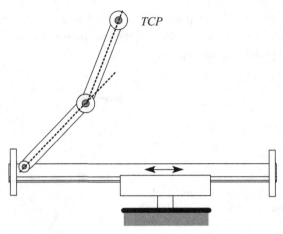

Figure 5.11. Mechanism with one prismatic and two revolute joints.

5.8. Consider the mechanism illustrated in Figure 5.11. The two links with the revolute joints may be assumed to be of equal length. Derive expressions for the boundary of the reachable workspace for this mechanism.

6

Grasping: Mechanics and Constraints

6.1 Forces and Moments

In the analyses of robot mechanisms considered in the preceding chapters, the influence of forces on the components was not considered. However, any analyses of the mechanics of such tasks as grasping a body must include an analysis of the forces acting on the components, the reactions to these applied forces, and the other consequences of the application of the force system.

Forces may be construed to have primary effects. In the first instance, forces acting on stationary rigid bodies generate internal reactions, and together they ensure that the body is in static equilibrium or in a state of rest. A state of static equilibrium is a state in which the rigid body experiences no acceleration. For a rigid body idealized as a single free particle so its orientation is unimportant, static equilibrium implies that there is no acceleration in any direction. For a general rigid body, static equilibrium not only implies that there is no acceleration in any direction but also that there is no angular acceleration in any direction.

The second and probably the more important of the two effects of forces acting on a rigid body is the production of acceleration. The moment of a force about any line is the product of the magnitude of a component of the force normal to the line and the perpendicular distance of this component from the line. The moment of two equal but opposite forces not acting at the same point about any line is a couple. A torque acting on a body with no net force is the resultant of one or more couples acting about the same line.

The influence of a system of forces is to generate acceleration in the direction of the resultant of these forces whereas the influence of a torque is to generate angular acceleration about the line of action of the torque or couple. Thus forces and moments are responsible for the generation of translational and angular acceleration, respectively. In this chapter we restrict our attention to the first of these effects and assume that the applied forces and moments do not generate any translational or angular acceleration, respectively.

For static equilibrium under the influence of a system of forces and moments, the resultant of the forces and the resultant of the moments must each be equal to zero. The resultants are both vectors with three components representing the forces and moments acting in each of the three directions.

6.2 Definition of a Wrench

For the purposes of robot mechanism or manipulator analysis, it is often quite expedient to consider both the forces and moments together. Hence a force acting along an axis and a moment about the same axis are both expressed as components of a single vector known as a *wrench*, **W**, where,

$$\mathbf{W} = \begin{bmatrix} \mathbf{M} \ \mathbf{F} \end{bmatrix}^T. \tag{6.1}$$

Given a force vector **F** acting at a point in space defined by a position vector **r**, from the origin of a *Cartesian reference frame*, the moment of the force about the origin is given by the vector cross product,

$$\mathbf{M} = \mathbf{r} \times \mathbf{F} = \begin{bmatrix} 0 & -z & y \\ z & 0 & -x \\ -y & x & 0 \end{bmatrix} \mathbf{F} = \hat{\mathbf{r}} \mathbf{F}, \tag{6.2}$$

where

$$\hat{\mathbf{r}} = \begin{bmatrix} 0 & -z & y \\ z & 0 & -x \\ -y & x & 0 \end{bmatrix}.$$

The general relationship between the moment of the force and force may be expressed as

$$\mathbf{M} = \mathbf{r} \times \mathbf{F} + p\mathbf{F} = \hat{\mathbf{r}}\mathbf{F} + p\mathbf{F} = [p\mathbf{I} + \hat{\mathbf{r}}]\,\mathbf{F}, \tag{6.3}$$

where the latter term $p\mathbf{F}$ is the component of the moment in the direction of the force. Hence it follows that the wrench is given by

$$\mathbf{W} = \begin{bmatrix} p\mathbf{I} + \hat{\mathbf{r}} \\ \mathbf{I} \end{bmatrix} \mathbf{F}. \tag{6.4}$$

The *pitch p* of the wrench is,

$$\frac{[\mathbf{F}\,0] \cdot \mathbf{W}}{\mathbf{F} \cdot \mathbf{F}} = \frac{[\mathbf{F}\,0] \cdot \begin{bmatrix} p\mathbf{I} + \hat{\mathbf{r}} \\ \mathbf{I} \end{bmatrix} \mathbf{F}}{\mathbf{F} \cdot \mathbf{F}} = \frac{\mathbf{M} \cdot \mathbf{F}}{\mathbf{F} \cdot \mathbf{F}} = p + \frac{\mathbf{F} \cdot \hat{\mathbf{r}}\mathbf{F}}{\mathbf{F} \cdot \mathbf{F}} = p. \tag{6.5}$$

6.3 Mechanics of Gripping

Robotic grasping systems in the real world are often faced with unknown objects in unknown poses. In the engineering world, the use of an appropriate end-effector to effectively grasp and then manipulate the grasped object to complete a given task involves, among others, task planning, custom design of the end-effector, a strategy for grasping, an active control system to achieve a controlled grasp, as well as a number of sensing devices. The grasping strategy plays an important role in the design of the gripper. Although the design ultimately must depend on the object shape and the type of grasp that is desired, for most applications, it is important that the object be held precisely and securely and that the surfaces that must be worked on remain accessible. Ideally the methods for planning grasps must perform well on completely unknown objects. However, most of the planning techniques are applicable to only grasps of known or partially known objects.

A grasp-planning method is essential to properly locate the contact points at which the body is held in the jaws of a gripper. The mechanics of gripping involves the general selection of a suitable gripping method from a range of such methods, the mechanics of contact and compliance, as well as the need to ensure the quality of the grasp, stability, and resistance to slipping. When a body is held in the jaws of a gripper, the body exerts on a number of forces the surface of the gripper. These forces depend to a larger extent on the nature of the surface contact, the requirements of no-slip and no-roll, and on the type of friction forces that are generated between the surfaces of the gripper and body. Surface contact between surfaces leads to forces that depend on the nature of the contact, which may be perfectly smooth or rough. It is assumed that the surfaces are not deliberately lubricated and hence are dry. Such surfaces behave consistently insofar as the frictional force is concerned, which is directly proportional to the normal force of contact over a small contacting surface. Contact forces in a contact task may also vary, depending on the nature of contact. Rigid contact refers to a situation in which the rigid gripping element comes into contact with the rigid object. It is the simplest case of contact task and is often solved in the mechanics of rigid bodies. Deformation – rigid contact is one of two types involving either contact of the rigid gripping element with the elastic object or an elastic gripping element with a rigid object. Fully deformation contact refers to situations in which there is real contact of two elastic bodies. It allows for dealing with physically meaningful situations with representative boundary conditions. Thus the forces and moments arising from the body being held in the jaws of the gripper are transmitted through the chain of components in a serial manipulator. Forces and torques must be applied at the joints to ensure that equilibrium is maintained. It is essential that the magnitudes and directions of these forces and moments are estimated so appropriate actuators can be installed to actuate the joints as required and provide the necessary forces and moments to maintain the manipulator in static equilibrium.

The condition for static equilibrium in the presence of externally applied forces and moments and internal reactions may be expressed in terms of a wrench as

$$\sum_i \mathbf{W}_i = \mathbf{0}. \tag{6.6}$$

A simple example illustrates the application of these conditions for static equilibrium. Consider an object held in the jaws of a gripper, as illustrated in Figure 6.1.

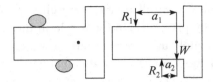

Figure 6.1. Object held in the jaws of a gripper and a free-body diagram of it.

Applying the condition for equilibrium in wrench form, we have

$$
\begin{bmatrix} 0 & 0 & 0 \\ 0 & 0 & -a_1 \\ 0 & a_1 & 0 \\ 0 & 0 & 0 \\ 0 & 1 & 0 \\ 0 & 0 & 0 \end{bmatrix}
\begin{bmatrix} 0 \\ R_1 \\ 0 \end{bmatrix}
-
\begin{bmatrix} 0 & 0 & 0 \\ 0 & 0 & -a_2 \\ 0 & a_2 & 0 \\ 0 & 0 & 0 \\ 0 & 1 & 0 \\ 0 & 0 & 0 \end{bmatrix}
\begin{bmatrix} 0 \\ R_2 \\ 0 \end{bmatrix}
+
\begin{bmatrix} 0 & 0 & 0 \\ 0 & 0 & 0 \\ 0 & 0 & 0 \\ 0 & 0 & 0 \\ 0 & 1 & 0 \\ 0 & 0 & 0 \end{bmatrix}
\begin{bmatrix} 0 \\ W \\ 0 \end{bmatrix}
=
\begin{bmatrix} 0 \\ 0 \\ 0 \\ 0 \\ 0 \\ 0 \end{bmatrix},
$$

$$(6.7)$$

which reduces to the two scalar equilibrium equations,

$$R_2 - R_1 = W, \quad a_2 R_2 - a_1 R_1 = 0. \tag{6.8}$$

Real links of a mechanism deform under externally applied forces and moments. This feature complicates the analysis and determination of the unknown reactions. Yet, for certain types of structures, known as *statically indeterminate structures*, it is the effect of the deformation that permits one to solve for the reaction forces acting on the links in the structure. A statically determinate structure is one in which there is only one distribution of internal forces and reactions that satisfies equilibrium. In a statically determinate structure, internal forces and reactions can be determined by considering nothing more than the equations of equilibrium.

The term statically determinate is used in the static analysis of mechanisms because, in such mechanisms, forces can be determined by the conditions of static equilibrium alone. If it is assumed that all link connections are revolute joints or pinned connections so they cannot resist bending moments, mechanisms are statically determinate when the following equation is satisfied:

$$m + r = n \cdot j, \tag{6.9}$$

where m is the number of links in the mechanism, r is the number of reactions, and j is the number of joints. In the preceding equation, the link forces and the reactions are the unknowns, and the number of these unknowns is $m + r$. The number of equations is equal to n times the number of joints, where n is equal to two for a planar mechanism when there are two force equilibrium equations at each joint and three for a spatial mechanism when there are three force equilibrium equations at each joint. The moments at each joint are determined by the applied joint torques.

6.4 Transformation of Forces and Moments

The transformation of the forces and moments from one joint to the other on a link may be determined by a formulation similar to the homogeneous transformation of displacements based on the DH convention and the associated reference frames. The forces and moments acting on a link in the $(i + 1)$th joint frame may be expressed in the ith joint frame as

$$\mathbf{F}_i = \mathbf{R}_{i,i+1}\mathbf{F}_{i+1}, \quad \mathbf{M}_i = \mathbf{R}_{i,i+1}\mathbf{M}_{i+1}. \tag{6.10}$$

The position vector of the origin of the joint $(i + 1)$ relative to the origin of the ith joint is

$$\mathbf{p}_{i,i+1} = \begin{bmatrix} \mathbf{I} & 0 \end{bmatrix} \mathbf{T}_{i,i+1} \begin{bmatrix} 0 & 0 & 0 & 1 \end{bmatrix}^T = \mathbf{d}_{i,i+1}. \tag{6.11}$$

The equilibrium equations are then given by

$$\bar{\mathbf{F}}_i + \mathbf{R}_{i,i+1}\mathbf{F}_{i+1} = 0,$$

$$\bar{\mathbf{M}}_i + \mathbf{R}_{i,i+1}\mathbf{M}_{i+1} + \mathbf{p}_{i,i+1} \times \mathbf{R}_{i,i+1}\mathbf{F}_{i+1} = 0, \tag{6.12}$$

where $\bar{\mathbf{F}}_i$ and $\bar{\mathbf{M}}_i$ are the force and moment vectors acting at the ith joint on the same link, which could be expressed as the single equation,

$$\begin{bmatrix} \bar{\mathbf{M}}_i \\ \bar{\mathbf{F}}_i \end{bmatrix} = - \begin{bmatrix} \mathbf{R}_{i,i+1} & \mathbf{p}_{i,i+1} \times \mathbf{R}_{i,i+1} \\ 0 & \mathbf{R}_{i,i+1} \end{bmatrix} \begin{bmatrix} \mathbf{M}_{i+1} \\ \mathbf{F}_{i+1} \end{bmatrix}. \tag{6.13}$$

Expressing it in terms of wrenches and adding any external forces and moments at the joints, we have

$$\bar{\mathbf{W}}_i = -\mathbf{W}_{ie} - \begin{bmatrix} \mathbf{R}_{i,i+1} & \mathbf{p}_{i,i+1} \times \mathbf{R}_{i,i+1} \\ 0 & \mathbf{R}_{i,i+1} \end{bmatrix} \mathbf{W}_{i+1}, \tag{6.14}$$

where $\mathbf{W}_{ie}$ is the external wrench acting at the joint on the link. Hence the reaction wrench on the next link in the chain connected to the ith joint is

$$\mathbf{W}_i = -\bar{\mathbf{W}}_i = \mathbf{W}_{ie} + \begin{bmatrix} \mathbf{R}_{i,i+1} & \mathbf{p}_{i,i+1} \times \mathbf{R}_{i,i+1} \\ 0 & \mathbf{R}_{i,i+1} \end{bmatrix} \mathbf{W}_{i+1}. \tag{6.15}$$

6.5 Compliance

Compliance refers to the ability of a robot component such as a gripper to comply with the surface of an obstacle rather than stubbornly grasp the body without any consideration of its characteristic features. Thus a compliant robot component is one that modifies its motions in such a way as to minimize a particular force or moment or a collection of forces and moments. When a component is compliant, the external and internal forces that are due to a given load cannot be calculated from the equations of equilibrium alone as the forces and moments are statically indeterminate.

Although the analysis of statically indeterminate assemblies is more involved than that of statically determinate assemblies, they offer some major advantages. The maximum stresses and deflections in a statically indeterminate assembly are smaller than those of its statically determinate counterpart. In an assembly, the presence of statically indeterminate forces has the tendency to redistribute the loading to redundant supports, which is beneficial when overloading occurs.

To determine the unknown reactions in a statically indeterminate assembly, a number of constraints must be imposed so that the number of unknowns is equal to the total number of equations. To do this it is essential to consider the elastic deformations of the links under the loading. However, the load distribution between components of a statically indeterminate assembly changes markedly because of the relative stiffnesses of the components. One approach is for the forces and moments generated by a set of compatible displacements to be included in the analysis as external forces and moments along with appropriate displacement constraints. The other approach is for the reaction forces and moments to be eliminated in terms of a fewer number of compatible displacements. By either method the number of unknown reaction forces and moments is made equal to the number of equations. The equations are then solved to determine all the forces and moments acting on a particular component.

6.5.1 Passive and Active Compliance

Compliance normally refers to passive compliance. In a passively compliant assembly the components of the assembly deform in a manner so as to comply with the applied forces and moments. In an actively compliant assembly the forces or moments acting on the assembly are sensed and compliance is actively induced by altering the displacements of the components. Thus active compliance involves not only the use of appropriate sensors to detect the compliant forces and moments but also a feedback loop and control strategy to control the forces of compliance.

6.5.2 Constraints: Natural and Artificial

Compliance to any task implies that the mechanism must satisfy certain constraints that may be associated with the tasks. Kinematic constraints, because of speed limitations, geometric constraints, or force constraints may have an influence on the types of paths generated by an end-effector.

A manipulator designed to write on a blackboard with a piece of chalk, for example, not only has to grip the chalk with the acceptable amount of force but also must be able to position it correctly relative to a set of coordinates fixed on the blackboard and then apply the correct force on the chalk normal to the blackboard so as to maintain the appropriate contact force. The manipulator then moves the chalk to write on the blackboard without altering the contact pressure. The geometric constraints describe the surface with which the end-effector has to maintain contact; they are referred to as geometric because they describe only the geometry of the surface without explicitly giving information about the path of the contact point on this surface. In fact, what is more useful is a description of the constraints as functions of time, known as kinematic constraints, in which a restricted path on the constraining surface is specified. Thus the end-effector's TCP must satisfy a set of position and force constraints in the task- or constraint-related frame that are naturally associated with the task. The frame is therefore known as a *constraint* frame, and the constraints are said to be *natural* as they arise naturally from the task-related mechanics of contact. To complete the specification of the robot-control problem, additional kinematics in the form of position and orientation constraints or force constraints may have to be specified. Such additional constraints imposed in order to complete the specification of the robot-control problem, without running the risk of overconstraining the problem, are known as *artificial constraints*. In the case of the problem of writing on the blackboard with a piece of chalk, the artificial constraints relate to the orientation, the unspecified components of the forces, and the moments acting on the chalk in the constraint frame at the point of contact with the blackboard.

6.5.3 Hybrid Control

From the preceding analysis of task compliance and the associated constraints, it is apparent that control of the end-effector's TCP involves both the control of position and orientation and the control of force and moment. Thus there is need for the simultaneous control of forces and moments and position and orientation.

When the forces and moments and position and orientation are independent of each other, simultaneous control of forces and moments and position and orientation is a relatively easy task. In the case of situations involving compliance, the forces and position are not completely independent of each other, and it may become important to establish optimal trade-offs to meet both the position and force constraints. In these cases, the normal approach is to exploit the orthogonality between the motion and force directions and control the forces or moments in certain directions for which force or moment control is more important while the position or orientation is controlled along the remaining axes. This is the concept of *hybrid control*. A practical gripper with a high degree of dexterity requires a hybrid controller including a mixed position–force gripper-control algorithm based on the gripper's functioning as a closed kinematic system that is able to keep grasping forces at a desired value in the presence of varying external exerted forces and is usually designed based on both the mechanics of contact and control system theory.

EXERCISES

6.1. Consider the slider–crank mechanism in Figure 6.2. The only force acting on the slider is a horizontal force, as shown. Obtain expressions for the wrench acting at the revolute joint to the fixed base.

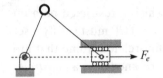

Figure 6.2. Slider–crank mechanism.

6.2. Consider the two-link manipulator of Figure 6.3, which has one revolute joint and one prismatic joint. A wrench with both a three-dimensional force and a three-dimensional moment acts at the point E. Obtain expressions for the wrenches acting at each of the joints.

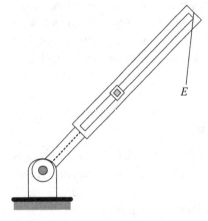

Figure 6.3. Two-link manipulator of Exercise 6.2.

6.3. Consider the three-link planar manipulator shown in Figure 6.4. A wrench with both a three-dimensional force and a three-dimensional moment acts at the point E. Obtain expressions for the wrenches acting at each of the three revolute joints.

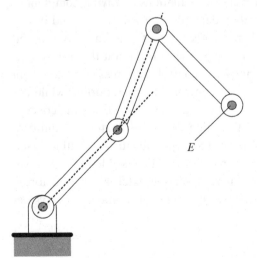

Figure 6.4. Three-link planar arm of Exercise 6.3.

6.4. Consider the two-link Cartesian manipulator shown in Figure 6.5. A wrench with both a three-dimensional force and a three-dimensional moment acts at the point E. Obtain expressions for the wrenches acting at both the prismatic joints.

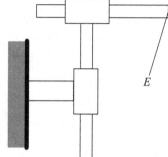

Figure 6.5. Two-link Cartesian manipulator of Exercise 6.4.

6.5. A manipulator end-effector turns a flat-headed screwdriver to drive a wood screw into a workpiece. The right-handed constraint frame is located at the point of contact between the screwdriver head and the screw, with the z axis pointing into the screwdriver along its axis and the x axis normal to the face. Briefly explain why the natural and artificial constraints in Table 6.1 apply in this case.

Table 6.1. *Natural and artificial constraints relating to a flat-headed screwdriver driving a wood screw into a workpiece*

(i)	Natural	Constraints	(ii)	Artificial	Constraints
$v_x = 0$	$\omega_x = 0$	$f_y = 0$	$v_y = 0$	$f_x = 0$	$m_x = 0$
$v_z = 0$	$\omega_y = 0$	$m_z = 0$	$\omega_z = \alpha_2$	$f_z = \alpha_3$	$m_y = 0$

6.6. A cylindrical dowel is smoothly driven into a matching hole by a manipulator end-effector. The right-handed constraint frame is located at the point of contact between the driver head and the dowel, with the z axis pointing up, along the dowel axis, and the x and y axes being diametrical axes of the circular cross section of the dowel. Briefly explain why the natural and artificial constraints in Table 6.2 apply in this case.

Table 6.2. *Natural and artificial constraints relating to a dowel smoothly driven into a matching hole*

(i)	Natural	Constraints	(ii)	Artificial	Constraints
$v_x = 0$	$\omega_x = 0$	$f_z = 0$	$v_z = v_{z0}$	$f_x = f_{x0}$	$m_x = m_{x0}$
$v_y = 0$	$\omega_y = 0$	$m_z = 0$	$\omega_z = \omega_{z0}$	$f_y = f_{y0}$	$m_y = m_{y0}$

7

Jacobians

7.1 Differential Motion

In Chapters 4 and 5, direct kinematics and inverse kinematics of manipulators, relating the position of the end-effector in base coordinates to the joint coordinates and vice versa, were considered. These relationships represent transformations from one set of coordinates to the other. However, in the context of forces acting at the joints, the definition of these transformations is incomplete. Although the work done by the joint forces is a scalar, it may be expressed as a surface or volume integral in the space defined by the coordinates. Thus it is important to relate the volume of an element in the Cartesian frame to the volume of an element in the frame defined by the joint coordinates. This relationship was first demonstrated by Carl Gustav Jacob Jacobi (1804–1851). Jacobi, who hailed from a family of Prussian bankers, worked at the University of Königsberg. He arrived there in May 1826 and pursued an academic career in pure mathematics. There he worked on, among other topics, determinants and studied the functional determinant now called the Jacobian. Although Jacobi was not the first to study the functional determinant that now bears his name, as it appears that the functional determinant was mentioned in 1815 by Cauchy, Jacobi wrote a paper on *De determinantibus functionalibus* in 1841 devoted to this determinant. He proved, among many other things, that if a set of n functions in n variables is functionally unrelated or independent, then the Jacobian cannot be identically zero.

Let the base coordinates x, y, and z be related to three joint variables, θ_1, θ_2, and θ_3, by the functional relations

$$x = f_1(\theta_1, \theta_2, \theta_3),$$

$$y = f_2(\theta_1, \theta_2, \theta_3),$$

$$z = f_3(\theta_1, \theta_2, \theta_3).$$

At this stage it is assumed that the number of base coordinates is equal to the number of joint variables. In general, it is quite possible that the number of joint variables is much larger than the number of base coordinates or the number of DOFs.

120

Consider small variations in the base coordinates x, y, and z, and the joint variables, θ_1, θ_2, and θ_3, given by, δx, δy, and δz, and the joint variables $\delta\theta_1$, $\delta\theta_2$, and $\delta\theta_3$. The relations among δx, δy, and δz, and $\delta\theta_1$, $\delta\theta_2$, and $\delta\theta_3$ are given by

$$\delta x = \frac{\partial}{\partial\theta_1} f_1 \times \delta\theta_1 + \frac{\partial}{\partial\theta_2} f_1 \times \delta\theta_2 + \frac{\partial}{\partial\theta_3} f_1 \times \delta\theta_3,$$

$$\delta y = \frac{\partial}{\partial\theta_1} f_2 \times \delta\theta_1 + \frac{\partial}{\partial\theta_2} f_2 \times \delta\theta_2 + \frac{\partial}{\partial\theta_3} f_2 \times \delta\theta_3,$$

$$\delta z = \frac{\partial}{\partial\theta_1} f_3 \times \delta\theta_1 + \frac{\partial}{\partial\theta_2} f_3 \times \delta\theta_2 + \frac{\partial}{\partial\theta_3} f_3 \times \delta\theta_3.$$

These relations can be expressed in matrix form:

$$\begin{bmatrix} \delta x \\ \delta y \\ \delta z \end{bmatrix} = \mathbf{J} \begin{bmatrix} \delta\theta_1 \\ \delta\theta_2 \\ \delta\theta_3 \end{bmatrix}, \tag{7.1}$$

where

$$\mathbf{J} = \begin{bmatrix} \dfrac{\partial}{\partial\theta_1} f_1 & \dfrac{\partial}{\partial\theta_2} f_1 & \dfrac{\partial}{\partial\theta_3} f_1 \\ \dfrac{\partial}{\partial\theta_1} f_2 & \dfrac{\partial}{\partial\theta_2} f_2 & \dfrac{\partial}{\partial\theta_3} f_2 \\ \dfrac{\partial}{\partial\theta_1} f_3 & \dfrac{\partial}{\partial\theta_2} f_3 & \dfrac{\partial}{\partial\theta_3} f_3 \end{bmatrix}. \tag{7.2}$$

Further, the volume of a parallelepiped, having the orthogonal increments δx, δy, and δz as three of its edges, may be obtained from the triple scalar product:

$$\delta x \times \delta y \times \delta z = |\mathbf{J}| \, \delta\theta_1 \times \delta\theta_2 \times \delta\theta_3, \tag{7.3}$$

where $|\mathbf{J}|$ is the determinant of $\mathbf{J}$. The matrix $\mathbf{J}$ is the Jacobian and $|\mathbf{J}|$ is the Jacobian determinant. When the Jacobian determinant $|\mathbf{J}| \neq 0$, the transformation

$$\delta\mathbf{x} \equiv \begin{bmatrix} \delta x \\ \delta y \\ \delta z \end{bmatrix} = \mathbf{J} \begin{bmatrix} \delta\theta_1 \\ \delta\theta_2 \\ \delta\theta_3 \end{bmatrix} \tag{7.4}$$

has an inverse. The quantities δx, δy, and δz represent differential translational motions and $\delta\theta_1$, $\delta\theta_2$ and $\delta\theta_3$ represent differential motions in the joint variables.

A similar relationship may be established for small differential rotational motions. Considering the *yaw–pitch–roll* Euler angle sequence, the rotational transformation is

$$\mathbf{R}_{\text{YPR}} = \begin{bmatrix} \cos\psi & -\sin\psi & 0 \\ \sin\psi & \cos\psi & 0 \\ 0 & 0 & 1 \end{bmatrix} \begin{bmatrix} \cos\theta & 0 & \sin\theta \\ 0 & 1 & 0 \\ -\sin\theta & 0 & \cos\theta \end{bmatrix} \begin{bmatrix} 1 & 0 & 0 \\ 0 & \cos\phi & -\sin\phi \\ 0 & \sin\phi & \cos\phi \end{bmatrix}. \tag{7.5}$$

If we assume the rotations to be small and equal to $\delta\psi$, $\delta\theta$, $\delta\phi$ corresponding to yaw, pitch, and roll, the trigonometric functions may be approximated as

$$\cos\,(\text{angle}) \approx 1 \text{ and } \sin\,(\text{angle}) \approx \text{angle (in radians)}.$$

Hence it follows that for small rotations about an initial set of attitudes all equal to zero the rotation matrix is

$$\mathbf{I}+\delta\mathbf{R}_{\text{YPR}} \approx \begin{bmatrix} 1 & -\delta\psi & 0 \\ \delta\psi & 1 & 0 \\ 0 & 0 & 1 \end{bmatrix} \begin{bmatrix} 1 & 0 & \delta\theta \\ 0 & 1 & 0 \\ -\delta\theta & 0 & 1 \end{bmatrix} \begin{bmatrix} 1 & 0 & 0 \\ 0 & 1 & -\delta\phi \\ 0 & \delta\phi & 1 \end{bmatrix}.$$

Performing the multiplication and ignoring all second-order terms, we obtain

$$\mathbf{I}+\delta\mathbf{R}_{\text{YPR}} \approx \begin{bmatrix} 1 & -\delta\psi & 0 \\ \delta\psi & 1 & 0 \\ 0 & 0 & 1 \end{bmatrix} \begin{bmatrix} 1 & 0 & \delta\theta \\ 0 & 1 & 0 \\ -\delta\theta & 0 & 1 \end{bmatrix} \begin{bmatrix} 1 & 0 & 0 \\ 0 & 1 & -\delta\phi \\ 0 & \delta\phi & 1 \end{bmatrix}$$

$$\approx \begin{bmatrix} 1 & -\delta\psi & \delta\theta \\ \delta\psi & 1 & -\delta\phi \\ -\delta\theta & \delta\phi & 1 \end{bmatrix}.$$

Thus any differential rotation about an arbitrary direction is the direct composition of three differential rotations about three orthogonal directions, which may be performed in any order.

It can be shown that the differential homogeneous transformation corresponding to a general differential rotation and a differential translation about any initial attitude and position is

$$\delta\mathbf{T} = \begin{bmatrix} \mathbf{R}_{\text{YPR}}+\delta\mathbf{R}_{\text{YPR}} & \mathbf{x}+\delta\mathbf{x} \\ \mathbf{0} & 1 \end{bmatrix} - \begin{bmatrix} \mathbf{R}_{\text{YPR}} & \mathbf{x} \\ \mathbf{0} & 1 \end{bmatrix} = \begin{bmatrix} \delta\mathbf{R}_{\text{YPR}} & \delta\mathbf{x} \\ \mathbf{0} & 0 \end{bmatrix},$$

where

$$\mathbf{R}_{\text{YPR}}+\delta\mathbf{R}_{\text{YPR}} = \mathbf{R}_{\text{YPR}}\,(\phi+\delta\phi,\theta+\delta\theta,\psi+\delta\psi).$$

Using the relationships for trigonometric functions of the sum of any angle and another small angle δ in radians,

$$\cos\,(\text{angle}+\delta) \approx \cos\,(\text{angle}) - \delta\sin\,(\text{angle}),$$

$$\sin\,(\text{angle}+\delta) \approx \sin\,(\text{angle}) + \delta\cos\,(\text{angle}),$$

we may express the differential rotational transformation in terms of the differential rotations as

$$\delta\mathbf{R}_{\text{YPR}} = \begin{bmatrix} 1 & -\delta\tilde{\psi} & \delta\tilde{\theta} \\ \delta\tilde{\psi} & 1 & -\delta\tilde{\phi} \\ -\delta\tilde{\theta} & \delta\tilde{\phi} & 1 \end{bmatrix} \mathbf{R}_{\text{YPR}} - \mathbf{R}_{\text{YPR}} = \begin{bmatrix} 0 & -\delta\tilde{\psi} & \delta\tilde{\theta} \\ \delta\tilde{\psi} & 0 & -\delta\tilde{\phi} \\ -\delta\tilde{\theta} & \delta\tilde{\phi} & 0 \end{bmatrix} \mathbf{R}_{\text{YPR}},$$

where $\delta\tilde{\psi},\delta\tilde{\theta},\delta\tilde{\phi}$ are functions of ψ,θ,ϕ and linear in $\delta\psi,\delta\theta,\delta\phi$.

7.1.1 Velocity Kinematics

One of the direct consequences of the relation between differential translation motions and the differential joint variables,

$$\delta \mathbf{x} \equiv \begin{bmatrix} \delta x \\ \delta y \\ \delta z \end{bmatrix} = \mathbf{J} \begin{bmatrix} \delta \theta_1 \\ \delta \theta_2 \\ \delta \theta_3 \end{bmatrix},$$

is

$$\lim_{\delta t \to 0} \frac{\delta \mathbf{x}}{\delta t} \equiv \frac{d\mathbf{x}}{dt} \equiv \begin{bmatrix} \dot{x} \\ \dot{y} \\ \dot{z} \end{bmatrix} = \mathbf{J} \begin{bmatrix} \dot{\theta}_1 \\ \dot{\theta}_2 \\ \dot{\theta}_3 \end{bmatrix}. \tag{7.6}$$

Further, it follows that

$$\dot{\mathbf{T}} = \begin{bmatrix} \dot{\mathbf{R}}_{\text{YPR}} & \dot{\mathbf{x}} \\ \mathbf{0} & 0 \end{bmatrix}, \tag{7.7}$$

where the rate of change of the rotational transformation is expressed in terms of the angular velocity components as

$$\dot{\mathbf{R}}_{\text{YPR}} = \lim_{\delta t \to 0} \frac{1}{\delta t} \begin{bmatrix} 0 & -\delta\tilde{\psi} & \delta\tilde{\theta} \\ \delta\tilde{\psi} & 0 & -\delta\tilde{\phi} \\ -\delta\tilde{\theta} & \delta\tilde{\phi} & 0 \end{bmatrix} \mathbf{R}_{\text{YPR}} = \begin{bmatrix} 0 & -\omega_3 & \omega_2 \\ \omega_3 & 0 & -\omega_1 \\ -\omega_2 & \omega_1 & 0 \end{bmatrix} \mathbf{R}_{\text{YPR}}.$$

Here the differentiation of the rotation matrix is equivalent to the application of the chain rule of differentiation.

Thus, consider a general homogeneous transformation relating the position vector in base coordinates to the position vector in the end-effector coordinates:

$$\begin{bmatrix} x_0 \\ y_0 \\ z_0 \\ 1 \end{bmatrix} = \begin{bmatrix} \mathbf{R} & \mathbf{d} \\ \mathbf{0} & 1 \end{bmatrix} \begin{bmatrix} x_e \\ y_e \\ z_e \\ 1 \end{bmatrix}. \tag{7.8}$$

Differentiating both sides with respect to time, we obtain

$$\begin{bmatrix} \dot{x}_0 \\ \dot{y}_0 \\ \dot{z}_0 \\ 0 \end{bmatrix} = \begin{bmatrix} \dot{\mathbf{R}} & \dot{\mathbf{d}} \\ \mathbf{0} & 0 \end{bmatrix} \begin{bmatrix} x_e \\ y_e \\ z_e \\ 1 \end{bmatrix} + \begin{bmatrix} \mathbf{R} & \mathbf{d} \\ \mathbf{0} & 1 \end{bmatrix} \begin{bmatrix} \dot{x}_e \\ \dot{y}_e \\ \dot{z}_e \\ 0 \end{bmatrix},$$

which is equivalent to

$$\begin{bmatrix} \dot{x}_0 \\ \dot{y}_0 \\ \dot{z}_0 \end{bmatrix} = \mathbf{R} \begin{bmatrix} \dot{x}_e \\ \dot{y}_e \\ \dot{z}_e \end{bmatrix} + \dot{\mathbf{R}} \begin{bmatrix} x_e \\ y_e \\ z_e \end{bmatrix} + \dot{\mathbf{d}}. \tag{7.9}$$

Consider a single screw motion of the end-effector axes about another axis that does pass through the origin. In this case,

$$
\begin{bmatrix} \mathbf{R} & \mathbf{d} \\ \mathbf{0} & 1 \end{bmatrix} = \begin{bmatrix} \mathbf{R} & \mathbf{p} - \mathbf{Rp} + \dfrac{\varphi}{2\pi}\mathbf{e}_\varphi p \\ \mathbf{0} & 1 \end{bmatrix}
$$

and

$$
\mathbf{d} = [\mathbf{I} - \mathbf{R}]\mathbf{p} + \frac{\varphi}{2\pi}\mathbf{e}_\varphi p.
$$

For a body fixed in the end-effector frame,

$$
[\dot{x}_e \quad \dot{y}_e \quad \dot{z}_e] = [0 \quad 0 \quad 0].
$$

Assuming that the pitch of the screw motion p and the position of the axis of rotation $\mathbf{p}$ are fixed,

$$
\dot{\mathbf{d}} = -\dot{\mathbf{R}}\mathbf{p} + \frac{\dot{\varphi}}{2\pi}\mathbf{e}_\varphi p
$$

and

$$
\dot{\mathbf{R}} = [\omega_1 \quad \omega_2 \quad \omega_3]^T \times \mathbf{R} = \dot{\varphi}\mathbf{e}_\varphi \times \mathbf{R}.
$$

It follows that, in this case, the velocity in base coordinates of a point in the end-effector frame is

$$
\begin{bmatrix} \dot{x}_0 \\ \dot{y}_0 \\ \dot{z}_0 \end{bmatrix} = \dot{\varphi}\left\{ (\mathbf{e}_\varphi \times \mathbf{R})\left(\begin{bmatrix} x_e \\ y_e \\ z_e \end{bmatrix} - \mathbf{p} \right) + \frac{p}{2\pi}\mathbf{e}_\varphi \right\}, \tag{7.10}
$$

which may be interpreted as the sum of two components, the first being the contribution of joint motion about the screw axis and the second being the contribution of the angular velocity of the screw motion to the translational velocity along the screw axis.

7.1.2 Translational Velocities and Acceleration

Rigid body motion of an object involves in general both translation of its mass center and rotation about its mass center. The motion could be with reference to a set of axes fixed in space or relative to a set of axes fixed in another object. Translation along a straight line in a fixed set of reference axes is considered first. The straight line could represent the ground track of a rigid body moving in space, i.e., the projection of a flight path of the body on a level plane. For the translation of a point A at constant speed V along a fixed straight line in Figure 7.1, the instantaneous distance from a fixed reference point is

$$
x_A = x_0 + V(t - t_0). \tag{7.11}
$$

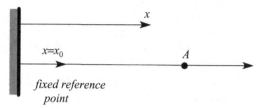

Figure 7.1. Motion along a straight line.

On the other hand, if the velocity is not constant, we have

$$x_A = x_0 + \int_{t_0}^{t} V dt. \tag{7.12}$$

When both sides are differentiated, the velocity can be expressed by

$$V = \frac{d}{dt} x_A. \tag{7.13}$$

Generalizations can now be made to two- and three-dimensional curves, in which the displacements in two or three directions are functions of time. With (7.12) applied to two displacements x, y, in two directions,

$$x_A = x_0 + \int_{t_0}^{t} V_x dt, \quad y_A = y_0 + \int_{t_0}^{t} V_y dt. \tag{7.14}$$

When both equations are differentiated with respect to time, the two equations may be written as

$$V_x = \frac{d}{dt} x_A, \quad V_y = \frac{d}{dt} y_A. \tag{7.15}$$

The next generalization we wish to consider is the effect of rotation about one of the reference axes perpendicular to the plane of the paper.

In Figure 7.2, a typical position vector at times t_0 and t_1 and the corresponding set of reference axes are shown. The position vector $\mathbf{r}$ at time t_0 is $\mathbf{r} + \delta\mathbf{r}$ and at the indicated position at time t_1. The reference frame itself has rotated by an angle $\delta\theta$ during this period of time. The components of the position vector in the rotated reference frame consist of two contributions. The first is due to the change in the position vector itself, and the second is due to the rotation of the reference frame. The change in the position vector that is due to the rotation of the reference frame

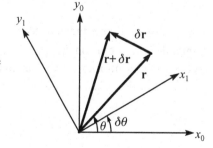

Figure 7.2. Position vector in a rotating reference frame.

is given by

$$\delta x = x\left(\cos \delta \theta - 1\right) - y \sin \delta \theta, \quad \delta y = x \sin \delta \theta + y\left(\cos \delta \theta - 1\right), \quad (7.16)$$

where x, y, and z are the components of the three-component position vector, $\mathbf{r} \equiv x\mathbf{i} + y\mathbf{j} + z\mathbf{k}$.

In the limit as time t_1 tends to t_0, i.e., as $t_1 - t_0 = \delta t \to \infty$,

$$\mathbf{V} = \frac{\delta \mathbf{r}}{\delta t} = \left[-y\frac{\delta \theta}{\delta t} \quad x\frac{\delta \theta}{\delta t} \quad 0 \right]^T, \quad (7.17)$$

where $\mathbf{V}$ is the three-component velocity vector. This can be represented in compact form:

$$\mathbf{V} = \frac{\delta \mathbf{r}}{\delta t} = \boldsymbol{\omega} \times \mathbf{r}, \quad (7.18)$$

where $\boldsymbol{\omega}$ is the rotation vector and $\boldsymbol{\omega} \times \mathbf{r}$ is the vector cross product or curl of the two vectors. The first contribution to the velocity that is due to change in the position vector can be expressed by Equation (7.13), assuming that the reference frame is "frozen" in time. The derivative is equivalent to a partial derivative with respect to time as the variables parametcrizing the rotating frame, the components of its attitude θ_i, are assumed fixed. Therefore the total velocity vector is given by

$$\mathbf{V} = \frac{d\mathbf{r}}{dt} = \frac{\partial \mathbf{r}}{\partial t} + \sum_i \frac{d\theta_i}{dt}\frac{\partial \mathbf{r}}{\partial \theta_i} = \frac{\partial \mathbf{r}}{\partial t} + \boldsymbol{\omega} \times \mathbf{r}. \quad (7.19)$$

Equation (7.19) represents the general relationship between the velocity and position vectors in a rotating reference frame. Similarly the equation relating the acceleration vector $\mathbf{A}$ to the velocity vector is given by

$$\mathbf{A} = \frac{d\mathbf{V}}{dt} = \frac{\partial \mathbf{V}}{\partial t} + \boldsymbol{\omega} \times \mathbf{V}. \quad (7.20)$$

Equation (7.19) can be substituted into Equation (7.20) to relate the acceleration vector directly to the position vector, giving

$$\mathbf{A} = \frac{\partial \mathbf{V}}{\partial t} + \boldsymbol{\omega} \times \mathbf{V} = \frac{\partial^2 \mathbf{r}}{\partial t^2} + \frac{\partial}{\partial t}\boldsymbol{\omega} \times \mathbf{r} + 2\boldsymbol{\omega} \times \frac{\partial \mathbf{r}}{\partial t} + \boldsymbol{\omega} \times \left(\boldsymbol{\omega} \times \mathbf{r}\right). \quad (7.21)$$

The vector $\mathbf{A}$ is the inertial acceleration vector referred to a rotating reference frame. The acceleration referred to the rotating frame is obtained by resolving the inertial acceleration into components in the rotating frame. One important special case of Equation (7.21) is when the position vector is constant. In this case,

$$\mathbf{A} = \frac{\partial}{\partial t}\boldsymbol{\omega} \times \mathbf{r} + \boldsymbol{\omega} \times \left(\boldsymbol{\omega} \times \mathbf{r}\right).$$

7.1.3 Angular Velocities

For the *yaw–roll–pitch* Euler angle sequence, we have already shown in Chapter 2 that

$$
\begin{bmatrix} \omega_1 \\ \omega_2 \\ \omega_3 \end{bmatrix} = \begin{bmatrix} 1 & 0 & -\sin\theta \\ 0 & \cos\phi & \sin\phi\cos\theta \\ 0 & -\sin\phi & \cos\phi\cos\theta \end{bmatrix} \begin{bmatrix} \dot\phi \\ \dot\theta \\ \dot\psi \end{bmatrix}.
$$

For a spherical wrist these may be expressed in terms of the joint angular velocities instead of the Euler angle rates.

It has already been shown in the previous section that, for a differential rotation, the rotational part of a homogeneous transformation satisfies

$$
\dot{\mathbf{R}} = \begin{bmatrix} 0 & -\omega_3 & \omega_2 \\ \omega_3 & 0 & -\omega_1 \\ -\omega_2 & \omega_1 & 0 \end{bmatrix} \mathbf{R}.
$$

The rotation matrix may be considered to be a fixed vector, and the preceding result follows when Equation (7.19) is applied to it. Thus, in general, the components of the angular velocity vector may always be expressed in terms of the joint angular velocities because

$$
\dot{\mathbf{R}} = \boldsymbol{\omega}\times\mathbf{R} = [\omega_1\ \ \omega_2\ \ \omega_3]^T\times\mathbf{R} = \begin{bmatrix} 0 & -\omega_3 & \omega_2 \\ \omega_3 & 0 & -\omega_1 \\ -\omega_2 & \omega_1 & 0 \end{bmatrix}\mathbf{R},
$$

which can be solved for the components of the angular velocity vector.

Thus, in the general case, this is a nonlinear relationship of the form

$$
\begin{bmatrix} \omega_1 \\ \omega_2 \\ \omega_3 \end{bmatrix} = \begin{bmatrix} g_1\left(\dot\theta_r,\ \dot\theta_{r+1},\ \dot\theta_{r+2}\right) \\ g_2\left(\dot\theta_r,\ \dot\theta_{r+1},\ \dot\theta_{r+2}\right) \\ g_3\left(\dot\theta_r,\ \dot\theta_{r+1},\ \dot\theta_{r+2}\right) \end{bmatrix}.
$$

However, in terms of a single screw motion about an axis $\mathbf{e}_\varphi$, the angular velocity equation is not only linear but also considerably simpler:

$$
\boldsymbol{\omega} = [\omega_1\ \ \omega_2\ \ \omega_3]^T = \mathbf{e}_\varphi\dot\varphi. \tag{7.22}
$$

7.2 Definition of a Screw Vector: Instantaneous Screws

Reconsider a single screw motion of the end-effector axes about another axis that does pass through the origin. In this case, from Equation (7.10),

$$
\begin{bmatrix} \dot x_0 \\ \dot y_0 \\ \dot z_0 \end{bmatrix} = \dot\varphi\left\{ (\mathbf{e}_\varphi\times\mathbf{R})\left(\begin{bmatrix} x_e \\ y_e \\ z_e \end{bmatrix}-\mathbf{p}\right)+\frac{p}{2\pi}\mathbf{e}_\varphi \right\}
$$

$$= \dot{\varphi} \left\{ \left(\begin{bmatrix} x_e \\ y_e \\ z_e \end{bmatrix} - \mathbf{p} \right) \frac{p}{2\pi} \right\} \begin{bmatrix} \mathbf{e}_\varphi \times \mathbf{R} \\ \mathbf{e}_\varphi \end{bmatrix} \equiv \mathbf{v},$$

and the quantity $\mathbf{v}$ may be expressed as

$$\mathbf{v} = \left[(\mathbf{r}_{ce} \times \mathbf{e}_\varphi) + \frac{p}{2\pi} \mathbf{e}_\varphi \right] \dot{\varphi}. \tag{7.23}$$

In Equation (7.23),

$$\mathbf{r}_{ce} = \mathbf{r}_{c0} - \mathbf{r}_e, \quad \mathbf{r}_{c0} = \mathbf{R}\mathbf{p}, \quad \mathbf{r}_e = \mathbf{R} \begin{bmatrix} x_e & y_e & z_e \end{bmatrix}^T,$$

and

$$\mathbf{r}_{ce} = \mathbf{r}_{c0} - \mathbf{r}_e = \mathbf{R}(\mathbf{p} - \begin{bmatrix} x_e & y_e & z_e \end{bmatrix}^T)$$

is the position vector of the center of rotation relative to the end-effector in base coordinates. Thus if the center of rotation is at the origin of the end-effector frame, $\mathbf{r}_{ce} = \mathbf{0}$. Thus the vector $\mathbf{v}$ is nothing but the translational velocity vector that is due to the screw motion.

Together with the angular velocity vector $\boldsymbol{\omega}$ about $\mathbf{e}_\varphi$, the axis of rotation, the vector

$$\mathbf{s} = \begin{bmatrix} \boldsymbol{\omega} \\ \mathbf{v} \end{bmatrix} = \begin{bmatrix} \mathbf{e}_\varphi \\ \mathbf{r}_{ce} \times \mathbf{e}_\varphi + \frac{p}{2\pi} \mathbf{e}_\varphi \end{bmatrix} \dot{\varphi} \tag{7.24}$$

is known as the *instantaneous screw vector* and plays a key role in the dynamics of the motion. The first three components are the components of the angular velocity vector whereas the last three are the components of the translational velocity vector associated with the screw motion. Physically it represents the components of the rotational and translational velocity vectors of the end-effector that are due to a joint-induced screw motion, assuming all other joints are fixed.

Consider the situation in which the number of base coordinates is fewer than the number of joint variables. In general, the number of joint variables is much more than the number of base coordinates or the number of DOFs. In a general situation, when all joints are revolute, prismatic, or screw joints, the homogeneous transformation may be considered to be a composition of successive relative rotations about each of the joint axes. Thus, considering a sequence of screw motions about each of the joint axes successively, we may write

$$\begin{bmatrix} \omega_1 \\ \omega_2 \\ \omega_3 \\ \dot{x} \\ \dot{y} \\ \dot{z} \end{bmatrix} = \sum_{i=1}^{n} \begin{bmatrix} \mathbf{e}_i h(r) \\ \mathbf{r}_i \times \mathbf{e}_i h(r) + \frac{p_i}{2\pi} \mathbf{e}_i \end{bmatrix} \dot{\theta}_i = \begin{bmatrix} \mathbf{s}_1 & \mathbf{s}_2 & \mathbf{s}_3 \cdots \mathbf{s}_{M-1} & \mathbf{s}_M \end{bmatrix} \begin{bmatrix} \dot{\theta}_1 \\ \dot{\theta}_2 \\ \dot{\theta}_3 \\ \dot{\theta}_4 \\ \cdots \\ \dot{\theta}_{M-1} \\ \dot{\theta}_M \end{bmatrix},$$

where

$$\mathbf{s}_i = \begin{bmatrix} \mathbf{e}_i h\left(r\right) \\ \mathbf{r}_i \times \mathbf{e}_i h\left(r\right) + \dfrac{p_i}{2\pi} \mathbf{e}_i \end{bmatrix}$$

is an instantaneous screw vector per unit joint variable rate. Thus,

$$\begin{bmatrix} \omega_1 \\ \omega_2 \\ \omega_3 \\ \dot{x} \\ \dot{y} \\ \dot{z} \end{bmatrix} \equiv \dot{\mathbf{x}} = \begin{bmatrix} \mathbf{s}_1 \ \mathbf{s}_2 \ \mathbf{s}_3 \cdots \mathbf{s}_{M-1} \ \mathbf{s}_M \end{bmatrix} \begin{bmatrix} \dot{\theta}_1 \\ \dot{\theta}_2 \\ \dot{\theta}_3 \\ \dot{\theta}_4 \\ \cdots \\ \dot{\theta}_{M-1} \\ \dot{\theta}_M \end{bmatrix} \equiv \mathbf{J}\dot{\Theta}, \qquad (7.25)$$

where

$$\mathbf{J} \equiv \begin{bmatrix} \mathbf{s}_1 \ \mathbf{s}_2 \ \mathbf{s}_3 \dots \mathbf{s}_{M-1} \ \mathbf{s}_M \end{bmatrix}, \qquad (7.26)$$

and $h\left(r\right)$ is equal to 1 for revolute and screw joints and to 0 for prismatic joints. For a prismatic joint, $p_i = 2\pi$ and θ_i represents the joint's translational displacement. This is just a convenient way of explicitly including the case of a prismatic, which may be treated as a rotation about a point at infinity.

Thus one may be easily assemble the Jacobian matrix $\mathbf{J}$ by associating an instantaneous screw vector with each joint, the components of which depend on the position, the orientation, and the pitch of the joint. Furthermore one may assemble the Jacobian matrix without computing any derivatives of the direct kinematic equations.

7.2.1 Duality with the Wrench

The instantaneous screw vector enjoys a special dual relationship with the wrench introduced in the preceding chapter. The rate at which work is done or the power is

$$P = \mathbf{W}^T \cdot \mathbf{s},$$

where $\mathbf{W}$ is the wrench that is responsible for generating the instantaneous screw $\mathbf{s}$.

The duality between the wrench and the instantaneous screw vector goes beyond the pairing operation of the two in the evaluation of the power. In Section 6.4 we observed how a wrench may be associated with each joint and that the wrench propagates from joint to joint, starting at the end-effector and propagating backward. A similar recursive propagation relation may be associated with instantaneous screw vectors at the joints. However, unlike the wrench, the instantaneous screw vector propagates forward from the fixed link with the base coordinate frame attached to it toward the end-effector. As one proceeds from the fixed link forward, one must progressively "undo" the DH link transformations introduced in Chapter 4. We assume that there are only revolute and prismatic joints and hence only the joint angles θ_{i+1} and the joint offsets d_{i+1} will be assumed to be functions of time. Thus the angular velocity vector of the $(i+1)$th link may be expressed in terms of the angular velocity vector of the ith link, the ith link-to-$(i + 1)$th link rotation matrix, and the relative angular velocity of the $(i + 1$th) link to the ith link as

$$\boldsymbol{\omega}_{i+1} = \mathbf{R}_{i+1,i} \left(\boldsymbol{\omega}_i + \dot{\theta}_{i+1} \begin{bmatrix} 0 \\ 0 \\ 1 \end{bmatrix} \right).$$

Similarly the translational velocity vector of the $(i + 1)$th link may be expressed in terms of the translational velocity vector of the ith link, the ith link-to-$(i + 1)$th link rotation matrix, and the relative translational velocity of the $(i + 1)$th link to the ith link as

$$\mathbf{v}_{i+1} = \mathbf{R}_{i+1,i} \left(\mathbf{v}_i + \dot{d}_{i+1} \begin{bmatrix} 0 \\ 0 \\ 1 \end{bmatrix} \right) + \mathbf{p}_{i,i+1} \times \boldsymbol{\omega}_{i+1},$$

where $\mathbf{p}_{i,i+1}$ is the position vector of the origin of the ith joint relative to the origin of the $(i + 1)$th joint. The equations for the angular and translational velocities of the $(i + 1)$th joint may be expressed as

$$\boldsymbol{\omega}_{i+1} = \mathbf{R}_{i+1,i} \left(\boldsymbol{\omega}_i + \boldsymbol{\omega}_{ie} \right),$$

$$\mathbf{v}_{i+1} = \mathbf{R}_{i+1,i} \left(\mathbf{v}_i + \mathbf{v}_{ie} \right) + \mathbf{P}_{i,i+1} \mathbf{R}_{i+1,i} \left(\boldsymbol{\omega}_i + \boldsymbol{\omega}_{le} \right),$$

where

$$\boldsymbol{\omega}_{ie} = \dot{\theta}_{i+1} \begin{bmatrix} 0 \\ 0 \\ 1 \end{bmatrix}, \mathbf{v}_{ie} = \dot{d}_{i+1} \begin{bmatrix} 0 \\ 0 \\ 1 \end{bmatrix}, \mathbf{P}_{i,i+1} = \begin{bmatrix} 0 & -p_3 & p_2 \\ p_3 & 0 & -p_1 \\ -p_2 & p_1 & 0 \end{bmatrix} \& \begin{bmatrix} p_1 \\ p_2 \\ p_3 \end{bmatrix} = \mathbf{p}_{i,i+1}.$$

Hence, for the forward propagation of the instantaneous screw vector, we have

$$\begin{bmatrix} \boldsymbol{\omega}_{i+1} \\ \mathbf{v}_{i+1} \end{bmatrix} = \begin{bmatrix} \mathbf{I} & \mathbf{0} \\ \mathbf{P}_{i,i+1} & \mathbf{I} \end{bmatrix} \begin{bmatrix} \mathbf{R}_{i+1,i} & \mathbf{0} \\ \mathbf{0} & \mathbf{R}_{i+1,i} \end{bmatrix} \left\{ \begin{bmatrix} \boldsymbol{\omega}_i \\ \mathbf{v}_i \end{bmatrix} + \begin{bmatrix} \boldsymbol{\omega}_{ie} \\ \mathbf{v}_{ie} \end{bmatrix} \right\}.$$

7.2.2 Transformation of a Compliant Body Wrench

In the previous chapter we discussed the process of transforming the wrench acting on the end-effector along with the torques acting on the joints to equivalent forces and moments in the base coordinates. It is also important to be able to transform the wrench acting in the base coordinates to equivalent joint torques.

When a wrench acts on a mechanism, it does work as the mechanism moves in response to the applied wrench. We find the total amount of work done by summing the products of all the forces acting on each body and the corresponding displacements in the directions of the force as well as the products of all the moments acting on each body and the corresponding rotations along the axes of the moments. Alternatively, we can also find the amount of work done by applying the principle of virtual work to the applied wrench and the corresponding virtual screw displacement to each body. It is also equal to the sum of the products of each of the joint torques

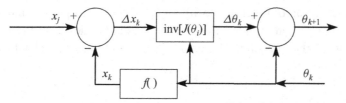

Figure 7.3. Computation of the inverse kinematics by the Newton–Raphson procedure.

or forces and the corresponding virtual joint displacements. This fact allows us to express the joint torques in terms of the wrench in base coordinates. Thus the virtual work done is

$$W = \mathbf{W}^T \cdot \delta \mathbf{x} = \mathbf{T}^T \cdot \delta \Theta,$$

where $\mathbf{T}$ is the vector of joint torques. However, the Jacobian satisfies

$$\delta \mathbf{x} = \mathbf{J} \delta \Theta,$$

and it follows that

$$W = \mathbf{W}^T \mathbf{J} \cdot \delta \Theta = \mathbf{T}^T \cdot \delta \Theta.$$

Thus the joint torques $\mathbf{T}$ may be expressed as

$$\mathbf{T} = \mathbf{J}^T \mathbf{W}. \qquad (7.27)$$

In the case in which the end-effector grasps a compliant body, the wrench may be expressed as

$$\mathbf{W} = -\mathbf{K}\mathbf{x} = -\mathbf{K}\mathbf{J}\Theta,$$

where $\mathbf{K}$ is the compliant body stiffness matrix and the associated joint torques are

$$\mathbf{T} = \mathbf{J}^T \mathbf{W} = -\mathbf{J}^T \mathbf{K}\mathbf{x} = -\mathbf{J}^T \mathbf{K}\mathbf{J}\Theta. \qquad (7.28)$$

7.3 The Jacobian and the Inverse Jacobian

The Jacobian is a matrix function and is the vector version of the first derivative of a scalar function. This Jacobian or Jacobian matrix is one of the most important quantities in the analysis and control of manipulator kinematics. It arises in almost every aspect of manipulation of the robot mechanism such as the on-line recursive computation of the inverse position kinematics by the Newton–Raphson method, as illustrated in block-diagram form in Figure 7.3, the computation of inverse velocity kinematics, the determination of singular configurations, in the implementation of coordinated anthropomorphic motion, in the derivation of the dynamic equations of motion, in the transformation of a wrench, and in the planning and execution of smooth trajectories.

The computation of inverse velocity kinematics is conceptually and apparently simpler than the computation of inverse position kinematics. Considering the relationship between the rates in the base coordinates and the joint velocities,

$$\dot{\mathbf{x}} = \mathbf{J}\dot{\Theta},$$

inverse velocity kinematics is conceptually given by

$$\dot{\Theta} = \mathbf{J}^{-1}\dot{\mathbf{x}},$$

where $\mathbf{J}^{-1}$ is the inverse Jacobian. Computing the inverse Jacobian by direct matrix inversion is probably not only the worst approach to this problem but also not practically feasible.

In general, the Jacobian matrix is not square, and as a consequence the method is not feasible even conceptually. Now, because

$$\mathbf{J}\mathbf{H}\mathbf{J}^T \left(\mathbf{J}\mathbf{H}\mathbf{J}^T\right)^{-1} = \mathbf{I},$$

we have

$$\mathbf{J}\mathbf{H}\mathbf{J}^T \left(\mathbf{J}\mathbf{H}\mathbf{J}^T\right)^{-1} = \mathbf{J}\mathbf{J}^{+} = \mathbf{I},$$

where

$$\mathbf{J}^{+} = \mathbf{H}\mathbf{J}^T \left(\mathbf{J}\mathbf{H}\mathbf{J}^T\right)^{-1}. \qquad (7.29)$$

A solution to the inverse problem is

$$\dot{\Theta} = \mathbf{J}^{+}\dot{\mathbf{x}} + \left(\mathbf{I} - \mathbf{J}^{+}\mathbf{J}\right)\dot{\mathbf{x}}_0, \qquad (7.30)$$

because the result of multiplying both sides by $\mathbf{J}$ is the relationship between the rates in the base coordinates and the joint velocities:

$$\dot{\mathbf{x}} = \mathbf{J}\dot{\Theta}.$$

where $\dot{\mathbf{x}}_0$ is an arbitrary vector that must be chosen with some care. The matrix $\mathbf{J}^{+}$ is known as the *right pseudoinverse* of the matrix $\mathbf{J}$ and sometimes conceptually referred to as the inverse Jacobian. However, given two different inverse solutions, "sufficiently close" to each other,

$$\dot{\Theta}_1 = \mathbf{J}^{+}\dot{\mathbf{x}}_1 + \left(\mathbf{I} - \mathbf{J}^{+}\mathbf{J}\right)\dot{\mathbf{x}}_0, \quad \dot{\Theta}_2 = \mathbf{J}^{+}\dot{\mathbf{x}}_2 + \left(\mathbf{I} - \mathbf{J}^{+}\mathbf{J}\right)\dot{\mathbf{x}}_0,$$

we have,

$$\Delta\dot{\Theta} = \dot{\Theta}_2 - \dot{\Theta}_1 = \mathbf{J}^{+}\left(\dot{\mathbf{x}}_2 - \dot{\mathbf{x}}_1\right) = \mathbf{J}^{+}\Delta\dot{\mathbf{x}}. \qquad (7.31)$$

This relationship is often used in estimating the errors in joint angles and velocities.

The special case in which the number of joint variables is equal to the number of equations is of some importance as in this case the Jacobian matrix is square, and it follows that

$$\mathbf{J}^{+} = \mathbf{J}^{-1}. \qquad (7.32)$$

Table 7.1. *Connectivities of typical kinematic joints*

Joint type	Rigid	Revolute	Prismatic	Screw	Cylindric	Planar	Spherical
c_i	0	1	1	1	2	3	3

It is possible to identify the number of joint variables in a manipulator as this corresponds to the number of DOFs of the kinematic chain.

7.3.1 The Mobility Criterion: Overconstrained Mechanisms

Any free link possesses six DOFs with respect to a fixed link. Thus the two-link system has a mobility of six. Every further connection imposes constraints, and there is a corresponding reduction in the DOFs. Alternatively, if the two links are rigidly connected, the mobility is zero. A kinematic joint provides the links with a certain number of additional DOFs. Thus we define the *connectivity* of a joint as the number of DOFs of a rigid body connected to another fixed rigid body through the joint.

Given a manipulator with n number of physical links and j number of joints, the *mobility* or the number of DOFs of the chain is defined by the Grübler or Kutzbach mobility criterion,

$$M = L(n - j) + \sum_{i=1}^{j} c_i,$$

where c_i is the connectivity of each joint, $L = 6$ for a spatial manipulator, and $L = 3$ for a planar manipulator. The connectivities c_i of four of the typical joints are listed in Table 7.1. Thus the number of joint variables is equal to the mobility M.

In a spatial or three-dimensional manipulator the Jacobian matrix is square when $M = 6$, and in a two-dimensional manipulator it is square when $M = 3$. When the mobility is greater than zero one could expect the system to be mobile, and when $M = 0$ the system would be immobile and consequently act as a structural framework under normal circumstances. However, there are linkages that have full range mobility, and therefore they are mechanisms even though they should be structures according to the mobility criterion. These linkages are called overconstrained mechanisms. A typical example is the so-called six-bar Bricard mechanism, which may, in principle, be used to transfer revolute motion to another revolute joint when the two joint axes are any skew lines in space, thus eliminating the need to use various types of expensive and heavy gears such as conic, spur, bevel, or worm gears. Some overconstrained mechanisms, with at least one revolute and one prismatic joint, may be used to transfer rotational motion to linear and vice versa when the revolute joint axis and the prismatic joint axis are any skew lines in three-dimensional space. The mobility of overconstrained mechanisms is due to the existence of special geometric conditions between the mechanism joint axes that are called overconstraint conditions. Although we obtain these conditions by solving the inverse kinematics of the closed-loop kinematic chain, which leads to a homogeneous system of equations and an associated characteristic equation, their spatial kinematic characteristics make them good candidates in modern linkage designs for which spatial motion is needed. In the case of some overconstrained mechanisms the mobility may even be negative, implying that even the removal of certain constraints would not make it "mobile" in the sense of the mobility criterion, although they physically realizable mechanisms.

7.3.2 Singularities: Physical Interpretation

When the determinant of the square Jacobian matrix is zero it is not invertible except in a generalized sense. The situations in which this happens correspond to singularities of the Jacobian, when the manipulator loses one or more DOFs and consequently matrix inversion is not possible. Before the computation of the inverse Jacobian is discussed, an in-depth analysis of the singularities of the Jacobian matrix and the relationship of these singularities to singular configurations or configurations in which the manipulator loses one or more DOFs are presented. The presence of singular configurations is not limited to serial manipulators alone and is a serious shortcoming of several parallel manipulators.

The Jacobian is a function of the joint variables, and together they determine the configuration of the manipulator. The set of joint variables is therefore denoted as the configuration set Q. Because the Jacobian is a function of the configuration set, those configurations for which the Jacobian matrix loses its rank are of special importance. Such configurations are called singularities or singular configurations. Identifying the singular configurations of a manipulator is essential to gaining an understanding of the limitations of the end-effector in terms of the directions of motion, joint velocities, joint torques, and the reachable workspace. Thus certain singularities represent configurations from which certain directions of motion may be unattainable or situations in which bounded end-effector velocities or torques correspond to unbounded joint velocities or torques, respectively. In some cases they would also represent transformations to unreachable points outside the workspace. Normally the inverse kinematic joint velocities are nonunique in the vicinity of the singularities.

The singularities associated with a manipulator may be broadly classified into two groups: (1) Singularities associated with the manipulator links or arms and (2) singularities associated with the wrist. The former are sometimes further classified as shoulder and elbow singularities. Considering the PUMA 560 configuration, the shoulder singularity refers to a situation in which the upper armed is locked along the axis of the body rotation. The elbow singularity refers to the situation in which the forearm and the upper arm behave like a single rigid link. To illustrate the wrist singularity, consider a spherical wrist based on the 3–2–3 Euler angle sequence. When the second rotation about the pitch axis is zero, it is impossible to distinguish between the first and the third rotations in the sequence. In this situation the wrist has just one DOF rather than three.

An alternative approach to singularities classification is based on the location of the singular point relative to the workspace. Singularities within the workspace are generally more serious than singularities on the workspace boundary.

Table 7.2. *The five-bar mechanism and associated singular configurations*

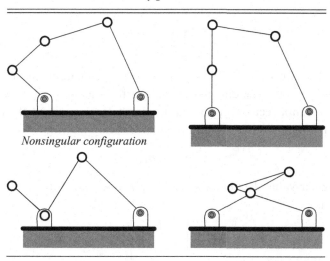

Nonsingular configuration

For a typical example of a kinematic chain with singular configurations, we consider the planar five-bar mechanism. The mechanism has a mobility of 2. Table 7.2 illustrates the five-bar mechanism and three associated singular configurations.

The principal method of identifying the singularities is *singular-value decomposition* of the Jacobian matrix. For every Jacobian matrix of size $N \times M$ and rank r, one can obtain the singular-value decomposition,

$$\mathbf{J} = \mathbf{U} \Sigma \mathbf{V}^*, \tag{7.33}$$

where $\mathbf{V}^*$ is the conjugate transpose of $\mathbf{V}$ and $\mathbf{U}$ and $\mathbf{V}$ are unitary matrices of size $N \times N$ and $M \times M$, respectively, so that their determinants are equal to unity and

$$\Sigma = \begin{bmatrix} \Sigma_r & 0 \\ 0 & 0 \end{bmatrix},$$

where $\Sigma_r = \mathrm{diag}\,[\sigma_1, \sigma_2, \ldots, \sigma_{r-1}, \sigma_r]$ and $\sigma_1 \geq \sigma_2 \geq \cdots \geq \sigma_{r-1} \geq \sigma_r > 0$. The diagonal elements of Σ_r are called singular values, and the columns of $\mathbf{U}$ and $\mathbf{V}$ are referred to as the left and right singular vectors, respectively. When plotted in hyperspace, the singular values trace an ellipsoid as the end-effector traverses its workspace. The lowest singular value is a measure of how close the Jacobian is to becoming singular. It is also a measure of the *manipulability* of the mechanism that is the ability to change position or orientation at a given configuration. The maximum singular value is a measure of the maximum amplification, and the minimum singular value is also a measure of the maximum attenuation. A singular value is unlikely to be precisely equal to zero. When it is less than a small measure of machine precision (say 2×10^{-16}) it is taken to be zero.

A major reason for computing the singular value is to be able to compute its *norm*. Norms are nonnegative numbers that are measures of the length of a vector.

Given p, where p is a positive real number, the p norm or the $\mathbf{L}_p$ norm of an $n \times 1$ vector x is defined as

$$\|x\|_p = \left[\sum_{i=1}^{n} |x_i|^p \right]^{1/p}.$$

In particular, the two-norm is also known as the Euclidian norm or the ordinary vector length. The corresponding matrix norm is the maximum of the vector norms of each of the column vectors. From singular-value decomposition, the two-norm of the Jacobian is equal to the maximum singular value:

$$\|\mathbf{J}\|_2 = \sigma_1. \tag{7.34}$$

From the singular-value decomposition, it is possible to define a pseudoinverse of the Jacobian as

$$\mathbf{J}^{\dagger} = \mathbf{V} \begin{bmatrix} \Sigma_r^{-1} & 0 \\ 0 & 0 \end{bmatrix} \mathbf{U}^*, \tag{7.35}$$

which satisfies the relations $\mathbf{J} \times \mathbf{J}^{\dagger} \times \mathbf{J} = \mathbf{J}$ and $\mathbf{J}^{\dagger} \times \mathbf{J} \times \mathbf{J}^{\dagger} = \mathbf{J}^{\dagger}$. When $r = M = N$,

$$\mathbf{J}^{\dagger} = \mathbf{J}^{-1}. \tag{7.36}$$

The pseudoinverse provides a restricted solution to the inverse velocity kinematics. If we let $\dot{\Theta} = \mathbf{J}^{\dagger} \mathbf{v}$ then

$$\mathbf{J}^{\dagger} \dot{\mathbf{x}} = \mathbf{J}^{\dagger} \mathbf{J} \dot{\Theta} = \mathbf{J}^{\dagger} \mathbf{J} \mathbf{J}^{\dagger} \mathbf{v} = \mathbf{J}^{\dagger} \mathbf{v}, \tag{7.37}$$

which implies that only a component of $\mathbf{v}$ equals a corresponding component of $\dot{\mathbf{x}}$.

Although the singular values are the square roots of the nonzero eigenvalues of $\mathbf{J}^T \mathbf{J}$, computing the eigenvalues of it is just another bad way of computing the singular values. The standard way of computing the singular-value decomposition is the *QR factorzation*, and most commercial software tools such as MATLAB provide this facility. The singular-value decomposition of the Jacobian matrix also permits the right pseudoinverse to be computed relatively easily.

7.3.3 Manipulability: Putting Redundant Mechanisms to Work

Many robot manipulators are designed to have redundant DOFs so as to facilitate their performing an array of tasks. However, to physically achieve this objective, it is necessary to recognize the goal of introducing redundancy and its utilization and to select a goal that satisfies trajectory.

In designing the mechanism of a robot manipulator or in planning the postures for performing a task, it is essential to be able to change the position and orientation of the end-effector with ease and dexterity. The renowned Japanese engineer Tsuneo Yoshikawa proposed using a measure based on the volume of the manipulability ellipsoid, as derived from a manipulator's kinematic properties, i.e., the Jacobian. This concept of manipulability is often used to obtain a quantitative measure of the end-effector's dexterity. It is a dynamic measure that is based on the idea that, when we consider the set of all end-effector velocities that are realizable by joint motions under a magnitude constraint, it is desirable that this set be not only large but also within a spherically bounded region. Measures of dynamic manipulability summarize a manipulator's capacity to generate accelerations for arbitrary tasks, and such measures are useful tools for the design and control of general-purpose robot mechanisms. For example, considering the case of a two-joint, two-link arm moving in a plane, then the aforementioned set of all end-effector velocities that are realizable by joint motions under a magnitude constraint becomes an ellipse (called the manipulability ellipsoid). Regarding the area of this ellipse as a measure of manipulation ability of the arm (called manipulability measure), we can show that this measure is a maximum when the elbow angle is a right angle and when the lengths of upper arm and forearm are the same if the total arm length is constant. The major and minor axes of the ellipse may be shown to be proportional to the maximum and minimum singular values of the Jacobian matrix. Thus the ratio of the minimum and maximum singular values of the Jacobian is also often used as a measure of *dynamic manipulability*.

A manipulator is kinematically redundant if the number of active joints is greater than the number of DOFs of its end-effector. Generally a higher degree of kinematic redundancy, involving more joints than DOFs, improves the manipulability, i.e., the end-effector dexterity. On the other hand, because of the nonuniqueness of inverse kinematics, path planning is complex, particularly when one seeks for the optimum usage of the available freedom in distributing the commanded motion to the full set of redundant joints.

The extra freedom that is due to kinematic redundancy also offers other advantages over conventional nonredundant manipulators in robot planning and control. Considering the case of a manipulator that executes a pointing task, tracing the trajectory of the object is given higher priority than avoiding obstacles in the workspace. The object's trajectory must be tracked exactly, whereas loose tolerances are typically sufficient for avoiding obstacles. Thus the DOFs associated with orientation of the camera located at the end-effector's tip must be tracked with greater fidelity than the remaining DOFs and is the basis for goal-satisfying optimization. This results in a tricky weighted optimization problem that one can effectively solve by harnessing the additional freedom in a kinematically redundant manipulator.

7.3.4 Computing the Inverse Kinematics: The Lyapunov Approach

It is often of interest to compute inverse kinematics without explicitly inverting the Jacobian. Such a technique would bypass the need to invert the Jacobian matrix. The method is based on the so-called Lyapunov approach pioneered by Aleksandr Mikhailovich Lyapunov (1857–1918) and a student Pafnuty L. Chebyshev (1821–1894), who was himself a contemporary of one of Jacobi's students. Chebyshev, who taught kinematics at the University of St. Petersburg, invented several straight-line

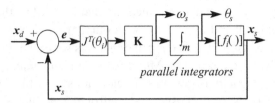

Figure 7.4. Computation of inverse kinematics by an approach based on Lyapunov's method.

mechanisms that were important in metal-planing machines and textile manufacturing. Although Lyapunov's method is not presented here in full, elements of the Lyapunov approach to inverse kinematics computation are briefly presented as the method itself plays a central role in manipulator and robot control.

The computation of inverse kinematics by an approach based on Lyapunov's method is illustrated as a multi-input multi-output block diagram in Figure 7.4. In the figure the direct kinematic equations are denoted as

$$\mathbf{x}_s = [x_{is}] = [f_i(\theta_{is})],$$

where $\mathbf{J}^T(\theta_i)$ is the transpose of the corresponding Jacobian matrix and $\int_m$ denotes a string of m parallel integrators.

Given a desired Cartesian trajectory x_d and assuming that the corresponding solution to the direct kinematic equations for the joint variables is θ_d, consider the system illustrated in Figure 7.4. The given Cartesian trajectory is assumed to be such that the time rate of change $\dot{x}_d$ is bounded from above. Given also that $\mathbf{K}$ is a positive-definite matrix and also that the Jacobian is singularity free along the particular trajectory under consideration, then there exists a positive scalar δ and a time $T > 0$, such that for all time, $t > T$,

$$\|\theta_s - \theta_d\|_2 \le \delta.$$

Furthermore, δ can be made arbitrarily small by increasing the minimum eigenvalue of $\mathbf{K}$.

To justify the preceding assertion, define a joint configuration error signal,

$$\varepsilon(t) = \theta_s(t) - \theta_d(t), \tag{7.38}$$

so it follows that

$$\theta_s(t) = \theta_d(t) + \varepsilon(t).$$

From the connected blocks in Figure 7.4 the following equations may be identified:

$$\mathbf{e}(t) = \mathbf{x}_d(t) - \mathbf{x}_s(t) = [f_i(\theta_{id}) - f(\theta_{is})],$$

$$\omega_s(t) = \mathbf{K}\mathbf{J}^T \mathbf{e}(t),$$

and the parallel integrators satisfy

$$\dot{\theta}_s(t) = \omega_s(t).$$

Hence,

$$\dot{\theta}_s(t) = -\mathbf{K}\mathbf{J}^T \left[f_i(\theta_{is}) - f(\theta_{id}) \right]. \tag{7.39}$$

For notational convenience we denote $[f_i(\theta_{is}) - f(\theta_{id})]$ as $\mathbf{E}$. Hence,

$$\dot{\varepsilon}(t) = -\mathbf{K}\mathbf{J}^T \left[f_i(\theta_{is}) - f(\theta_{id}) \right] - \dot{\theta}_d(t) = -\left[\mathbf{K}\mathbf{J}^T \mathbf{E} + \dot{\theta}_d(t) \right]. \tag{7.40}$$

To show that the error vector $\varepsilon(t)$ can be made arbitrarily small after some time T, we consider a positive-definite functional of $\varepsilon(t)$ in the form

$$V(\varepsilon, t) = \frac{1}{2} \left[f_i(\theta_{is}) - f(\theta_{id}) \right]^T \left[f_i(\theta_{is}) - f(\theta_{id}) \right] = \frac{1}{2} \mathbf{E}^T \mathbf{E}, \tag{7.41}$$

and $V(\varepsilon, t) > 0$ when $\varepsilon(t) \neq 0$.

Differentiating $V(\varepsilon, t)$ along the trajectories of $\varepsilon(t)$ that satisfy Equation (7.40), we have

$$\dot{V}(\varepsilon, t) = \frac{\partial V(\varepsilon, t)}{\partial \varepsilon} \frac{\partial \varepsilon}{\partial t} + \frac{\partial V(\varepsilon, t)}{\partial t} = \left(\frac{\partial \varepsilon^T}{\partial t} \frac{\partial \mathbf{E}^T}{\partial \varepsilon^T} + \frac{\partial \mathbf{E}^T}{\partial t} \right) \mathbf{E},$$

and it follows that,

$$\dot{V}(\varepsilon, t) = -\mathbf{E}^T \mathbf{J}\mathbf{K}\mathbf{J}^T \mathbf{E} - \dot{\theta}_d^T(t) \mathbf{J}^T \mathbf{E}. \tag{7.42}$$

Given that the minimum eigenvalue of the positive-definite matrix $\mathbf{K}$ is $\lambda_{\mathbf{K}}$ and that $\theta_d^T(t)$ and $\mathbf{J}$ are bounded, an upper bound for $\dot{V}(\varepsilon, t)$ is

$$\dot{V}(\varepsilon, t) \leq -2\lambda_{\mathbf{K}}\sigma_r^2 \|V(\varepsilon, t)\|_2 - \dot{\theta}_d^T(t) \mathbf{J}^T \mathbf{E} \leq -2\lambda_{\mathbf{K}}\sigma_r^2 \|V(\varepsilon, t)\|_2 - c\mathbf{E},$$

where c is a constant.

Hence

$$\dot{V}(\varepsilon, t) \leq -\lambda_{\mathbf{K}}\sigma_r^2 \|\mathbf{E}\|_2^2 - c \|\mathbf{E}\|_2, \tag{7.43}$$

which may be expressed as the negative of a sum of squares and

$$\dot{V}(\varepsilon, t) \leq -\left(\sqrt{\lambda_{\mathbf{K}}} \times \sigma_r \|\mathbf{E}\|_2 - \frac{c}{2\sqrt{\lambda_{\mathbf{K}}} \times \sigma_r} \right)^2 - \left(\frac{c}{2\sqrt{\lambda_{\mathbf{K}}} \times \sigma_r} \right)^2. \tag{7.44}$$

Thus the positive-definite function $V(\varepsilon, t)$ can be made to be always decreasing with increasing time along the trajectories of $\varepsilon(t)$ that satisfy Equation (7.40). Hence it follows that the error vector $\varepsilon(t)$ can be made arbitrarily small after some time T by an appropriate choice of the positive-definite matrix $\mathbf{K}$, and as a consequence it follows that

$$\theta_s(t) \to \theta_d(t)$$

sufficiently fast as the time $t \to \infty$. Thus the result allows one to recursively compute the inverse kinematics without inverting the Jacobian.

In fact, the method could be extended to compute the joint accelerations as well. However, the computation of joint accelerations from first principles is generally preferred. Because

$$\dot{\mathbf{x}} = \mathbf{J}\dot{\Theta}, \quad \ddot{\mathbf{x}} = \dot{\mathbf{J}}\dot{\Theta} + \mathbf{J}\ddot{\Theta}.$$

Hence,

$$\mathbf{z} = \ddot{\mathbf{x}} - \dot{\mathbf{J}}\dot{\Theta} = \mathbf{J}\ddot{\Theta},$$

and conceptually it follows that

$$\ddot{\Theta} = \mathbf{J}^{-1}\mathbf{z} = \mathbf{J}^{-1}(\ddot{\mathbf{x}} - \dot{\mathbf{J}}\dot{\Theta}).$$

The computation of the joint accelerations may be carried out by a recursive process similar to the one adopted for the joint velocities.

EXERCISES

7.1. Consider the two-link planar manipulator illustrated in Figure 4.2. Use the instantaneous screw vector approach and show that the 6×2 Jacobian matrix for the transformation from the end-effector to the base coordinates is of the form

$$\mathbf{J} = \begin{bmatrix} \mathbf{z}_0 & \mathbf{z}_0 \\ \mathbf{z}_0 \times \mathbf{r}_2 & \mathbf{z}_0 \times (\mathbf{r}_2 - \mathbf{r}_1) \end{bmatrix},$$

where

$$\mathbf{z}_0 = \begin{bmatrix} 0 \\ 0 \\ 1 \end{bmatrix}, \ \mathbf{r}_1 = \begin{bmatrix} a_1 c_1 \\ a_1 s_1 \\ 0 \end{bmatrix}, \ \text{and} \ \mathbf{r}_2 = \begin{bmatrix} a_1 c_1 + a_2 c_{12} \\ a_1 s_1 + a_2 s_{12} \\ 0 \end{bmatrix}.$$

Confirm your results by differentiation with respect to the joint velocities of the direct kinematic equations in Chapter 4.

7.2. Obtain the determinant of the 2×2 matrix product $\mathbf{J}^T\mathbf{J}$ and determine the singularities, if any, associated with the two-link planar manipulator.

7.3. Show that for a kinematically decoupled manipulator the Jacobian may be expressed in block-triangular form as

$$\mathbf{J} = \begin{bmatrix} \mathbf{J}_{11} & \mathbf{J}_{12} \\ \mathbf{0} & \mathbf{J}_{22} \end{bmatrix}.$$

Hence show that the singularities may be partitioned into two independent sets.

7.4. Obtain the Jacobian matrix for the transformation from the end-effector to the base coordinates of the three-link planar manipulator in Figure 4.8.

7.5. Show that for a typical three-link spherical wrist the Jacobian is of the form

$$\mathbf{J} = \begin{bmatrix} \mathbf{z}_3 & \mathbf{z}_4 & \mathbf{z}_5 \\ \mathbf{z}_3 \times \mathbf{r} & \mathbf{z}_4 \times \mathbf{r} & \mathbf{z}_5 \times \mathbf{r} \end{bmatrix}.$$

7.6. Obtain the Jacobian matrix for the transformation from the end-effector to the base coordinates of the two-link planar manipulators in Figures 4.9 and 4.10.

7.7. Obtain the Jacobian matrix for the transformation from the end-effector to the base coordinates of a $3R$ spherical wrist (3–2–3 Euler angle sequence).

7.8. Obtain the Jacobian matrix for the transformation from the end-effector to the base coordinates of the three-link planar manipulator in Figure 4.11.

7.9. Consider the four-DOF SCARA manipulator (Figure 1.18) and show that the 6×4 Jacobian matrix for the transformation from the end-effector to the base coordinates is of the form

$$
\mathbf{J} = \begin{bmatrix} \mathbf{z}_0 & \mathbf{z}_1 & \mathbf{0} & \mathbf{z}_3 \\ \mathbf{z}_0 \times \mathbf{r}_4 & \mathbf{z}_1 \times (\mathbf{r}_4 - \mathbf{r}_1) & \mathbf{z}_2 & \mathbf{0} \end{bmatrix}.
$$

Hence obtain the singularities associated with the SCARA manipulator.

7.10. Consider the six-DOF Stanford manipulator and obtain the 6×6 Jacobian matrix for the transformation from the end-effector to the base coordinates.

7.11. Consider the PUMA 560 manipulator (Figure 1.19) and obtain the Jacobian matrix for the transformation from the end-effector to the base coordinates.

  (a) Hence write a computer program to recursively compute the inverse velocity kinematics based on the Lyapunov approach.

  (b) Write a computer program to perform the singular-value decomposition of the Jacobian matrix corresponding to the PUMA 560 manipulator and determine all the singular configurations associated with it.

7.12. A Hooke's joint is used to connect to two shafts whose axes intersect at $150°$. The driving shaft rotates uniformly at 240 rpm.

  (a) If the inclination of the driven shaft to the driving shaft is α, show that the rotation θ of the driving shaft is related to the rotation ϕ of the driven shaft by

$$
\tan \theta = \cos \alpha \tan \phi.
$$

  (b) Hence deduce a general expression for the angular velocity of the driven shaft in terms of the angle of intersection between the two shafts and the angular position of one arm of the crosspiece.

  (c) Hence determine the maximum and the minimum speeds of the driven shaft.

7.13. Consider an axisymmetric three-leg-actuated fully parallel platform. The actuated legs are connected to two equilateral triangular platforms of unequal size by two-DOF universal joints. The base platform, which is the larger of the two self-similar triangles, is fixed in space. The linear actuators of the legs are in the form of the controlled prismatic joints.

Show that by placing the universal joints such that within one leg their axes are always parallel to each other, the motion of the moving platform may be restricted to translation only, relative to the base platform.

8

Newtonian, Eulerian, and Lagrangian Dynamics

8.1 Newtonian and Eulerian Mechanics

The kinematic synthesis of manipulators pointedly ignores one important aspect – size. It must be said that size matters! The parameter that is probably one of the best measures of size is inertia. The power required to drive a manipulator largely depends on the inertia of the manipulator. Thus the forces and moments that must be generated to ensure that the manipulator is not only mobile but performs the motions it is expected to take on a special significance. Yet there is one feature of the kinetics of mechanisms that sets it apart from the kinetics of free bodies. In the kinetic analysis of free bodies one is provided with the forces and moments acting on the body, and the objectives of the analysis are to determine the ensuing motions of the bodies. In the case of mechanisms the situation is just the opposite. The motions of all the kinematic links may be determined first, and it is then necessary to determine the forces and moments acting on each one of them. However, the basic principles of kinetic analysis remain the same. They all form a part of the field of *dynamics*, the study of the action of forces and moments on bodies and their relationship of the resulting motions. The study of the action of the forces and moments and the motions they generate relates to the *kinetics* whereas the relationships among the motion attributes, position, velocity, and acceleration relate to the *kinematics*.

Probably the primary underpinning principles of kinetic analysis are enshrined in Sir Isaac Newton's three laws of motion, which were enunciated in 1687 in the *Principia Mathematica*. The three laws may be stated as follows:

1. Every body continues in its state of rest or of uniform motion in a straight line unless it is made to change that state by an applied force (law of inertia).
2. The rate of change of momentum of a body is proportional to the net applied force and takes place in the direction of the net applied force.
3. To every action there is an equal and opposite reaction.

Of these three, it is the second that is most relevant to the study of the kinetics of rigid bodies such as the links in a manipulator. Because Newton's second law states that the rate of change of momentum is equal to the applied force, it may be expressed in vector form as

$$\mathbf{F} = \frac{d}{dt}\mathbf{p}, \tag{8.1}$$

where the translational momentum is $\mathbf{p} = m\mathbf{v}$. The law applies to a single particle of mass. However, a rigid body may be considered to be a collection of particles held together such that the distances between one particle and the others remain constant for all time. We obtain the total mass of the body by summing the masses of each particle that makes up the body. Each of the particles is characterized by a volume element and an associated material density. Hence, to estimate the total mass, each element of volume is multiplied by the density and the product integrated over the entire volume of the body. The general relationship defining the total mass is

$$M = \int_V \rho \, dV, \tag{8.2}$$

where ρ is the local density of material of the body. The center of mass of a rigid body is the point in the body where the sum of the moments of the masses of each of the individual particles is equal to zero. Thus, if the position vector of the center of mass is defined by $\mathbf{c}$ and $\mathbf{r}$ is the position vector of a particle, then the integral

$$\int_V \rho \, (\mathbf{r} - \mathbf{c}) \, dV = 0. \tag{8.3}$$

However, because the center of mass is a fixed point in the body,

$$\mathbf{c} \int_V \rho \, dV = \int_V \rho \mathbf{r} \, dV. \tag{8.4}$$

Hence the position of the center of mass is given by the vector

$$\mathbf{c} = \int_V \rho \mathbf{r} \, dV \bigg/ \int_V \rho \, dV = \frac{1}{M} \int_V \rho \mathbf{r} \, dV. \tag{8.5}$$

Moreover, the total linear momentum of all the particles that make the rigid body is

$$\mathbf{p} = \int_V \rho \mathbf{v} \, dV = \int_V \rho \frac{d\mathbf{r}}{dt} dV = \frac{d}{dt} \int_V \rho \mathbf{r} \, dV = \int_V \rho \, dV \frac{d\mathbf{c}}{dt} = M \frac{d\mathbf{c}}{dt}. \tag{8.6}$$

Summing the forces acting on all the particles that make up the rigid body, we obtain

$$\sum_i \mathbf{F}_i = \mathbf{F}_{\text{total}} = \frac{d}{dt}\mathbf{p} = M \frac{d^2}{dt^2}\mathbf{c}. \tag{8.7}$$

Let the position vector of the center of mass of the ith link from the origin of the frame fixed in the link and located at the joint center be $\mathbf{c}_i$. Furthermore, let the velocity of the joint center be $\mathbf{v}_i$ and the angular velocity of the link be $\boldsymbol{\omega}_i$. Then the Newtonian force equation for the motion of the ith link with mass M_i is

$$\sum_i \mathbf{F}_i = \mathbf{F}_{\text{total}} = M_i \frac{d}{dt}(\mathbf{v}_i + \boldsymbol{\omega}_i \times \mathbf{c}_i). \tag{8.8a}$$

These may also be expressed as

$$\sum_i \mathbf{F}_i = \mathbf{F}_{\text{total}} = M_i \frac{d}{dt} \left(\mathbf{v}_i + \mathbf{C}_i^T \boldsymbol{\omega}_i \right), \tag{8.8b}$$

where

$$\mathbf{C}_i \equiv \begin{bmatrix} 0 & -c_{3i} & c_{2i} \\ c_{3i} & 0 & -c_{1i} \\ -c_{2i} & c_{1i} & 0 \end{bmatrix}, \quad \begin{bmatrix} c_{1i} \\ c_{2i} \\ c_{3i} \end{bmatrix} \equiv \mathbf{c}_i.$$

The moment of momentum of a particle is defined as

$$\mathbf{h} = \mathbf{r} \times \mathbf{p}. \tag{8.9}$$

The rate of change of the moment of momentum is given by

$$\frac{d}{dt}\mathbf{h} = \frac{d}{dt} (\mathbf{r} \times \mathbf{p}) = \frac{d}{dt}\mathbf{r} \times \mathbf{p} + \mathbf{r} \times \frac{d}{dt}\mathbf{p} = \mathbf{r} \times \mathbf{F}, \tag{8.10}$$

because

$$\frac{d}{dt}\mathbf{r} \times \mathbf{p} = \frac{d}{dt}\mathbf{r} \times m\frac{d}{dt}\mathbf{r} = 0. \tag{8.11}$$

Thus the rate of change of moment of momentum is equal to the moment of the applied force:

$$\frac{d}{dt}\mathbf{h} = \mathbf{r} \times \mathbf{F}. \tag{8.12}$$

Considering a rigid body that can be thought of as collection of particles that are constrained not to move relative to each other, we have

$$\mathbf{h}_{\text{body}} = \sum \mathbf{r} \times \mathbf{p} = \int_V \mathbf{r} \times \frac{d\mathbf{r}}{dt}\rho dV. \tag{8.13}$$

Using the fact that, for the *i*th link,

$$\frac{d\mathbf{r}}{dt} = \mathbf{v}_i + \boldsymbol{\omega}_i \times \mathbf{r},$$

we obtain

$$\mathbf{h}_{\text{body}} = \sum \mathbf{r} \times \mathbf{p} = \int_V \mathbf{r} \times (\mathbf{v}_i + \boldsymbol{\omega}_i \times \mathbf{r}) \rho dV. \tag{8.14}$$

Evaluating the integral, we have

$$\mathbf{h}_{\text{body}} = \sum \mathbf{r} \times \mathbf{p} = M_i (\mathbf{c}_i \times \mathbf{v}_i) + \int_V \mathbf{r} \times (\boldsymbol{\omega}_i \times \mathbf{r}) \rho dV. \tag{8.15}$$

The integral of the vector triple product is linear in the angular velocity components and is defined as

$$\int_V \mathbf{r} \times (\boldsymbol{\omega}_i \times \mathbf{r}) \rho dV = \mathbf{I}_i \boldsymbol{\omega}_i, \tag{8.16}$$

where $\mathbf{I}_i$ is a 3×3 matrix, known as the moment of inertia matrix. With this substitution,

$$\mathbf{h}_{\text{body}} = \sum \mathbf{r} \times \mathbf{p} = M_i \left(\mathbf{c}_i \times \mathbf{v}_i \right) + \mathbf{I}_i \boldsymbol{\omega}_i. \tag{8.17}$$

However,

$$\frac{d}{dt} \mathbf{h}_{\text{body}} = \frac{d}{dt} \left(M_i \left(\mathbf{c}_i \times \mathbf{v}_i \right) + \mathbf{I}_i \boldsymbol{\omega}_i \right) = \sum \mathbf{r} \times \mathbf{F} = \mathbf{M}_{\text{total}}. \tag{8.18}$$

The equation governs the rotational equilibrium of a rigid body and is complementary to the Newtonian force equation. It may also be expressed as

$$\mathbf{M}_{\text{total}} = \sum \mathbf{r} \times \mathbf{F} = \frac{d}{dt} \left(M_i \mathbf{C}_i \mathbf{v}_i + \mathbf{I}_i \boldsymbol{\omega}_i \right). \tag{8.19}$$

The equation was independently enunciated in the context of the motion of rigid bodies in 1765 by the Swiss mathematician Leonard Euler almost 80 years after Sir Isaac Newton published his three laws of motion. Hence it is often referred to as Euler's equation of motion.

8.1.1 Kinetics of Screw Motion: The Newton–Euler Equations

Together, the Newtonian equation for force equilibrium and the Eulerian equation for moment equilibrium may be expressed as a single matrix differential equation:

$$\begin{bmatrix} \mathbf{M}_{\text{total}} \\ \mathbf{F}_{\text{total}} \end{bmatrix} = \sum_i \begin{bmatrix} \mathbf{r}_i \times \mathbf{F}_i \\ \mathbf{F}_i \end{bmatrix} = \frac{d}{dt} \left(\begin{bmatrix} \mathbf{I}_i & M_i \mathbf{C}_i \\ M_i \mathbf{C}_i^T & M_i \mathbf{I} \end{bmatrix} \times \begin{bmatrix} \boldsymbol{\omega}_i \\ \mathbf{v}_i \end{bmatrix} \right), \tag{8.20}$$

which could then be written in terms of the applied wrench and the instantaneous screw vector as

$$\mathbf{W}_{\text{total}} = \frac{d}{dt} \left(\mathbf{N}_i \mathbf{s}_i \right), \tag{8.21}$$

where

$$\mathbf{N}_i = \begin{bmatrix} \mathbf{I}_i & M_i \mathbf{C}_i \\ M_i \mathbf{C}_i^T & M_i \mathbf{I} \end{bmatrix}.$$

The combined single matrix equation is often referred to as the Newton–Euler matrix equation and is the equations of motion of a rigid body in a fixed frame of reference.

It is important to observe that $\mathbf{N}_i$ is not a constant matrix and that its elements could be functions of time.

Following d'Alembert's interpretation of the inertial terms in Newton's law, we introduce the inertial wrench defined as

$$\mathbf{W}_{\text{inertial}} = -\frac{d}{dt} \left(\mathbf{N}_i \mathbf{s}_i \right). \tag{8.22}$$

Now the dynamic equations of motion may be stated as equivalent dynamic equilibrium equations in the same form as that of the static equilibrium equations by including the inertial wrench as yet another external wrench acting on the body. The dynamic equilibrium equations in a fixed frame of reference are

$$\sum \mathbf{W} = \mathbf{W}_{\text{total}} + \mathbf{W}_{\text{inertial}} = \mathbf{0}. \tag{8.23}$$

8.1.2 Moments of Inertia

The relative moment of momentum of a system of particles is defined by

$$\mathbf{h} = \int_V (\mathbf{r} \times \mathbf{v}) \, dm = \int_V \rho \, (\mathbf{r} \times \mathbf{v}) \, dV, \tag{8.24}$$

where $\mathbf{r}$ is the position vector of a mass particle and $\mathbf{v}$ is the particle velocity relative to a moving frame of reference. Assuming that the relative motion is purely rotational, we may define the moment of momentum vector of a rigid body as

$$\mathbf{h} = \int_V \mathbf{r} \times (\boldsymbol{\omega} \times \mathbf{r}) \, dm = \int_V [\boldsymbol{\omega} \, (\mathbf{r} \cdot \mathbf{r}) - \mathbf{r} \, (\mathbf{r} \cdot \boldsymbol{\omega})] \, dm = \mathbf{I}\boldsymbol{\omega}, \tag{8.25}$$

where $\mathbf{I}$ is the moment of inertia matrix and $\boldsymbol{\omega}$ is the angular velocity vector of the reference frame given by

$$\boldsymbol{\omega} = [\omega_1 \quad \omega_2 \quad \omega_3]^T \tag{8.26a}$$

or

$$\boldsymbol{\omega} = \omega_1 \, i + \omega_2 \, j + \omega_3 \, k \tag{8.26b}$$

in terms of the mutually perpendicular unit vectors i, j, and k in the three body axes, where ω_1, ω_2, and ω_3 are the three Cartesian components of angular velocity.

If the axes along which $\mathbf{h}$ is resolved are defined to be coincident with the physical principal axes of the body, then $\mathbf{I}$ is a diagonal matrix. Thus, when $\mathbf{h}$ is not resolved along principal body axes, we get

$$\mathbf{h} = \mathbf{I}\boldsymbol{\omega} = \begin{bmatrix} I_{xx} & -I_{xy} & -I_{xz} \\ -I_{xy} & I_{yy} & -I_{yz} \\ -I_{xz} & -I_{yz} & I_{zz} \end{bmatrix} \begin{bmatrix} \omega_1 \\ \omega_2 \\ \omega_3 \end{bmatrix}. \tag{8.27}$$

Assuming that the particulate position vector $\mathbf{r}$ is given by

$$\mathbf{r} = xi + yj + yk, \tag{8.28}$$

we find that the vector triple product

$$\mathbf{r} \times (\boldsymbol{\omega} \times \mathbf{r}) = [\boldsymbol{\omega} \, (\mathbf{r} \cdot \mathbf{r}) - \mathbf{r} \, (\mathbf{r} \cdot \boldsymbol{\omega})] = (\mathbf{r} \cdot \mathbf{r}) \, \boldsymbol{\omega} - \mathbf{r}\mathbf{r} \cdot \boldsymbol{\omega} = [(\mathbf{r} \cdot \mathbf{r}) \, \mathbf{I} - \mathbf{r}\mathbf{r}] \cdot \boldsymbol{\omega}$$

is

$$\mathbf{r} \times (\boldsymbol{\omega} \times \mathbf{r}) = \left[(x^2 + y^2 + z^2) \begin{bmatrix} 1 & 0 & 0 \\ 0 & 1 & 0 \\ 0 & 0 & 1 \end{bmatrix} - \begin{bmatrix} x^2 & xy & xz \\ xy & y^2 & yz \\ xz & yz & z^2 \end{bmatrix} \right] \begin{bmatrix} \omega_1 \\ \omega_2 \\ \omega_3 \end{bmatrix}. \tag{8.29}$$

Thus the 3×3 inertia matrix is given by

$$\mathbf{I} = \int_V \rho \left[(\mathbf{r} \cdot \mathbf{r}) \mathbf{I} - \mathbf{rr} \right] dV \tag{8.30a}$$

or by

$$\mathbf{I} = \int_V \rho \left[(x^2 + y^2 + z^2) \begin{bmatrix} 1 & 0 & 0 \\ 0 & 1 & 0 \\ 0 & 0 & 1 \end{bmatrix} - \begin{bmatrix} x^2 & xy & xz \\ xy & y^2 & yz \\ xz & yz & z^2 \end{bmatrix} \right] dV. \tag{8.30b}$$

If we rotate the body by a rotational transformation $\mathbf{R}$ then the transformed moment of inertia matrix is

$$\mathbf{I}' = \int_V \rho \mathbf{R} \left[(\mathbf{r} \cdot \mathbf{r}) \mathbf{I} - \mathbf{rr} \right] \mathbf{R}^T dV = \mathbf{R} \mathbf{I} \mathbf{R}^T. \tag{8.31}$$

Thus it follows that, for the ith link, the effect of a rotational transformation $\mathbf{R}_i$ is

$$\mathbf{I}'_i = \mathbf{R}_i \mathbf{I}_i \mathbf{R}_i^T. \tag{8.32}$$

To see the effect of a translational displacement $\mathbf{d}$ of the position vector $\mathbf{r}$, on the moment of inertia, we set

$$\mathbf{r}' = \mathbf{r} + \mathbf{d} \tag{8.33}$$

and first note that

$$\mathbf{r}' \times (\boldsymbol{\omega} \times \mathbf{r}') = (\mathbf{r} + \mathbf{d}) \times \left[\boldsymbol{\omega} \times (\mathbf{r} + \mathbf{d}) \right]. \tag{8.34}$$

Consequently, the vector triple product

$$(\mathbf{r} + \mathbf{d}) \times \left[\boldsymbol{\omega} \times (\mathbf{r} + \mathbf{d}) \right] = \mathbf{r} \times (\boldsymbol{\omega} \times \mathbf{r}) + \mathbf{d} \times (\boldsymbol{\omega} \times \mathbf{r}) + \mathbf{r} \times (\boldsymbol{\omega} \times \mathbf{d}) + \mathbf{d} \times (\boldsymbol{\omega} \times \mathbf{d}),$$

and it follows that, for the ith link, the moment of inertia is

$$\mathbf{I}'_i = \mathbf{I}_i - M_i \mathbf{T}_i \mathbf{C}_i - M_i \mathbf{C}_i \mathbf{T}_i - M_i \mathbf{T}_i^2, \tag{8.35}$$

where

$$\mathbf{C}_i \equiv \begin{bmatrix} 0 & -c_{3i} & c_{2i} \\ c_{3i} & 0 & -c_{1i} \\ -c_{2i} & c_{1i} & 0 \end{bmatrix}, \begin{bmatrix} c_{1i} \\ c_{2i} \\ c_{3i} \end{bmatrix} \equiv \mathbf{c}_i, \quad \mathbf{T}_i \equiv \begin{bmatrix} 0 & -d_{3i} & d_{2i} \\ d_{3i} & 0 & -d_{1i} \\ -d_{2i} & d_{1i} & 0 \end{bmatrix}, \begin{bmatrix} d_{1i} \\ d_{2i} \\ d_{3i} \end{bmatrix} \equiv \mathbf{d}_i,$$

and $\mathbf{d}_i$ is the translation displacement of the ith link.

Under the effect of a combined translational and rotational transformation, the position vector of the center of mass is

$$\mathbf{c}' = \frac{1}{M} \int_V \rho \mathbf{r}' dV = \frac{1}{M} \int_V \rho \left(\mathbf{R} \mathbf{r} + \mathbf{d} \right) dV = \mathbf{R} \mathbf{c} + \mathbf{d}. \tag{8.36}$$

8.1.3 Dynamics of a Link's Moment of Inertia

In general the translational and rotational transformations are both functions of time. Thus the time rate of change of the moment of inertia is directly related to the time rates of change of the translational and rotational transformations. To identify this relationship explicitly, assuming that only the rotational transformation is a function of time, we note that

$$\frac{d}{dt}\mathbf{I}'_i = \frac{d}{dt}\mathbf{R}_i \times \mathbf{I}_i\mathbf{R}_i^T + \mathbf{R}_i\mathbf{I}_i \times \frac{d}{dt}\mathbf{R}_i^T. \tag{8.37}$$

Because

$$\frac{d}{dt}\mathbf{R}_i = \boldsymbol{\omega}_i \times \mathbf{R}_i,$$

$$\frac{d}{dt}\mathbf{I}'_i\boldsymbol{\omega}_i = \mathbf{I}'_i\frac{\partial}{\partial t}\boldsymbol{\omega}_i + \boldsymbol{\omega}_i \times \mathbf{R}_i\mathbf{I}_i\mathbf{R}_i^T\boldsymbol{\omega}_i + \mathbf{R}_i\mathbf{I}_i\left(\boldsymbol{\omega}_i \times \mathbf{R}_i\right)^T\boldsymbol{\omega}_i.$$

However, because

$$\mathbf{R}_i\mathbf{I}_i\left(\boldsymbol{\omega}_i \times \mathbf{R}_i\right)^T\boldsymbol{\omega}_i = \mathbf{R}_i\mathbf{I}_i\mathbf{R}_i^T\left(\boldsymbol{\omega}_i \times \boldsymbol{\omega}_i\right) = \mathbf{0},$$

it follows that

$$\frac{d}{dt}\mathbf{I}'_i\boldsymbol{\omega}_i = \mathbf{I}'_i\boldsymbol{\alpha}_i + \boldsymbol{\omega}_i \times \mathbf{I}'_i\boldsymbol{\omega}_i, \quad \boldsymbol{\alpha}_i = \frac{\partial}{\partial t}\boldsymbol{\omega}_i. \tag{8.38}$$

Further,

$$\frac{d}{dt}\mathbf{v}_i = \frac{\partial}{\partial t}\mathbf{v}_i + \boldsymbol{\omega}_i \times \mathbf{v}_i = \mathbf{a}_i + \boldsymbol{\omega}_i \times \mathbf{v}_i. \tag{8.39}$$

Thus, applying the chain rule of differentiation, we may express the Newton–Euler equations as

$$\begin{bmatrix} \mathbf{M}_{\text{total}} \\ \mathbf{F}_{\text{total}} \end{bmatrix} = \begin{bmatrix} \mathbf{I}_i & M_i\mathbf{C}_i \\ M_i\mathbf{C}_i^T & M_i\mathbf{I} \end{bmatrix}\begin{bmatrix} \boldsymbol{\alpha}_i \\ \mathbf{a}_i \end{bmatrix} + \begin{bmatrix} \boldsymbol{\omega}_i \times (\mathbf{I}_i\boldsymbol{\omega}_i + M_i\mathbf{C}_i\mathbf{v}_i) \\ \boldsymbol{\omega}_i \times M_i\mathbf{v}_i \end{bmatrix}. \tag{8.40}$$

These are the equations of motion of the rigid link in a fixed frame of reference so the inertia matrix and the position vector of the center of mass are not constants. However, the form of the equations is unchanged if we were to consider a moving reference frame fixed in the body.

To transform the equations to a joint centered reference frame fixed in the link, we assume that the fixed reference frame is colocated with the link-fixed moving frame at the joint center. Thus we need to transform the fixed frame through a sequence of rotations to align it with the link-fixed moving frame. Recall that the ith-link body coordinates are related to the fixed coordinates by the transformation

$$\begin{bmatrix} \boldsymbol{\omega}_k \\ \mathbf{v}_k \end{bmatrix}^B = \prod_{i=0}^{k}\begin{bmatrix} \mathbf{R}_{i,i-1} & \mathbf{0} \\ \mathbf{0} & \mathbf{R}_{i,i-1} \end{bmatrix}\begin{bmatrix} \boldsymbol{\omega}_k \\ \mathbf{v}_k \end{bmatrix}^F = \mathbf{R}_k\begin{bmatrix} \boldsymbol{\omega}_k \\ \mathbf{v}_k \end{bmatrix}. \tag{8.41}$$

When the transformation is applied to the Newton–Euler equations, the equations of motion remain the same in form, but are now valid in the ith-link body coordinates. Further the superscript B is not explicitly indicated for brevity. Thus the Newton–Euler equations in the ith-link body coordinates are

$$\begin{bmatrix} \mathbf{M}_{\text{total}} \\ \mathbf{F}_{\text{total}} \end{bmatrix} = \begin{bmatrix} \mathbf{I}_i & M_i\mathbf{C}_i \\ M_i\mathbf{C}_i^T & M_i\mathbf{I} \end{bmatrix}\begin{bmatrix} \boldsymbol{\alpha}_i \\ \mathbf{a}_i \end{bmatrix} + \begin{bmatrix} \boldsymbol{\omega}_i \times (\mathbf{I}_i\boldsymbol{\omega}_i + M_i\mathbf{C}_i\mathbf{v}_i) \\ \boldsymbol{\omega}_i \times M_i\mathbf{v}_i \end{bmatrix}. \tag{8.42}$$

Introducing a compact notation for the product of a screw and a wrench as

$$\left\{ \begin{bmatrix} \mathbf{a} \\ \mathbf{b} \end{bmatrix} \times \begin{bmatrix} \mathbf{A} \\ \mathbf{B} \end{bmatrix} \right\} = \begin{bmatrix} \mathbf{b} \times \mathbf{A} \\ \mathbf{a} \times \mathbf{A} + \mathbf{b} \times \mathbf{B} \end{bmatrix}, \tag{8.43}$$

we may express the Newton–Euler equations as

$$\mathbf{W}_{\text{total}} = \mathbf{N} \begin{bmatrix} \boldsymbol{\alpha}_i \\ \mathbf{a}_i \end{bmatrix} + \left\{ \begin{bmatrix} \boldsymbol{\omega}_i \\ \mathbf{v}_i \end{bmatrix} \times \mathbf{N} \begin{bmatrix} \boldsymbol{\omega}_i \\ \mathbf{v}_i \end{bmatrix} \right\}. \tag{8.44}$$

Again following d'Alembert's interpretation of the inertial terms in Newton's law, we define the inertial wrench as

$$\mathbf{W}_{\text{inertial}} = -\begin{bmatrix} \mathbf{I}_i & M_i \mathbf{C}_i \\ M_i \mathbf{C}_i^T & M_i \mathbf{I} \end{bmatrix} \begin{bmatrix} \boldsymbol{\alpha}_i \\ \mathbf{a}_i \end{bmatrix} - \begin{bmatrix} \boldsymbol{\omega}_i \times (\mathbf{I}_i \boldsymbol{\omega}_i + M_i \mathbf{C}_i \mathbf{v}_i) \\ \boldsymbol{\omega}_i \times M_i \mathbf{v}_i \end{bmatrix}. \tag{8.45}$$

Again, the dynamic equations of motion may be stated as equivalent dynamic equilibrium equations in the same form as that of the static equilibrium equations by including the inertial wrench as yet another external wrench acting on the body. The dynamic equilibrium equations in the ith-link body-fixed frame of reference are

$$\sum \mathbf{W} = \mathbf{W}_{\text{total}} + \mathbf{W}_{\text{inertial}} = \mathbf{0}. \tag{8.46}$$

In the special case of a serial manipulator with only revolute joints, the equations of motion of a typical link may be expressed in a fairly simple form. In this case,

$$\mathbf{M}_{\text{total}} = \mathbf{I}_i \boldsymbol{\alpha}_i + \boldsymbol{\omega}_i \times \mathbf{I}_i \boldsymbol{\omega}_i. \tag{8.47}$$

8.1.4 Recursive Form of the Newton–Euler Equations

We now discuss in detail the recursive formulation of robot dynamics. The idea behind the recursive formulation is a two-step iteration process.

The dynamic equilibrium equations are initially formulated for the entire serial manipulator. From Section 6.4, recursive equations relating the static wrench acting on the $(i + 1)$th link to that acting on the ith link are

$$\mathbf{W}_i = \mathbf{W}_e + \begin{bmatrix} \mathbf{R}_{i,i+1} & \mathbf{p}_{i,i+1} \times \mathbf{R}_{i,i+1} \\ \mathbf{0} & \mathbf{R}_{i,i+1} \end{bmatrix} \mathbf{W}_{i+1}. \tag{8.48}$$

In the dynamic case we may simply modify the external wrench component $\mathbf{W}_{ie}$ by adding to it the inertial wrench in the sense of d'Alembert. Hence it follows that

$$\mathbf{W}_i + \begin{bmatrix} \mathbf{I}_i & M_i \mathbf{C}_i \\ M_i \mathbf{C}_i^T & M_i \mathbf{I} \end{bmatrix} \begin{bmatrix} \boldsymbol{\alpha}_i \\ \mathbf{a}_i \end{bmatrix} + \begin{bmatrix} \boldsymbol{\omega}_i \times (\mathbf{I}_i \boldsymbol{\omega}_i + M_i \mathbf{C}_i \mathbf{v}_i) \\ \boldsymbol{\omega}_i \times M_i \mathbf{v}_i \end{bmatrix}$$

$$= \mathbf{W}_{ie} + \begin{bmatrix} \mathbf{R}_{i,i+1} & \mathbf{p}_{i,i+1} \times \mathbf{R}_{i,i+1} \\ \mathbf{0} & \mathbf{R}_{i,i+1} \end{bmatrix} \mathbf{W}_{i+1}. \tag{8.49}$$

These equations propagate backward starting at the end-effector, where the wrench acting on the workpiece at the TCP or tool tip is specified, toward the fixed link,

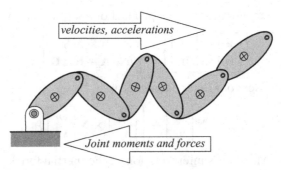

Figure 8.1. The Newton–Euler recursive formulation.

the base. On the other hand, they are now functions of the instantaneous screw vector and its time derivative. Thus we must necessarily augment the *backward-propagating* wrench equations with the *forward-propagating* instantaneous screw equations given by

$$\begin{bmatrix} \boldsymbol{\omega}_{i+1} \\ \mathbf{v}_{i+1} \end{bmatrix} = \begin{bmatrix} \mathbf{I} & \mathbf{0} \\ \mathbf{P}_{i,i+1} & \mathbf{I} \end{bmatrix} \begin{bmatrix} \mathbf{R}_{i+1,i} & \mathbf{0} \\ \mathbf{0} & \mathbf{R}_{i+1,i} \end{bmatrix} \left\{ \begin{bmatrix} \boldsymbol{\omega}_i \\ \mathbf{v}_i \end{bmatrix} + \begin{bmatrix} \boldsymbol{\omega}_{ie} \\ \mathbf{v}_{ie} \end{bmatrix} \right\}. \qquad (8.50)$$

The complete set of forward- and backward-propagating equations is similar to the so-called Luh, Walker, and Paul algorithm. However, there are indeed a number of versions of these based on the nature of the formulation. The wrench–screw-based formulation presented here is particularly suitable for the matrix analysis of the manipulator dynamics. The forward and backward recursive processes in the Newton–Euler equations are illustrated in Figure 8.1.

The two-step iterative process may now be explicitly stated. In the forward iteration the generalized velocities and accelerations of each link are propagated from the base to the tip, each quantity expressed in local reference frames attached at the joint of each link. In the backward iteration the generalized forces are propagated backward from the tip to the base, also expressed with respect to local reference frames attached at the joint of each link.

The dynamic equations of motion for a robot manipulator with only revolute and prismatic joints may be expressed in terms of the joint space variables. In this form it is known as the configuration state equation.

To express the equations in terms of joint space variables, it is essential to extract the externally applied joint force or torque from the external wrench $\mathbf{W}_{ie}$ acting on each link. In the binary links it can be shown that the joint wrench $\mathbf{W}_{\Gamma i}$ satisfies the relation

$$\mathbf{W}_{\Gamma i} + \mathbf{W}_{Ri} + \sum_{j=i}^{N} \mathbf{G}_j = \sum_{j=i}^{N} \mathbf{W}_{je}, \qquad (8.51)$$

where $\mathbf{W}_{Ri}$ is the reaction wrench and $\mathbf{G}_j$ is the gravity wrench acting on each link. Further, the reaction wrench $\mathbf{W}_{Ri}$ acts in a direction orthogonal to the *i*th-joint axis.

For a revolute joint, the ith-joint axis may be defined by the unit screw vector

$$\mathbf{S}_i = \frac{1}{\|\mathbf{v}_i\|}[\mathbf{0} \quad \mathbf{v}_i]^T. \tag{8.52}$$

Hence, by taking the dot product of the joint wrench $\mathbf{W}_{\Gamma i}$ and the unit screw vector $\mathbf{S}_i$, we obtain

$$\mathbf{W}_{\Gamma i} \cdot \mathbf{S}_i = \mathbf{W}_{\Gamma i} = \sum_{j=i}^{N} \mathbf{W}_{je} \cdot \mathbf{S}_i - \mathbf{W}_{Ri} \cdot \mathbf{S}_i - \sum_{j=i}^{N} \mathbf{G}_j \cdot \mathbf{S}_i. \tag{8.53}$$

However, because the dot product of the reaction wrench with the unit screw vector in the direction of the joint axis is zero,

$$\mathbf{W}_{\Gamma i} = \sum_{j=i}^{N} \mathbf{W}_{je} \cdot \mathbf{S}_i - \sum_{j=i}^{N} \mathbf{G}_j \cdot \mathbf{S}_i. \tag{8.54}$$

Assuming that gravity normally acts downward and is the negative Z direction in each link, the gravity wrench $\mathbf{G}_i$ may be shown to be

$$\mathbf{G}_i = \begin{bmatrix} \mathbf{I_i} & M_i \mathbf{C}_i \\ M_i \mathbf{C}_i^T & M_i \mathbf{I} \end{bmatrix} \begin{bmatrix} \mathbf{0} \\ -g\mathbf{k} \end{bmatrix}. \tag{8.55}$$

Further, the joint wrench may expressed as the sum of a control wrench and the end-effector wrench referred to base coordinates. Hence, starting from the dynamic equilibrium wrench equations, the governing equations may be expressed as

$$I_0(\mathbf{q}_0)\ddot{\mathbf{q}}_0 + \mathbf{C}_0(\mathbf{q}_0, \dot{\mathbf{q}}_0) - \Gamma_{g0}(\mathbf{q}_0) = \Gamma = \Gamma_{c0} + \mathbf{J}_0^T(\mathbf{q}_0)\mathbf{W}_N, \tag{8.56}$$

where $\mathbf{q}_0$ is the M vector of joint space variables in the frame of reference attached to the fixed link, $I_0(\mathbf{q}_0)$ is the $M \times M$ inertia matrix, $\mathbf{C}_0(\mathbf{q}_0, \dot{\mathbf{q}}_0)$ represents the Coriolis and centrifugal force vectors, Γ_{g0} is the vector of gravity torques acting at the joints, and Γ_{c0} is the vector of the generalized input control moments at the joints, and $\mathbf{W}_N$ is the wrench acting on the end-effector mapped to the joint space via the transpose of the Jacobian matrix $\mathbf{J}_0^T(\mathbf{q}_i)$. In particular, the Jacobian $\mathbf{J}_0(\mathbf{q}_0)$ expresses the relation between the instantaneous screw vector at the origin of the end-effector in the reference Cartesian frame and the joint velocities:

$$[\omega_{0e} \quad \mathbf{v}_{0e}]^T = \mathbf{J}_0(\mathbf{q}_0)\dot{\mathbf{q}}_0. \tag{8.57}$$

The coefficient matrices in Equation (8.57) correspond to the reference frame and are functions only of the joint space coordinates in this frame. The coefficient matrices in the configuration state equation, Equation (8.56), are a function of the entire configuration state vector. We determine the gravity torques at each joint recursively by assuming that the base of the manipulator is accelerating in a direction opposite to that of the gravity vector with an acceleration equal to $1g$, where g is the local acceleration that is due to gravity.

8.2 Lagrangian Dynamics of Manipulators

The *Lagrangian dynamic formulation* is an alternative approach to the Newton–Euler approach. Pioneered by the French mathematician Joseph-Louis Lagrange (1736–1813) and stated first in 1788, it is based on a energy characterization of a dynamic system in contrast to the force–moment-balance-based Newton–Euler approach. Although the governing equations of motion obtained by either approach are completely equivalent, the Lagrangian approach is considered more useful for establishing the governing equations of motion explicitly (i.e., symbolically), and the Newton–Euler approach is considered superior for purposes of computing the dynamics numerically. In the Lagrangian formulation, we typically derive the equations of motion of open-chain robotic systems by first establishing the Lagrangian. We follow this with deriving from the Lagrangian the Euler–Lagrange or simply Lagrange's equations.

In the Lagrangian framework, once a suitable set of generalized coordinates $\mathbf{q}_0 = [q_1 \, q_2 \, q_3 \ldots q_i \ldots q_{M-1} \, q_M]^T$ has been chosen (here the q_i represent joint angles for revolute joints and linear joint displacements for prismatic joints), the equations of motion are generated by means of Euler–Lagrange or simply Lagrange's equations:

$$\frac{d}{dt}\frac{\partial L}{\partial \dot{q}_i} - \frac{\partial L}{\partial q_i} = Q_i, \tag{8.58}$$

where $L = L(q_i, \dot{q}_i)$ is a scalar function called the Lagrangian that may be defined as

$$L = T(q_i, \dot{q}_i) - V(q_i). \tag{8.59}$$

Here $T(q_i, \dot{q}_i)$ denotes the total kinetic energy of the system, $V(q_i)$ is the total potential energy, and Q_i (typically denoted by τ in robot dynamics literature) is a vector representing the generalized forces and moments acting on the system other than those accounted for by the potential energy function.

For a robotic system the kinetic energy can be expressed as

$$T(q_i, \dot{q}_i) = \frac{1}{2}\sum_{i=1}^{M}\sum_{j=1}^{M} I_{ij}(\mathbf{q})\,\dot{q}_i\dot{q}_j, \tag{8.60}$$

where $I_{ij}(\mathbf{q})$ is a symmetric positive-definite matrix known as the *inertia matrix*. On substitution of the Lagrangian L into the Euler–Lagrange equations, it can be easily verified that the equations of motion take the following standard form:

$$\sum_{j=1}^{M} I_{kj}(\mathbf{q})\,\ddot{q}_j + \sum_{i=1}^{M}\sum_{j=1}^{M}\Gamma_{ijk}(\mathbf{q})\,\dot{q}_i\dot{q}_j + \phi_k(\mathbf{q}) = \tau_k, \quad k = 1, 2, 3, \ldots, M, \tag{8.61}$$

where τ_k denotes the applied torque or force at joint k,

$$\Gamma_{ijk} = \frac{1}{2}\left(\frac{\partial I_{kj}(\mathbf{q})}{\partial q_i} + \frac{\partial I_{ki}(\mathbf{q})}{\partial q_j} - \frac{\partial I_{ij}(\mathbf{q})}{\partial q_k}\right) \tag{8.62}$$

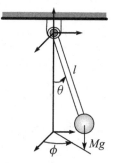

Figure 8.2. A typical spherical pendulum.

represents the Coriolis and centrifugal effects, and

$$\phi_k = \frac{\partial V(\mathbf{q})}{\partial q_k}. \tag{8.63}$$

The equations of motion may be written more compactly in matrix notation:

$$\mathbf{I}_0(\mathbf{q}_0)\ddot{\mathbf{q}}_0 + \mathbf{C}_0(\mathbf{q}_0, \dot{\mathbf{q}}_0) - \Gamma_{g0}(\mathbf{q}_0) = \Gamma, \tag{8.64}$$

where $\mathbf{q}_0$ is the M vector of joint space variables in the frame of reference attached to the fixed link, $\mathbf{I}_0(\mathbf{q}_0)$ is the $M \times M$ inertia matrix, $\mathbf{C}_0(\mathbf{q}_0, \dot{\mathbf{q}}_0)$ represents the Coriolis and centrifugal force vectors, and Γ_{g0} is the vector of gravity torques acting at the joints. When the vector of applied torques is expressed as

$$\Gamma = \Gamma_{c0} + \mathbf{J}_0^T(\mathbf{q}_0)\mathbf{W}_N, \tag{8.65}$$

where Γ_{c0} is the vector of the generalized input control moments at the joints and $\mathbf{W}_N$ is the wrench acting on the end-effector mapped to the joint space via the transpose of the Jacobian matrix $\mathbf{J}_0^T(\mathbf{q}_i)$, the equations are identical to those obtained by the Newton–Euler formulation.

A typical example illustrates the application of the Lagrangian formulation. Consider the idealized spherical pendulum illustrated in Figure 8.2.

This mechanical system consists of a body of mass m, which is attached to a spherical joint by a light rod that is assumed to be of negligible mass. Two angles θ and ϕ describe the position of the mass. All the initial conditions are assumed to be known, and the equations of motion of the mass must be derived under the influence of gravity.

As the first step, the system's Lagrangian L is established. If related to the origin of the pendulum, the position of the mass m is given by the vector $\mathbf{r}$:

$$\mathbf{r} = [l\sin\theta\cos\phi \quad l\sin\theta\sin\phi \quad l\cos\theta]^T. \tag{8.66}$$

Consequently the kinetic energy is given by

$$T(q_i, \dot{q}_i) = \frac{1}{2}\sum_{i=1}^{M}\sum_{j=1}^{M} I_{ij}(\mathbf{q})\dot{q}_i\dot{q}_j = \frac{1}{2}m\|\dot{\mathbf{r}}\|^2,$$

$$T(q_i, \dot{q}_i) = \frac{1}{2}m\|\dot{\mathbf{r}}\|^2 = \frac{1}{2}ml^2[\dot{\theta}^2 + (1 - \cos^2\theta)\dot{\phi}^2]. \tag{8.67}$$

The potential energy is given by

$$V(q_i) = -mgl\cos\theta. \tag{8.68}$$

Thus the Lagrangian is

$$L = T(q_i, \dot{q}_i) - V(q_i) = \frac{1}{2}ml^2[\dot{\theta}^2 + (1 - \cos^2\theta)\dot{\phi}^2] + mgl\cos\theta. \tag{8.69}$$

Here the generalized coordinates are $\mathbf{q} = (\theta, \phi)$. Substituting the Lagrangian into the Euler–Lagrange equations of motion, we find the following individual terms:

$$\frac{d}{dt}\frac{\partial L}{\partial \dot{q}_1} = \frac{d}{dt}\frac{\partial L}{\partial \dot{\theta}} = ml^2\ddot{\theta}, \quad \frac{\partial L}{\partial q_1} = \frac{\partial L}{\partial \theta} = ml^2\sin\theta\cos\theta\dot{\phi}^2 - mgl\sin\theta,$$

$$\frac{d}{dt}\frac{\partial L}{\partial \dot{q}_2} = \frac{d}{dt}\frac{\partial L}{\partial \dot{\phi}} = ml^2\frac{d}{dt}(\sin^2\theta\dot{\phi}) = ml^2(\sin^2\theta\ddot{\phi} + 2\sin\theta\cos\theta\dot{\theta}\,\dot{\phi}), \tag{8.70}$$

$$\frac{\partial L}{\partial q_2} = \frac{\partial L}{\partial \phi} = 0, \quad Q_1 = Q_2 = 0.$$

The governing equation of motion may be expressed as

$$ml^2\begin{bmatrix} 1 & 0 \\ 0 & \sin\theta \end{bmatrix}\begin{bmatrix} \ddot{\theta} \\ \ddot{\phi} \end{bmatrix} + ml^2\sin\theta\cos\theta\begin{bmatrix} -\dot{\phi}^2 \\ 2\dot{\theta}\dot{\phi} \end{bmatrix} + mgl\begin{bmatrix} \sin\theta \\ 0 \end{bmatrix} = \begin{bmatrix} 0 \\ 0 \end{bmatrix}. \tag{8.71}$$

Given the initial conditions, the initial coordinate position vector $\mathbf{q}(0) = [\theta(0), \phi(0)]$ and the initial coordinate velocity vector $\dot{\mathbf{q}}(0) = [\dot{\theta}(0), \dot{\phi}(0)]$, we may determine the position and velocity of the center of mass for all time $t > 0$ by numerically integrating the preceding governing equation of motion.

8.2.1 Forward and Inverse Dynamics

In the last section it was shown that the equations of link motion may be written more compactly in matrix notation as follows:

$$\mathbf{I}_0(\mathbf{q}_0)\ddot{\mathbf{q}}_0 + \mathbf{C}_0(\mathbf{q}_0, \dot{\mathbf{q}}_0) - \Gamma_{g0}(\mathbf{q}_0) = \Gamma, \tag{8.72}$$

where $\mathbf{q}_0$ is the M vector of joint space variables in the frame of reference attached to the fixed link, $\mathbf{I}_0(\mathbf{q}_0)$ is the $M \times M$ inertia matrix, $\mathbf{C}_0(\mathbf{q}_0, \dot{\mathbf{q}}_0)$ represents the Coriolis and centrifugal force vectors, and Γ_{g0} is the vector of gravity torques acting at the joints. These equations may be applied in a variety of ways to manipulator problems, and this aspect is briefly discussed here.

Two basic formulations arise in the dynamic analysis of multibody manipulator systems. The *forward dynamics* problem is to determine the accelerations of the system when given the initial positions and velocities of the system and the applied forces. Thus the forward dynamics problem is as follows:

> Given the vectors of joint positions $\mathbf{q}_0$, joint velocities $\dot{\mathbf{q}}_0$, and joint torques Γ, as well as the mass distribution of each link, find the resulting acceleration of the end-effector's pose.

Hence forward dynamics is used for *simulation* of the end-effector's motion, to find what the manipulator does when known joint torques are applied.

The *inverse dynamics* problem is to determine the applied forces required for producing a specified motion of the system. Whereas forward dynamics uses forces and moments to create motion, inverse dynamics uses the motion to attempt to find the forces or moments responsible for the motion. The inverse dynamics problem is then as follows:

> Given the vectors of joint positions $\mathbf{q}_0$, joint velocities $\dot{\mathbf{q}}_0$, and desired joint accelerations $\ddot{\mathbf{q}}_0$ (or the end-effector's motion), as well as the mass matrix of each link, find the vector of joint torques Γ required for generating the desired pose.

Inverse dynamics is generally used for two main purposes:

Controller set-point synthesis: When one desires the manipulator to follow a specified trajectory, one has to convert the desired motion into the joint forces that will generate this motion. These are then used to generate the command signals or set points to the joint control actuators.

Actuator sizing: When generating a desired motion for the manipulator end-effector, one can use the inverse dynamics of the manipulator to check whether the manipulator's actuators are capable of generating the joint forces needed to execute the trajectory. Thus inverse dynamics may be used for sizing the desired actuator in the design of new manipulator.

Developing application-specific algorithms to calculate either forward or inverse dynamics is much more important for *serial* manipulators than for *parallel* manipulators. The dynamics of parallel manipulators may be reasonably approximated by the dynamics of *one single* rigid body. Parallel manipulators have light links, and all actuators are in, or close to, the base, such that the contributions of the manipulator inertias themselves are limited. Furthermore, the parallel connection of the links allows one to express the Euler–Lagrange or Lagrangian dynamical equations of motion as

$$\mathbf{I}_0\left(\mathbf{q}_0\right)\ddot{\mathbf{q}}_0 + \mathbf{C}_0\left(\mathbf{q}_0, \dot{\mathbf{q}}_0\right) - \Gamma_{g0}\left(\mathbf{q}_0\right) = \mathbf{J}_0^T\left(\mathbf{q}_0\right)\mathbf{W}_N, \qquad (8.73)$$

where the vector of applied torques is expressed in terms of $\mathbf{W}_N$, which is the wrench acting on the end-effector mapped to the joint space via the transpose of the Jacobian matrix $\mathbf{J}_0^T\left(\mathbf{q}_i\right)$.

Inverse dynamics is easily found when the chain is an *open chain*, with no resistance to motion at the terminal segment, as all the kinematic variables are known from motion analysis. When there is contact of a link with another object, such as the ground or a previous link in the chain, the forces between the two links in this *closed chain* must be measured. Although this complicates the formulation of the inverse dynamics, a number of practical algorithms exist for determining inverse dynamics of serial and parallel manipulators involving both open and closed kinematic chains. The recursive Newton–Euler equations are well suited for solving the inverse dynamics problem.

In the field of robotics, in which simulation is a critical part of the analysis and evaluation of manipulator systems, it is also of paramount importance to be able to solve the forward dynamics problem, i.e., given the initial state of the system in terms of the initial values of the joint variables, the joint velocities and joint accelerations, the applied generalized force acting at the tip, and the torques applied at the joints, one must determine the subsequent state of the system in terms of the joint variables, the joint velocities, and joint accelerations. Moreover, for applications such as the testing of advanced control algorithms and terrestrial-based teleoperations of remote space robots, the development of computationally efficient forward dynamics algorithms is also absolutely vital.

8.3 The Principle of Virtual Work

An important principle that is extremely handy in estimating the applied forces or torques is the principle of virtual work. In using this principle, the inertial forces and torques are first estimated based on the linear and angular velocities of each of the bodies. Then the whole of the manipulator is considered to be in dynamic equilibrium, and the principle of virtual work is applied to derive the input force or torque necessary to drive the motion without having to estimate the reaction forces and torques.

The concepts of *virtual displacement* and *virtual work* are extremely useful abstractions and are essential to the elucidation of the principles of dynamics. Suppose the particles that constitute a dynamical system undergo small instantaneous displacements, which are independent of time and consistent with constraints on the system and such that all internal and external forces remain unchanged in magnitude and direction during the displacements. Such displacements are said to be *virtual* because of their hypothetical nature.

Let the ith particle of mass m_i at position $\mathbf{r}_i$ at time t experience a virtual displacement to position $\mathbf{r}_i + \delta\mathbf{r}_i$. Let $\mathbf{F}_i$ and $\mathbf{R}_i$ be the external and internal forces acting on m_i, respectively. The virtual work done on m_i in the displacement is $(\mathbf{F}_i + \mathbf{R}_i) \cdot \delta\mathbf{r}_i$, and so the total virtual work done on all particles of the system when similar displacements are made is

$$\delta W = \sum_{i=1}^{N} (\mathbf{F}_i + \mathbf{R}_i) \cdot \delta\mathbf{r}_i = \sum_{i=1}^{N} \mathbf{F}_i \cdot \delta\mathbf{r}_i + \sum_{i=1}^{N} \mathbf{R}_i \cdot \delta\mathbf{r}_i. \qquad (8.74)$$

Now

$$\delta W_r = \sum_{i=1}^{N} \mathbf{R}_i \cdot \delta\mathbf{r}_i \qquad (8.75)$$

is the total work done by the internal forces of the system. In most practical manipulator systems this is zero, and we shall assume this to be true, unless otherwise stated. When the internal forces do no work in a virtual displacement,

$$\delta W = \sum_{i=1}^{N} \mathbf{F}_i \cdot \delta\mathbf{r}_i = \sum_{i=1}^{N} X_i \delta x_i + Y_i \delta y_i + Z_i \delta z_i, \qquad (8.76)$$

where $\mathbf{F}_i = [X_i, Y_i, Z_i]$ and $\delta\mathbf{r}_i = [\delta x_i, \delta y_i, \delta z_i]$.

δW is termed the *virtual work function*, and we note that the coefficients in it of δx_i, δy_i, and δz_i are the external force components, X_i, Y_i, and Z_i.

Considering the case of mechanical systems in static equilibrium, *the principle of virtual work* may be stated as follows: If a system with workless constraints is in static equilibrium, the total virtual work done on all the virtual displacements must be equal to zero. Hence,

$$\delta W = \sum_{i=1}^{N} X_i \delta x_i + Y_i \delta y_i + Z_i \delta z_i = 0. \tag{8.77}$$

When the constraints are frictionless, the reactive forces at the constraints do no work, as their directions are orthogonal to the directions of the virtual displacements. Frictionless constraints are a typical example of workless constraints. Thus the forces at workless constraints need not be considered in evaluating virtual work.

It is essential to restate that the principle of virtual work requires the following:

1. internal forces to do no work unless the "internal" forces that do work are treated as "external" forces,
2. reactions to be frictionless unless friction forces are explicitly included as external forces,
3. virtual displacements be compatible with geometric constraints, and
4. the system to be in static equilibrium.

The last requirement may be somewhat relaxed, and the principle of virtual work may be extended to the case of dynamic equilibrium following the application of d'Alembert's principle. D'Alembert's principle may be restated as follows:

Every state of motion may be considered at any instant as a state of equilibrium if the inertial forces are also considered as external forces. The inertial forces may be obtained as the negative of the product of the mass and the relevant acceleration vectors. Including the inertial forces, the principle of virtual work may be stated as

$$\delta W = \sum_{i=1}^{N} (X_i - m_i \ddot{x}_i)\, \delta x_i + (Y_i - m_i \ddot{y}_i)\, \delta y_i + (Z_i - m_i \ddot{z}_i)\, \delta z_i = 0. \tag{8.78}$$

The simultaneous solution of this equation along with the constraint equations gives the equations of motion of the system under consideration. It applies to a collection of particles and hence also to rigid bodies. Therefore it follows that

$$\delta W = \delta \mathbf{q}_0^T [\Gamma - \mathbf{I}_0 (\mathbf{q}_0)\, \ddot{\mathbf{q}}_0 - \mathbf{C}_0 (\mathbf{q}_0, \dot{\mathbf{q}}_0) + \Gamma_{g0} (\mathbf{q}_0)] = 0, \tag{8.79}$$

where $\delta \mathbf{q}_0$ is the vector of virtual displacements meeting the constraints and compatibility conditions. The vector of virtual displacements may usually be expressed in terms of a single input virtual displacement. Hence it has the form

$$\delta \mathbf{q}_0 = \mathbf{C} \delta u, \tag{8.80}$$

where **C** is vector constraining the virtual displacements $\delta\mathbf{q}_0$. The expression for the virtual work done then reduces to

$$\delta W = \mathbf{C}^T \left[\Gamma - \mathbf{I}_0 \left(\mathbf{q}_0 \right) \ddot{\mathbf{q}}_0 - \mathbf{C}_0 \left(\mathbf{q}_0, \dot{\mathbf{q}}_0 \right) + \Gamma_{g0} \left(\mathbf{q}_0 \right) \right] \delta u = 0. \tag{8.81}$$

It follows that the virtual work done by an actuator to provide the required input force is given by

$$\delta W_{\text{actuator}} = -\mathbf{C}^T \Gamma \delta u = \mathbf{C}^T \left[\Gamma_{g0} \left(\mathbf{q}_0 \right) - \mathbf{I}_0 \left(\mathbf{q}_0 \right) \ddot{\mathbf{q}}_0 - \mathbf{C}_0 \left(\mathbf{q}_0, \dot{\mathbf{q}}_0 \right) \right] \delta u. \tag{8.82}$$

Equation (8.82) provides a generic expression that is particularly useful in computing the total work done by the actuator in driving the input as well as the load and power requirements to drive the entire manipulator.

EXERCISES

8.1. Given the general inertia matrix of a rigid body, as

$$\mathbf{N} = \begin{bmatrix} \mathbf{I} & MC \\ MC^T & M\mathbf{I} \end{bmatrix},$$

show that, by moving the origin to the center of mass of the body, the general inertia matrix may be expressed as

$$\mathbf{N} = \begin{bmatrix} \mathbf{I} & C^T \\ 0 & \mathbf{I} \end{bmatrix} \begin{bmatrix} \mathbf{I} & MC \\ MC^T & M\mathbf{I} \end{bmatrix} \begin{bmatrix} \mathbf{I} & 0 \\ C & \mathbf{I} \end{bmatrix} = \begin{bmatrix} \mathbf{I} - MC^2 & 0 \\ 0 & M\mathbf{I} \end{bmatrix}.$$

Hence show that a general inertia matrix may be reduced to a diagonal form.

8.2. Show that, for a single link constrained to rotate about a single revolute joint, the Newton–Euler equations may be expressed as

$$\mathbf{W}_{\text{total}}^T \mathbf{S} = \mathbf{S}^T \mathbf{N S} \ddot{\theta} + \{ \mathbf{S} \times \mathbf{N S} \}^T \cdot \mathbf{S} \dot{\theta}^2 = \mathbf{S}^T \mathbf{N S} \ddot{\theta},$$

where **S** is a unit instantaneous screw vector in the direction of the joint axis.

8.3. Consider the two-link planar manipulator illustrated in Figure 4.2.

(a) Derive the forward-propagating equations relating the instantaneous screw vectors of each link.

(b) Derive the backward-propagating dynamic wrench equilibrium equations for each link. Assume each link to be uniform, homogeneous, of equal length, and of mass m.

(c) Hence derive the joint torque equations in terms of the joint space coordinates.

(d) Verify the joint torque equations by using an alternative energy method.

8.4. Consider the two-link planar manipulator illustrated in Figure 4.3.

(a) Derive the forward-propagating equations relating the instantaneous screw vectors of each link.

(b) Derive the backward-propagating dynamic wrench equilibrium equations for each link. Assume each link to be uniform, homogeneous, of equal length, and of mass m.

(c) Hence derive the appropriate joint force or torque equations in terms of the joint space coordinates.

(d) Verify the joint force or torque equations by using an alternative energy method.

8.5. Consider the two-link Cartesian manipulator illustrated in Figure 4.9.

(a) Derive the forward-propagating equations relating the instantaneous screw vectors of each link.

(b) Derive the backward-propagating dynamic wrench equilibrium equations for each link. Assume each link to be uniform, homogeneous, of equal length, and of mass m.

(c) Hence derive the joint force equations in terms of the joint space coordinates.

(d) Verify the joint force equations by using an alternative energy method.

8.6. Consider the two-link planar manipulator illustrated in Figure 4.10.

(a) Derive the forward-propagating equations relating the instantaneous screw vectors of each link.

(b) Derive the backward-propagating dynamic wrench equilibrium equations for each link. Assume each link to be uniform, homogeneous, of equal length, and of mass m.

(c) Hence derive the appropriate joint force or torque equations in terms of the joint space coordinates.

(d) Verify the joint force or torque equations by using an alternative energy method.

8.7. The bob of a simple pendulum is replaced with a heavy slender uniform rod, which is attached to a cord at one end, as shown in Figure 8.3. The length of the cord and that of the rod is each equal to L.

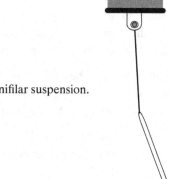

Figure 8.3. Unifilar suspension.

(a) Find the number of DOFs and a set of generalized coordinates.

(b) Determine the potential and kinetic energy functions for the system in terms of the generalized coordinates obtained in (a).

   (c) Obtain the equations of motion of the rod by the Euler–Lagrange method.

   (d) Obtain the linearized equations of motion for small displacements.

8.8. A uniform rod of mass m and length L is suspended by two symmetrically placed light cords of length h, as shown in Figure 8.4. The distance between the two cords is a. Initially the rod is assumed to be at rest.

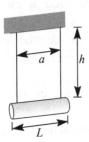

Figure 8.4. Bifilar suspension.

   (a) If the rod is given a slight rotation about a vertical axis passing through its center of mass and released, determine the kinetic energy, the potential energy, and the equation of motion. Hence determine the period of oscillations.

   (b) If the rod is given a slight horizontal displacement in the plane of the cords, while carefully avoiding any rotation or vertical displacement, and released, determine the kinetic energy, the potential energy, and the equation of motion. Hence determine the period of oscillations.

8.9. Consider a general rigid body, and show that the total kinetic energy may be expressed as

$$\mathbf{T} = \frac{1}{2}\mathbf{V}^T\mathbf{N}\mathbf{V},$$

where $\mathbf{V}$ is the instantaneous velocity screw of the rigid body and $\mathbf{N}$ is the general inertia matrix.

Assume that the potential energy is equal to zero and show that the Euler–Lagrange equations for the body are identical to the corresponding Newton–Euler equations.

8.10. Consider the cable-stayed square planar platform $ABCD$ of side a and of mass m as illustrated in Figure 5.6. Assume the weight to be negligible. Derive the equations of motion by the application of Lagrange's method.

8.11. Consider the four-DOF SCARA robot manipulator, discussed in Chapter 1 (Figure 1.18). The fourth coordinate rotates the gripper that represents a free rotational DOF.

   (a) Considering just the first three DOFs, obtain expressions for the kinetic energy and gravitational potential energy.

   (b) Hence show that the equilibrium equations are of the following form:

$$[\mathbf{I}_0(\mathbf{q}) + \cos(q_2)\,\mathbf{I}_1(\mathbf{q})]\,\ddot{\mathbf{q}} + \sin(q_2)\,\mathbf{C}_0(\mathbf{q},\dot{\mathbf{q}}) - g\Gamma_g = \Gamma,$$

where

$$\mathbf{q} = \begin{bmatrix} q_1 \\ q_2 \\ q_3 \end{bmatrix}, \quad \mathbf{I}_0(\mathbf{q}) = \begin{bmatrix} I_1 & I_3 & 0 \\ I_3 & I_3 & 0 \\ 0 & 0 & m_3 \end{bmatrix}, \quad \mathbf{I}_1(\mathbf{q}) = I_2 \begin{bmatrix} 2 & 1 & 0 \\ 1 & 0 & 0 \\ 0 & 0 & 0 \end{bmatrix}, \quad \Gamma_g = \begin{bmatrix} 0 \\ 0 \\ m_3 \end{bmatrix},$$

$$\mathbf{C}_0(\mathbf{q}, \dot{\mathbf{q}}) = I_2 \begin{bmatrix} -2\dot{q}_1\dot{q}_2 - \dot{q}_2^2 \\ \dot{q}_1^2 \\ 0 \end{bmatrix},$$

$$\begin{bmatrix} I_1 \\ I_2 \\ I_3 \end{bmatrix} = \begin{bmatrix} P_{C1}^2 & P_{C2}^2 + L_1^2 & L_1^2 + L_2^2 \\ 0 & L_1 P_{C2} & L_1 L_2 \\ 0 & P_{C2}^2 & L_2^2 \end{bmatrix} \begin{bmatrix} m_1 \\ m_2 \\ m_3 \end{bmatrix} + \begin{bmatrix} 1 & 1 & 1 \\ 0 & 0 & 0 \\ 0 & 1 & 1 \end{bmatrix} \begin{bmatrix} {}^C I_{zz1} \\ {}^C I_{zz2} \\ {}^C I_{zz3} \end{bmatrix},$$

m_i is the mass of link i, L_i is the length of link i, P_{Ci} is the vector from joint i to the center of mass of link i, and ${}^C I_{zzi}$ is the moment of inertia of link i with respect to an axis parallel to the z axis.

8.12. Consider the four-bar mechanism illustrated in Figure 8.5. The links AB, BC, and DC have lengths equal to l_i, $i = 1, 2$, and 3, with masses m_i, $i = 1, 2$, and 3, respectively. The links are respectively at the angles θ_i, $i = 1, 2$, and 3, to the horizontal directly toward the right at A, B, D. All angles are measured counterclockwise. The centers of mass of the links are at C_i, $i = 1, 2$, and 3, which are at the distances r_i, $i = 1, 2$, and 3 from A, B, and D, respectively. The vectorial distances of the centers of mass are denoted by $\mathbf{r}_i$, $i = 1, 2$, and 3, respectively, in reference frames with their origins at A, B, and D.

Figure 8.5. The four-bar mechanism.

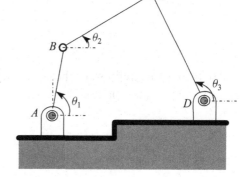

(a) Show that the coordinates of point D in the frame fixed at point A satisfy the following equations:

$$x_D = l_1 \cos\theta_1 + l_2 \cos\theta_2 - l_3 \cos\theta_3, \quad y_D = l_1 \sin\theta_1 + l_2 \sin\theta_2 - l_3 \sin\theta_3.$$

(b) Hence show that the angular rates satisfy the conditions

$$\begin{bmatrix} \dot{\theta}_1 \\ \dot{\theta}_2 \end{bmatrix} = \begin{bmatrix} -l_1 \sin\theta_1 & -l_2 \sin\theta_2 \\ l_1 \cos\theta_1 & l_2 \cos\theta_2 \end{bmatrix}^{-1} \begin{bmatrix} -l_3 \sin\theta_3 \\ l_3 \cos\theta_3 \end{bmatrix} \dot{\theta}_3.$$

(c) Differentiating the result obtained in (b), show that,

$$\begin{bmatrix} \ddot{\theta}_1 \\ \ddot{\theta}_2 \end{bmatrix} = \begin{bmatrix} -l_1 \sin\theta_1 & -l_2 \sin\theta_2 \\ l_1 \cos\theta_1 & l_2 \cos\theta_2 \end{bmatrix}^{-1} \mathbf{h},$$

where

$$\mathbf{h} = \dot{\mathbf{e}} - \dot{\mathbf{D}} \begin{bmatrix} \dot{\theta}_1 \\ \dot{\theta}_2 \end{bmatrix}, \quad \mathbf{e} = \begin{bmatrix} -l_3 \sin\theta_3 \\ l_3 \cos\theta_3 \end{bmatrix} \dot{\theta}_3, \quad \mathbf{D} = \begin{bmatrix} -l_1 \sin\theta_1 & -l_2 \sin\theta_2 \\ l_1 \cos\theta_1 & l_2 \cos\theta_2 \end{bmatrix}.$$

(d) Determine the accelerations of the centers of mass of each link and show that they may be expressed as

$$\mathbf{a}_1 = \ddot{\theta}_1 \mathbf{E} \mathbf{r}_1 - \dot{\theta}_1^2 \mathbf{r}_1, \quad \mathbf{a}_2 = \ddot{\theta}_1 \mathbf{E} \mathbf{l}_1 - \dot{\theta}_1^2 \mathbf{l}_1 + \ddot{\theta}_2 \mathbf{E} \mathbf{r}_2 - \dot{\theta}_2^2 \mathbf{r}_2,$$
$$\mathbf{a}_3 = \ddot{\theta}_3 \mathbf{E} \mathbf{r}_3 - \dot{\theta}_3^2 \mathbf{r}_3,$$

where

$$\mathbf{E} = \begin{bmatrix} 0 & -1 \\ -1 & 0 \end{bmatrix}, \quad \mathbf{r}_i = \begin{bmatrix} r_i \cos\theta_i \\ r_i \sin\theta_i \end{bmatrix}, \quad \mathbf{l}_1 = \begin{bmatrix} l_1 \cos\theta_1 \\ l_1 \sin\theta_1 \end{bmatrix}.$$

8.13. Reconsider the four-bar mechanism in the previous exercise.

(a) Show that the inertial forces, including the effect of gravity, and inertia torques at the center of mass of each moving link are given by

$$\mathbf{f}_i = -m_i (\mathbf{a}_i - \mathbf{g}), \quad i = 1, 2, 3,$$

and

$$\tau_i = -\mathbf{I}_{0i} \dot{\boldsymbol{\omega}}_i - \boldsymbol{\omega}_i \times (\mathbf{I}_{0i} \boldsymbol{\omega}_i), \quad i = 1, 2, 3,$$

where the vector $\mathbf{g}$ represents the acceleration that is due to gravity, $\mathbf{I}_{0i} = \mathbf{R}_i \mathbf{I}_i \mathbf{R}_i^T$, $\mathbf{I}_i$ is the moment of inertia tensor of each link about its center of mass,

$$\boldsymbol{\omega}_i = \begin{bmatrix} 0 \\ 0 \\ \dot{\theta}_i \end{bmatrix},$$

and the rotation matrices $\mathbf{R}_i$ are given by

$$\mathbf{R}_i = \begin{bmatrix} \cos\theta_i & -\sin\theta_i & 0 \\ \sin\theta_i & \cos\theta_i & 0 \\ 0 & 0 & 1 \end{bmatrix}.$$

(b) Show that the virtual linear displacement of each link in the mechanism is given by

$$\delta_1 = \delta\theta_1 \mathbf{Er}_1, \quad \delta_2 = \delta\theta_1 \mathbf{El}_1 + \delta\theta_2 \mathbf{Er}_2, \quad \delta_3 = \delta\theta_3 \mathbf{Er}_3.$$

(c) Then use the principle of virtual work and obtain an expression for the total input power requirements to drive the mechanism, assuming that link 1 is driven at a constant angular velocity by a servomotor.

9

Path Planning, Obstacle Avoidance,
and Navigation

9.1 Fundamentals of Trajectory Following

The motion of the end-effector of a manipulator is described by a trajectory in multidimensional space. It refers to a time history of the position, the velocity, and the acceleration of each DOF. Trajectory control is a fundamental problem in robotics and involves two distinct steps. The first of these steps is the planning of the desired trajectory or path of the end-effector. The second is the design and implementation of a suitable controller so as to ensure that the robot does indeed follow the planned path. This step is known as the trajectory or path tracking and is essentially a feedback-control problem. The path tracker is responsible for making the robot's actual position and velocity match the desired values of position and velocity provided to it by the path planner.

Path planning is one of the most vital issues to ensure autonomy of a robot, whether it is a robot manipulator or a mobile robot. It can be viewed as finding a safe and optimum path through an obstacle-filled environment from starting point to some destination in a collision-free manner. Once the desired path of an end-effector is planned there is also concern about the representation of the path for subsequent computational purposes. In the case of manipulators, a continuous time path either in the joint variable space or in a Cartesian base frame is essential. The trajectory controller then synthesizes an appropriate path following the control law based on this continuous time path.

Path planning in mobile robot applications is usually different from the path-planning problem for manipulators. In these situations it is either map based, in which a precompiled map is used to plan paths that guarantee collision-free motion, or sensor based, in which the sensory information is directly used by the trajectory controller to control the robot in its environment.

The problem of trajectory following involves feedback control of the robot to ensure that it follows the specified path as closely as is desired. Surprisingly, feedback control of either a manipulator or a mobile robot can cause some subtle problems. This is primarily due to the number of controls being less than the number of states of the system that must be controlled. Sometimes, when there are a large number of controls, they are not independent and the constraints between them are nonholonomic, thus making it impossible for them to be transformed to a minimal set.

9.1.1 Path Planning: Trajectory Generation

The basic problem is to move the manipulator configuration from an initial pose to some desired final pose. On the one hand, this implies moving the TCP from its current position to some desired final position. In doing so every point in the manipulator would move from its current position to a new position. Thus the manipulator takes on a new pose that, in the first instance, must be feasible before it can be achieved. Further, every intermediate pose that the manipulator takes on, as it moves from its current pose to the desired final pose, must be completely feasible and achievable in the sense that there are no obstacles not only to any point on the manipulator but also within a specified region around the manipulator. Assuming this is so, the problem of path planning reduces to moving the TCP, in prespecified incremental steps, from its current position to the desired final position.

Sometimes it becomes necessary to specify the desired trajectory by including, in addition to the desired final position and the specific incremental steps, a sequential set of desired intermediate or via points that must be traversed en route to the final position. Each of these points then represents the origin of the frame for specifying the position and orientation of the tool or the end-effector. Thus a certain region around the desired path must be available for manipulations of the tool, and this requirement can be stated in the form of spatial constraints. Further, the end-effector or the TCP may be required to be at a certain point on the desired path at a certain time with a certain time frame. Thus there is usually a set of temporal constraints associated with the desired trajectory.

A further requirement is that the entire desired trajectory from the current position through the sequence of via points to the final desired position be smooth. This implies that each segment of the desired trajectory from one via point to the next is continuous with continuous first derivatives and possibly with continuous second derivatives. In practice the requirement is usually relaxed in the sense that the continuity of derivatives is not enforced at a finite set of points in each segment; rather, it is only required that the directions of the local tangents be the same. This ensures that the desired trajectory is sufficiently smooth.

Given two points in multidimensional space, one method of specifying a continuous path is by assuming that it is a straight line connecting both points. Such an approach leads to a piecewise linear approximation to a curved trajectory. Unless the trajectory being approximated is also piecewise linear, large numbers of via point coordinates must be created and stored to generate a reasonably accurate path. An alternative approach is to use curves that are of a higher degree than are the straight lines. The higher-degree approximation is based on the use of curves that are represented explicitly, implicitly, or parametrically as nonlinear functions.

The parametric representation for curves $x = x(t)$, $y = y(t)$, $z = z(t)$ in terms of the parameter t, which is not to be confused with time, offers a number of advantages, particularly in relation to the specification of desired trajectory for a robot manipulator. Consider, for example, the parametric representation of a straight line connecting two points, P_0 and P_1. Any point P on the line may be expressed as

$$P = P_0 + t(P_1 - P_0), \quad 0 \le t \le 1. \tag{9.1}$$

When $t = 0$, $P = P_0$, and when $t = 1$, $P = P_1$. However, the equation may also be expressed as

$$P = P_0(1 - t) + t P_1. \tag{9.2}$$

The functions $(1 - t)$ and t serve as blending functions, and it follows that a much more general representation of any P may be expressed as

$$P = B_0(t) P_0 + B_1(t) P_1, \tag{9.3}$$

where $B_0(t)$ and $B_1(t)$ are general blending functions. If we let $t_0 = 0$ and $t_1 = 1$, P may be expressed as

$$P = P_0 \frac{(t - t_1)}{(t_0 - t_1)} + P_1 \frac{(t - t_0)}{(t_1 - t_0)} = \sum_{i=0}^{1} P_i B_i(t), \tag{9.4}$$

and the blending functions are

$$B_0(t) = \frac{(t - t_1)}{(t_0 - t_1)}, \quad B_1(t) = \frac{(t - t_0)}{(t_1 - t_0)}. \tag{9.5}$$

This is a case of the classic Lagrange linear interpolation formula that could be generalized to the case of a higher-degree polynomial in t. Thus the general nonlinear Lagrange interpolation formula for interpolating through $n + 1$ points is

$$P = \sum_{i=0}^{n} P_i B_i(t), \tag{9.6}$$

where the nth-order nonlinear blending functions in parameter t are

$$B_i(t) = \frac{\prod_{j=0}^{i-1}(t - t_j) \prod_{j=i+1}^{n}(t - t_j)}{\prod_{j=0}^{i-1}(t_i - t_j) \prod_{j=i+1}^{n}(t_i - t_j)}. \tag{9.7}$$

At each of the via points, all terms in the summation drop out except one. Thus the formula provides a smooth nth-order polynomial passing through each of the via points.

Although the Lagrange interpolation formula is extremely elegant, it has several shortcomings. It has a tendency to produce approximations to trajectories that are significantly wigglier than the original itself. Furthermore, the nature of the error depends to a very large extent on the choice of the interpolation points, $t = t_i$. The maximum error made in using an n-degree Lagrange interpolation polynomial to approximate a parametric trajectory over the interval $0 \leq t \leq 1$ is

$$|P_L(t) - P(t)| \leq \frac{\max\limits_{0 \leq \bar{t} \leq 1} \left| \dfrac{d^{n+1} P(\bar{t})}{d\bar{t}^{n+1}} \right|}{(n+1)!} \left| \prod_{j=0}^{n} (t - t_j) \right|. \qquad (9.8)$$

Thus it depends on the $(n+1)$th derivative of the original trajectory $P(t)$, and that is not a very desirable feature.

The problem of the error in the approximation to the original trajectory was studied extensively by Pafnuty Chebyshev (1821–1924), more than 100 years ago, who was able to establish conditions for the choice of the via points under which the error was uniform and a minimum over the entire length of the approximate trajectory. He showed that the best choices of these points were unevenly spaced and that they could be obtained as the zeros of the $(n+1)$th-degree Chebyshev polynomial as

$$t_i = \frac{1}{2} \left\{ 1 + \cos \left[\frac{(2i+1)}{n+1} \frac{\pi}{2} \right] \right\}, \quad i = 0, 1, 2, 3, \ldots, n. \qquad (9.9)$$

9.1.2 Splines, Bézier Curves, and Bernstein Polynomials

The term spline originates from the long flexible strips used by the professional draughtsman to draw smooth curves representing aerofoil shapes, automobile contours, and ship hull profiles. "Ducks" are weights attached to the ends of the spline used to pull the spline in various directions. The mathematical equivalent of the draughtsman's spline is the natural cubic spline with C^0, C^1, and C^2 continuity, a cubic polynomial that interpolates the two, three, or four control points that may or may not be identical to the via points.

Parametric curves replace geometric slopes that could be infinite with finite parametric tangent vectors. This is achieved by approximation of the original trajectory by a piecewise polynomial curve rather than by a single polynomial interpolating function. Thus we reconstruct the original trajectory by patching together a number of segments, each of which is a polynomial of the same degree. For patching together two segments smoothly we require that they not only pass through the end points but also that their tangent-vector directions match no matter what the tangent-vector magnitudes are. Each segment Q of the overall curve is usually approximated by a cubic polynomial in parameter t. With a cubic it is indeed possible to interpolate with a curve segment that not only passes through the end points but also matches the slopes at the two end points. The parametric cubics are the lowest-order polynomials that permit this. They are also the lowest-order polynomials that are nonplanar in three dimensions.

A curve segment $Q(t)$ is defined by constraints on the end points, the join points or knots, tangent vectors at the knots, and continuity requirements between curve segments. As each cubic polynomial has four coefficients, four constraints are used to formulate four equations for the four unknown coefficients that are then solved.

There are three major types of cubic polynomial curves used to model curve segments: *Hermite splines*, defined by two end points that are two via points and two end-point tangent vectors, *Bézier curves*, defined by two end points and two other points that control the end-point tangent vectors, and several kinds of other *splines*, each defined by four control points. These splines have C^1 and C^2 at the join points and come close to their control points but normally do not interpolate the points. Three important types of these splines are uniform **B** splines, nonuniform **B** splines, and β splines. The term uniform implies that the knots are spaced at equal intervals of parameter t. The spline passes through all the knots but not necessarily through all the control points. The letter **B** refers to the term *basis*.

Splines may be classified into two groups based on the *support*. Given a typical Lagrange interpolation polynomial that is a finite-degree polynomial, the function passes through all the control points. It is continuous through its $(n-1)$th derivative with respect to t. Furthermore, all the blending functions and hence all the coefficients of the polynomial in t are influenced when one of the control points is altered. This property is known as *infinite support*. Because of a long series of via points, the property of infinite support can easily lead to ill-conditioned linear systems, causing numerical instability. The second type of splines has coefficients that are calculated from only a limited subset of control points and therefore have a *finite support*. As the coefficients depend on just a few control points, moving a control point affects only a small part of the curve. This is called local control. Because of the limited support, this type of spline does not suffer from the disadvantage of possible numerical instability. Moreover, the time required to compute the coefficients is greatly reduced. Hermite splines and **B** splines are typical examples of splines with finite support.

A typical example of the Hermite spline is

$$P(t) = [2t^3 - 3t^2 + 1 \quad 3t^2 - 2t^3 \quad t(t-1)^2 \quad t^2(t-1)] \begin{bmatrix} P_1 \\ P_4 \\ R_1 \\ R_4 \end{bmatrix}, \quad (9.10)$$

where $P(t)$ is any point on the spline curve, P_i is the ith control point, and R_i is the tangent vector at that same point. It is customary to express this as

$$P(t) = [t^3 \quad t^2 \quad t \quad 1] \begin{bmatrix} 2 & -2 & 1 & 1 \\ -3 & 3 & -2 & -1 \\ 0 & 0 & 1 & 0 \\ 1 & 0 & 0 & 0 \end{bmatrix} \begin{bmatrix} P_1 \\ P_4 \\ R_1 \\ R_4 \end{bmatrix} = \mathbf{TMG}, \quad (9.11)$$

where $\mathbf{T} = [t^3 \quad t^2 \quad t \quad 1]$, **M** is a basis matrix, and **G** is a geometry vector. The blending functions are given by

$$\mathbf{B}(t) = [t^3 \quad t^2 \quad t \quad 1] \begin{bmatrix} 2 & -2 & 1 & 1 \\ -3 & 3 & -2 & -1 \\ 0 & 0 & 1 & 0 \\ 1 & 0 & 0 & 0 \end{bmatrix} = \mathbf{TM}. \quad (9.12)$$

One could also relate the Hermite geometry vector to four control points by defining two control points in addition to the two end points:

$$
\begin{bmatrix} P_1 \\ P_4 \\ R_1 \\ R_4 \end{bmatrix} = \begin{bmatrix} 1 & 0 & 0 & 0 \\ 0 & 0 & 0 & 1 \\ -3 & 3 & 0 & 0 \\ 0 & 0 & -3 & 3 \end{bmatrix} \begin{bmatrix} P_1 \\ P_2 \\ P_3 \\ P_4 \end{bmatrix}. \tag{9.13}
$$

Eliminating the Hermite geometry vector from the Hermite spline gives

$$
P(t) = \begin{bmatrix} t^3 & t^2 & t & 1 \end{bmatrix} \begin{bmatrix} 2 & -2 & 1 & 1 \\ -3 & 3 & -2 & -1 \\ 0 & 0 & 1 & 0 \\ 1 & 0 & 0 & 0 \end{bmatrix} \begin{bmatrix} 1 & 0 & 0 & 0 \\ 0 & 0 & 0 & 1 \\ -3 & 3 & 0 & 0 \\ 0 & 0 & -3 & 3 \end{bmatrix} \begin{bmatrix} P_1 \\ P_2 \\ P_3 \\ P_4 \end{bmatrix}, \tag{9.14}
$$

which reduces to

$$
P(t) = \begin{bmatrix} t^3 & t^2 & t & 1 \end{bmatrix} \begin{bmatrix} -1 & 3 & -3 & 1 \\ 3 & -6 & 3 & 0 \\ -3 & 3 & 0 & 0 \\ 1 & 0 & 0 & 0 \end{bmatrix} \begin{bmatrix} P_1 \\ P_2 \\ P_3 \\ P_4 \end{bmatrix}, \tag{9.15}
$$

and is a typical example of a *Bézier* curve.

B splines are approximating splines with minimal support. Two different types of **B** splines may be identified. The first are nonrational **B** splines, which are defined in terms of cubic polynomials. The term nonrational is used to distinguish these splines from rational **B** splines, which are simply the ratios of two different cubic polynomials. A typical example of a uniform nonrational **B** spline is

$$
P(t) = \begin{bmatrix} t^3 & t^2 & t & 1 \end{bmatrix} \frac{1}{6} \begin{bmatrix} -1 & 3 & -3 & 1 \\ 3 & -6 & 3 & 0 \\ -3 & 0 & 3 & 0 \\ 1 & 4 & 1 & 0 \end{bmatrix} \begin{bmatrix} P_1 \\ P_2 \\ P_3 \\ P_4 \end{bmatrix}, \tag{9.16}
$$

where P_i is the ith control point, which is not the same as the knots. Thus in a **B** spline there is less control over where the curve goes. However, it is very smooth, and for this reason nonrational **B** splines are extremely suitable for trajectory planning. The location of the control points for these applications can be quite restrictive as they define the convex hull, a convex polygon formed by the four control points for two-dimensional splines. It is often required that the convex hull be within a collision-free feasible space, and this is quite restrictive.

An example of a Bézier curve was given earlier; it may also be expressed as

$$
P(t) = \mathbf{TM}_B\mathbf{G}_B = \mathbf{B}_B\mathbf{G}_B, \tag{9.17}
$$

Table 9.1. *Bernstein polynomials of degrees 1, 2, and 3*

$i = \rightarrow$	0	1	2	3
$n = 1$	$(1-t)$	t		
2	$(1-t)^2$	$2t(1-t)$	t^2	
3	$(1-t)^3$	$3t(1-t)^2$	$3t^2(1-t)$	t^3

where the blending functions are given by

$$\mathbf{B}_B(t) = [t^3 \quad t^2 \quad t \quad 1] \begin{bmatrix} -1 & 3 & -3 & 1 \\ 3 & -6 & 3 & 0 \\ -3 & 3 & 0 & 0 \\ 1 & 0 & 0 & 0 \end{bmatrix} = \mathbf{TM}_B. \tag{9.18}$$

They may be expressed as

$$\mathbf{B}_B(t) = \mathbf{T} \begin{bmatrix} -1 & 3 & -3 & 1 \\ 3 & -6 & 3 & 0 \\ -3 & 3 & 0 & 0 \\ 1 & 0 & 0 & 0 \end{bmatrix} = [(1-t)^3 \quad 3t(1-t)^2 \quad 3t^2(1-t) \quad t^3]. \tag{9.19}$$

The four blending polynomials are known as *Bernstein polynomials* and play a pivotal role in the definition of various blending functions. Examining the four Bernstein polynomials, we observe that their sum is always unity and that each polynomial is nonnegative everywhere for $0 \leq t \leq 1$.

The Bernstein polynomials of degree n are defined by

$$B_{i,n}(t) = \binom{n}{i} t^i (1-t)^{n-i} = \sum_{k=i}^{n} (-1)^{k-i} \binom{n}{k} \binom{k}{i} t^k, \tag{9.20}$$

for $i = 0, 1, \ldots, n$, where

$$\binom{n}{i} = \frac{n!}{i!(n-i)!}.$$

There are $n+1$ nth-degree Bernstein polynomials. For mathematical convenience, we usually set $B_{i,n}(t) = 0$ if $i < 0$ or $i > n$. The polynomials are quite easy to write as the coefficients could be obtained from Pascal's triangle, and the exponents on the t term increase by one as i increases and the exponents on the $(1-t)$ term decrease by one as i increases. The Bernstein polynomials of degrees 1, 2, and 3 are shown in Table 9.1.

We can define the Bernstein polynomials of degree n by blending two Bernstein polynomials of degree $n-1$. That is, the kth nth-degree Bernstein polynomial satisfies a recurrence relation and could be written as

$$B_{k,n}(t) = (1-t)B_{k,n-1}(t) + tB_{k-1,n-1}(t). \tag{9.21}$$

Derivatives of the nth-degree Bernstein polynomials are polynomials of degree $n-1$. Using the definition of the Bernstein polynomial, we can show that this derivative can be written as a linear combination of Bernstein polynomials. In particular

$$\frac{d}{dt} B_{k,n}(t) = n \left[B_{k-1,n-1}(t) - B_{k,n-1}(t) \right],$$ (9.22)

for $0 \leq k \leq n$.

Thus in terms of the Bernstein polynomials the Bézier curve may also be expressed as

$$P(t) = \mathbf{TM}_B \mathbf{G}_B = \mathbf{B}_B \mathbf{G}_B,$$ (9.23)

where

$$\mathbf{B}_B(t) = \mathbf{T} \begin{bmatrix} -1 & 3 & -3 & 1 \\ 3 & -6 & 3 & 0 \\ -3 & 3 & 0 & 0 \\ 1 & 0 & 0 & 0 \end{bmatrix} = \begin{bmatrix} B_{0,3}(t) & B_{1,3}(t) & B_{2,3}(t) & B_{3,3}(t) \end{bmatrix}.$$

We could now write a single formula for the whole curve in terms of the curve segments, which are denoted by $Q_i(t)$, $i = 3, 4, 5, \ldots, m$. For $i < 3$ or when there are multiple knots, i.e., $t_i = t_{i+1}$, it is taken to be a single point. With this notion of the curve segment reducing to a point, the general equation for the whole curve is the weighted sum:

$$Q_i(t) = P_{i-3} B_{i-3,4}(t) + P_{i-2} B_{i-2,4}(t) + P_{i-1} B_{i-1,4}(t) + P_i B_{i,4}(t),$$ (9.24)

where $3 \leq i \leq m, t_i \leq t \leq t_{i+1}$.

For other types of splines and **B** splines in particular, there are no explicit formulae for the blending functions. Yet they may be expressed quite compactly in a recursive form with the jth-order blending function $B_{i,j}(t)$ for weighting the control point P_i:

$$B_{i,1}(t) = \begin{cases} 1, & t_i \leq t \leq t_{i+1} \\ 0, & \text{otherwise} \end{cases},$$ (9.25a)

$$B_{i,2}(t) = \frac{t - t_i}{t_{i+1} - t_i} B_{i,1}(t) + \frac{t_{i+2} - t}{t_{i+2} - t_{i+1}} B_{i+1,1}(t),$$ (9.25b)

$$B_{i,3}(t) = \frac{t - t_i}{t_{i+2} - t_i} B_{i,2}(t) + \frac{t_{i+3} - t}{t_{i+3} - t_{i+1}} B_{i+1,2}(t),$$ (9.25c)

$$B_{i,4}(t) = \frac{t - t_i}{t_{i+3} - t_i} B_{i,3}(t) + \frac{t_{i+4} - t}{t_{i+4} - t_{i+1}} B_{i+1,3}(t).$$ (9.25d)

The general equation for the curve segment can be expressed as was done in the case of the Bézier curve.

To conclude, there are a number of methods of defining a desired manipulator trajectory either in Cartesian space or in joint configuration space. The joint interpolated approach uses piecewise polynomial segments, such as splines and Bézier curves, to yield a smooth joint trajectory in joint configuration space. The power of the cubic spline as a trajectory design tool is significantly enhanced if it is used to design contours directly. If contours are found by another process, they could be accurately represented by splines; however, this would require a larger number of interpolation points than would be required otherwise.

9.2 Dynamic Path Planning

In the path-planning techniques considered in the preceding section, no dynamics or temporal constraints of any kind were considered. During path planning, information about dynamic objects with exactly known paths should be taken into account as well as position information of other dynamic objects. Path planning is a task that can benefit from the information derived about environmental dynamics. Motion planning in a dynamic environment is a difficult problem, as it requires planning in the state space, that is, simultaneously solving path-planning and velocity-planning problems. Path planning is a kinematic problem, involving the computation of a collision-free path from start to goal, whereas velocity planning requires the consideration of robot dynamics and actuator constraints. Furthermore, in practice, there are a number of temporal constraints that must be satisfied in addition to the fact that the manipulator itself must satisfy conditions of dynamic equilibrium. Thus motion planning in a dynamic context involves the determination of joint positions, velocities, and accelerations. Motion planning also involves the synthesis of algorithms to compute collision-free paths for a mechanical system, such as a mobile robot or a manipulator, moving amid obstacles.

Methods of dynamic path planning are composed of four steps:

1. path determination to meet the geometric constraints on the manipulator;
2. determination of timing rules and switching laws to meet the temporal and velocity constraints of the manipulator;
3. joint coordinate determination by use of the inverse kinematic model under the constraints;
4. determination of actuator torques while meeting the physical limitations imposed on them.

One class of timing constraints, for example, is when the manipulator is moved from one configuration to the next in minimum time. This is the problem of minimum-time path planning. In many robot-manipulator applications, time is an important resource, and the minimum-time path-planning problem is just a special case of the more general minimum-cost problem. In these minimum-cost optimization situations the path-planning and the path-tracking problems are not decoupled. The tracking control law has a significant influence on the optimal path and vice versa. A simplified statement of the minimum-cost path-planning problem is as follows: What input signals will drive a given robot from a given initial configuration to a given final configuration with as low a cost possible, given constraints on the magnitudes and derivatives on the control torques and constraints on the intermediate configurations of the robot determined by the requirement that the robot must not hit any obstacles? To retain the general parametric specification of the paths

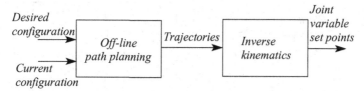

Figure 9.1. Off-line path planning and joint variable set-point computation.

planned, in the dynamic path-planning literature the optimal path is often specified parametrically in terms of a parameter that is now labeled as λ and not as t to avoid confusion with the time variable and substituted into the dynamic equations. Thus the optimal paths may be specified in terms of a geometric path in the joint variable space involving the path parameter λ and the time derivative of λ, which is labeled as $\mu = \dot{\lambda}$. There are then the other control torque and velocity constraints that must be imposed. This generally leads to the complex problem of constrained optimization with the dynamic equilibrium equations serving as differential equation constraints.

When the desired path is specified *a priori*, the problem reduces to determining the controls that will drive a given robot along a specified curve in joint space with minimum cost, given constraints on the input and final velocities and on the control signals and their derivatives. When the independent variable is taken as the path parameter λ, the state variables are reduced to a minimum of two states, μ and its derivative with respect to λ. There is therefore a reduction in the complexity of the problem, and the problem now reduces to finding a minimum-cost optimal control signal rather than a minimum-cost optimal path.

The control problem can be efficiently solved by the methodology of *dynamic programming* introduced in the 1950s by the eminent mathematician Richard Bellman. The algorithm essentially starts by solving the problem in stages, with the last stage taken up first and then working backward. The underlying principle was stated by Bellman:

> An optimal trajectory has the property that whatever the initial state and trajectory, the remaining part of the trajectory is always optimal with regard to the state resulting from that portion of it that is already known.

The methodology is quite computationally intensive and is well beyond the scope of this chapter.

For our part we shall restrict ourselves to the issue of continuous on-line or real-time updating of a planned path. The classical on-line computation of the desired trajectories for path planning and the subsequent inverse kinematic computations of the joint variables are illustrated in Figure 9.1. In an on-line or real-time situation the computations are carried out concurrently. The main routine performs the computation of the control signal driving the joint servomotors. When the computations are completed, a clock-controlled interrupt transfers control to an interrupt service routine that performs the path-planning computations. This is followed by another clock-controlled interrupt, and an associated interrupt service routine computes the

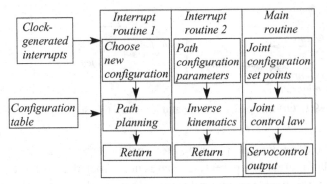

Figure 9.2. Concurrent computation of paths, inverse kinematics, and servocontrols.

inverse kinematics. Following the successful completion of these computations, the processor awaits the clock-controlled interrupt that transfers control back to the main routine, and the cycle is repeated.

A block diagram of the concurrent computation processes is illustrated in Figure 9.2. A typical example of the on-line process, often referred to as *Cartesian control,* requires that the calculation of the path points as well as the transformation to joint coordinates be performed sufficiently fast, i.e., in real time. However, this is possible only if one has fairly powerful computational resources both in terms of processor speed and memory.

An alternative strategy is to apply the Cartesian approach to only the initial and final configurations and interpolate the joint variables between these two extremes at equal intervals of time. A whole family of such techniques exists and is broadly known as *joint interpolated control schemes.* Some of these techniques make use of the fact that the second derivatives of the blending functions are linear functions of the parameter t, and for small changes in the control point locations, the increments in the segments may be computed by a finite-step numerical integration that is sufficiently accurate for cubic splines.

9.3 Obstacle Avoidance

Perhaps the most important driver for dynamic path planning is the continuous need to avoid obstacles in real time. Another important driver is the need to deal with uncertainty. In motion planning with uncertainty, the objective is to find a plan that is guaranteed to succeed even when the robot cannot execute it perfectly. Dealing with obstacles, particularly of the unexpected variety, is also a kind of uncertainty. Of course, the problems of dealing with obstacles are intimately related to compliant motion, provided motion-in-contact plans are acceptable. In what follows we exclude the possibility of motion-in-contact plans and focus primarily on obstacle avoidance.

Obstacle avoidance is of importance for both manipulators and mobile robots, although some of the problem constraints in the two situations can be significantly different from each other. However, in both situations one is interested in choosing

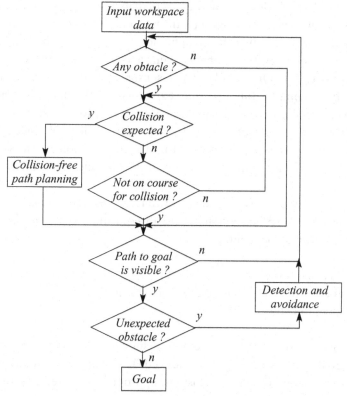

Figure 9.3. Basic collision-free path-planning flow chart.

a path that minimizes the risk of collision while reaching the goal. A general flow chart for path planning in the presence of obstacles is illustrated in Figure 9.3. Path-planning techniques in the presence of obstacles can be grouped into one of three major categories:

1. Map based, in which it is assumed that a precompiled map of the entire workspace including all the obstacles is available, and path planning reduces to finding a collision-free path in a workspace environment with a continuously known ensemble of obstacles. The map-based approach assumes that one has complete knowledge of the scene or operational environment.
2. Sensor based, in which a complete complement of obstacle detection and free-space sensors are used to guide the manipulator along a collision-free path to the goal.
3. Analogy based, in which various types of physical analogies are used for path-planning purposes. Typical examples of analogy-based methods are path-planning strategies based on (a) diffusion of a gaseous property in a planar obstacle-filled domain, (b) potential-flow-field methods based on a charge-distribution model, an electromagnetic-field model, or an optic-flow model, and (c) hydrodynamic wave propagation in the obstacle-filled region.

Knowledge about the location and shape of dynamic obstacles is in general hard to obtain, especially when the sensing platform is moving, and this is a drawback of the map-based approach. Sensor-based approaches can deal with incomplete knowledge of the scene or operational environment and are based on the principles of autonomous navigation. On the other hand, sensor- and analogy-based approaches are not only quite complex but also quite expensive to implement in practice. The optic-flow-based analogy method is particularly relevant for visual tracking control of a mobile robot. The concept of optic flow refers to the flow of objects in a visual scene as perceived by an observer moving toward a specific point in the scene. Optic flow is quite complementary to the visual control technique and is briefly discussed in Chapter 2.

In all these methods, however, the algorithm for obstacle avoidance is the most complex as well as the primary issue. Thus three distinct types of algorithms for obstacle avoidance may be identified:

1. hypothesize and test,
2. minimum-cost route with obstacle penalty metric,
3. navigating in the "free" space.

In the "hypothesize-and-test" method, a simple multisegment path from the start to the goal point is initially hypothesized. It is then tested for potential collisions. If there are any potential collisions, appropriate segments in the path are modified and a new path is proposed. The process is then repeated until an acceptable path is found.

Once such a path is found, it is "optimized" in the sense that it is a goal-satisfying trajectory. Although the resulting trajectory may never be optimal in the minimum-cost sense, it usually meets all the requirements and constraints imposed. The method is relatively simple in principle.

The "minimum-cost-route" method is based on associating a cost with any deviations from the planned path. A penalty function that encodes the presence of obstacles in the domain is also included in the cost metric. In general, the penalty is infinite if there are any potential collisions and drops off sharply as the minimum distance from the obstacle increases. The cumulative penalty function is obtained by summing together the penalties from individual obstacles. Once the total cost is found it is then minimized to find the optimal path.

The third category of obstacle-avoidance methods is based on an explicit representation of the robot's domain and the obstacles within it so as to elicit the free space wherein the robot is free of collisions. This involves estimating the area swept by the moving object in its planned path so the minimum requirements of the freeways can be established for a collision-free movement. This is done by ensuring that there is no overlap between the projected areas of the obstacles and the area swept by the robot. Obstacle avoidance is then reduced to the problem of finding a freeway that connects the initial and the final positions of the robot. Although computationally expensive and not normally used for robot manipulators, the method is quite

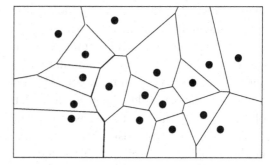

Figure 9.4. Example of a Voronoi diagram.

flexible in that there is the option of introducing a number of constraints and finding the "shortest" path.

The *Voronoi diagram* of a collection of geometric objects is a partition of space into cells, each of which consists of the points closer to one particular object than to any others. Figure 9.4 illustrates a typical Voronoi diagram. The Voronoi diagram has the property that, for each site, indicated by black circles, every point in the cell around that site is closer to that site than to any of the other sites. These diagrams, their boundaries, and their duals have been generalized to many situations including higher dimensions and applied to the motion-planning problem.

Voronoi diagrams tend to be involved in situations in which a space should be partitioned into "spheres of influence." They are modified to obtain a refined representation of the free space by using the concept of a generalized Voronoi diagram, which is the locus of points that are equidistant from two or more obstacle boundaries as well as from the workspace boundary. Thus, if a collision-free path exists, it must traverse one of the loci after joining it from its initial position, leaving it near only the goal point. The disadvantage of using the generalized diagram has always been that it is difficult to compute robustly and efficiently.

Voronoi diagram-based methods of path planning for mobile robots use a search graph, in which each of the paths between neighboring Voronoi regions is connected to form the graph. The graph is then searched for the desired optimum path.

The obstacle-avoidance problem can also be transformed to one of finding an optimum path traced by a robot represented as a point in a planar obstacle-filled region. To account for the finite dimension of the robot, the areas swept by the obstacles are increased into enlarged spaces. The problem is transformed to an equivalent but simpler problem of planning the motion of a point through an imaginary configuration space that excludes the enlarged configuration space obstacles.

When the configuration space obstacles are known, the connectivity of the free space is determined. Such connectivity information about the configuration space is used to determine a collision-free path from an initial to a final configuration. The *C-space approach*, as it is commonly known, provides an effective framework for investigating a range of robot motion-planning problems. Once transformed, the problem of planning the motion of a point through an imaginary configuration space that excludes the enlarged configuration space obstacles can be solved by one of a family of search techniques.

The C-space approach is computationally intensive and practically infeasible if one is interested in real-time computations of a trajectory. Several alternatives and generalizations have been proposed. The first of these is based on the use of probabilistic methods to solve the problem of determining the C-space and its dual, the obstacle space. The second is based on explicitly using velocity constraints in the formulation. Velocity constraints are not always integrable, and these are said to be nonholonomic. Although nonholonomic constraints are discussed in greater detail in the next chapter, they result in a whole class of nonholonomic motion-planning algorithms. Nonholonomic motion planning may be considered as the problem of planning open-loop control inputs to achieve the desired constraint-satisfying motion responses. Probably the most important development related the C-space approach is the realization that one does not need to compute the C-space to obtain the connectivity information within the workspace that is essential for motion planning.

An optimal path in the connected free space is obtained heuristically by searching the enlarged configuration space in a systematic manner. One could represent the sequence of moves in the search by a *tree*, as shown in Figure 9.5(a). All applicable moves in the search are arranged in order and represented in the tree, with the first choice on the left. One may regard the problem as growing some of this tree as it runs and thus exploring it. Two principal questions arise. First, there is the issue of the order in which to grow the tree. Second, we must consider the type of strategy to adopt to perform the search for the optimal solution as fast as is possible. Figure 9.5(b) illustrates a possible strategy that may be adopted by simulating a typical search. This is the basis of a *depth-first search*. That is, we keep going down, taking the leftmost branch at every point where the tree bifurcates or trifurcates, until there is need to back up one or more levels. We then explore the next branch after excluding all branches that have already been searched. Unfortunately this does not necessarily give us an optimum solution, as shown in Figure 9.5(c). Although we could find a suboptimum solution on the left-hand side of the tree, well before finding this we find an optimum solution to its right. A plausible remedy to this is to search the whole tree before making a choice of the optimum solution. Alternatively, we could explore the solutions in parallel so the optimum solution is found relatively faster. If by "optimum" it is implied that it corresponds to the minimum number of moves, we could advance one branch of the tree in one step and then go back and repeat the process. Such a search is said to be a *breadth-first search*.

In practice, there are a number of enhancements to the depth-first and breadth-first strategies designed to obtain the optimum solution as rapidly as possible. The A^* search algorithm is a tree-search algorithm to find a path from a given initial node to a given goal node. A "heuristic estimate" that ranks each node by an estimate of the best route that goes through that node is used in this search method. The A^* search algorithm begins at a selected initial node. The minimum "cost" of entering this node is estimated. The algorithm also estimates the minimum "distance" of this

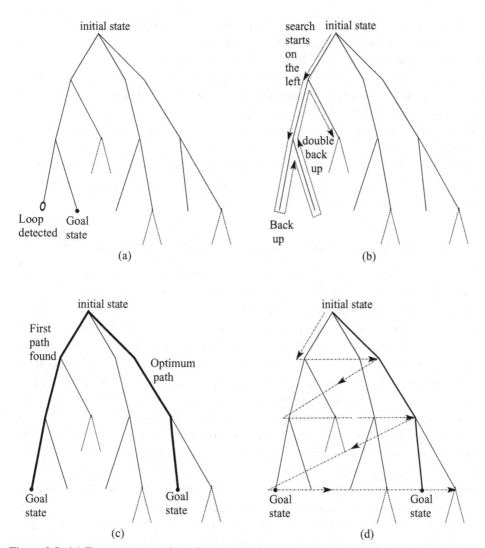

Figure 9.5. (a) Tree representation of the search for an optimum path, (b) depth-first search, (c) first solution found may not be optimum, (d) breadth-first search.

current node from the goal node. The two estimates are weighted appropriately and added together. This is the heuristic that is assigned to the path leading to this node. The search is performed of the nodes in the order of this heuristic estimate. The A^* algorithm is therefore an example of best-first or optimal search algorithm.

Other enhancements to the central search algorithms are based on the subject of *artificial intelligence* that deals with building computing machines capable of finding solutions to complex problems by incorporating characteristics from human reasoning in the algorithm. Genetic algorithms are one such family of paradigms that provide a powerful tool for automatic path planning. They are methods that find the best-fit solution by means of natural selection, a concept borrowed from genetics. A typical path in the free space is represented by a sequence of nodes, in which each node is identified by its spatial coordinates. Each path is represented as a *chromosome* or a string of data. Two paths are selected at random, and the fitter path is selected as the first parent. The same approach is adopted for the second parent. The parents form new chromosomes that are fitter than either of the parents by adopting one of a set of evolutionary operators. The operations are repeated for several iterations until the algorithm converges to an optimal path. Another artificial-intelligence-based paradigm is based on experiential learning. To learn a certain path, the robot must perceive an obstacle and deal with the situation. It does this from its past experiences that are encapsulated in fuzzy-logic-based reasoning schemes involving a set of if–then rules. Although a complete discussion of these methods is beyond the scope of this section, the reader is referred to the Bibliography for further information on the topics.

9.4 Inertial Measuring and Principles of Position and Orientation Fixing

Sensor-based dynamic path planning in the presence of obstacles often requires a continuous estimation of a robot's or platform's position and orientation. The principles for measuring orientation and position of a moving body by rigidly strapping a set of gyroscopes and accelerometers on the body have been well established in the field of inertial navigation systems (INSs) designed for aircraft and submarines. Navigation is not only the process of determining one's position and orientation but is also the process of determining and maintaining a course or trajectory to a goal location. Whereas local navigation requires the recognition of only one location, namely, the goal, robots are often involved in way finding that involves the recognition of several places and the representation of relations between places that may be outside the current range of perception. Although way finding allows the agent to find places that cannot be found by local navigation alone, it relies on position fixing and local navigation skills to move from one place to another. The primary skills in local navigation may be classified into four groups: search, direction following, aiming, and guidance. Position and orientation fixing, however, is basic to the whole process of navigation. A process related to way finding is motion planning, which was discussed in an earlier section. Yet motion planning differs from way finding in that, in motion planning, the task is to generate a collision-free path for a movable object among stationary, time-varying, constrained, or arbitrarily movable obstacles.

The original INSs were built around a gimballed platform that was stabilized to a particular navigation reference frame by gyros on the platform to drive the gimbal motors in a feedback loop. The platform-mounted accelerometers could then be individually double integrated to obtain position updating in each direction. Most recent systems eliminated the need for a platform. They are called strapped-down INSs, which eliminate the mechanical gimbals and measure the orientation of a vehicle by integrating the angular rates from three orthogonal angular rate-sensing gyroscopes (hereafter referred to as "rate gyros") strapped down to the frame of the vehicle.

The basic principle of an inertial measuring system is actually quite simple. To obtain position, three linear accelerometers, also rigidly strapped to the moving

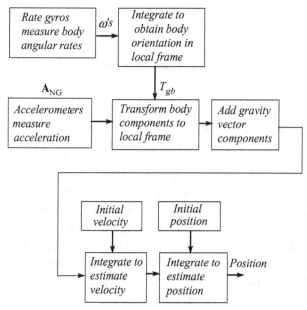

Figure 9.6. Principle of operation of an IMU.

body with their measuring axes orthogonal to each other, are used to measure all the components of the total acceleration vector of the body relative to inertial space. This acceleration vector can be converted from body coordinates to Earth coordinates by use of the known instantaneous orientation of the body determined by the gyros. Adding the gravity vector component to the measured acceleration vector and then performing double integration starting from a known initial position, one estimates the instantaneous position of the moving body. The inertial measuring system is encapsulated in a stand-alone unit known as an inertial measuring unit (IMU), which is a primary component of an autonomous vehicle navigation system. Whether in two dimensions or in three dimensions, the basic principles of inertial measurement are the same.

Figure 9.6 illustrates the principle of operation of an IMU. Together with software for determining the position and orientation of the body, the system is known as an inertial reference system (IRS). The technology is currently being adopted for inertial position and orientation fixing of mobile robots and platforms.

To set up the mathematical basis for the concept of inertial measurement, consider the total velocity vector in a rotating reference frame given by

$$\mathbf{V} = \frac{\partial \mathbf{R}}{\partial t} + \boldsymbol{\omega} \times \mathbf{R}. \tag{9.26}$$

Equation (9.26) represents the general relationship between the velocity and position vectors in a rotating reference frame. Similarly the equation relating the acceleration vector $\mathbf{A}$ to the velocity vector is given by

$$\mathbf{A} = \frac{\partial \mathbf{V}}{\partial t} + \boldsymbol{\omega} \times \mathbf{V}. \tag{9.27}$$

Equation (9.26) can be substituted into Equation (9.27) to relate the acceleration vector directly to the position vector, giving

$$\mathbf{A} = \frac{\partial \mathbf{V}}{\partial t} + \boldsymbol{\omega} \times \mathbf{V} = \frac{\partial^2 \mathbf{R}}{\partial t^2} + \frac{\partial}{\partial t} \boldsymbol{\omega} \times \mathbf{R} + 2\boldsymbol{\omega} \times \frac{\partial \mathbf{R}}{\partial t} + \boldsymbol{\omega} \times (\boldsymbol{\omega} \times \mathbf{R}), \quad (9.28)$$

which may be expressed as

$$\frac{\partial^2 \mathbf{R}}{\partial t^2} = \mathbf{A} - \frac{\partial}{\partial t} \boldsymbol{\omega} \times \mathbf{R} - 2\boldsymbol{\omega} \times \frac{\partial \mathbf{R}}{\partial t} - \boldsymbol{\omega} \times (\boldsymbol{\omega} \times \mathbf{R}). \quad (9.29)$$

The vector $\mathbf{A}$ is the inertial acceleration vector referred to a rotating reference frame. We obtain the acceleration referred to the rotating frame by resolving the inertial acceleration into components in the rotating frame. Equations (9.26)–(9.29) are the basic equations used in deriving the relationships between the acceleration and angular rate measurements and the estimated position and orientation outputs.

One approach for calculating the position vector of a body in arbitrary motion is to estimate it from measurements of the acceleration vector or velocity vector in a nonrotating reference frame fixed in space. Clearly in this case, the position vector may be written in terms of the velocity as

$$\mathbf{R}_A = \mathbf{R}_0 + \int_{t_0}^{t} \mathbf{V} dt. \quad (9.30)$$

This relation is valid for an arbitrary ground track and can also be written as

$$\mathbf{R}_A = \mathbf{R}_0 + \mathbf{V}_0 (t - t_0) + \int_{t_0}^{t} \int_{t_0}^{\tau} (\mathbf{A}_I + \mathbf{G}_I) \, dt \, d\tau, \quad (9.31)$$

where $\mathbf{G}_I$ and $\mathbf{A}_I$ denote the gravitational and nongravitational components, respectively, of the body accelerations. These equations can easily be extended to the case in which the body is tracking another whose acceleration is measured as well. If this acceleration is $\mathbf{Z}_t$, Equation (9.31) may be written as

$$\mathbf{R}_T = \mathbf{R}_{0T} + \mathbf{V}_{0T} (t - t_0) + \int_{t_0}^{t} \int_{t_0}^{\tau} (\mathbf{A}_I + \mathbf{G}_I - \mathbf{Z}_t) \, dt \, d\tau, \quad (9.32)$$

where $\mathbf{R}_T$ is the relative position vector of the tracked body and $\mathbf{R}_{0T}$ and $\mathbf{V}_{0T}$ are the initial relative position and velocity vectors, respectively.

If the acceleration measurements are slightly biased or slightly erroneous, as is to be expected in all practical measurements, it is quite obvious from Equations (9.31) and (9.32) that the position errors would increase quadratically with time. Hence such an approach is generally not acceptable from a practical point of view. It is therefore essential to choose a rotating reference frame in which the position errors that are due to errors in the measurements made in a reference frame fixed in the moving body would be within specific limits. A number of reference frames are used to define the position and orientation estimation problem. These are briefly summarized.

Earth-Centered Inertial (i_e frame) and Earth-Fixed Reference Frames (e frame).
The inertial reference frame used in inertial measuring systems is usually an orthogonal frame consisting of three mutually perpendicular axes fixed in space and with

Figure 9.7. Earth coordinates, latitudes, and longitudes.

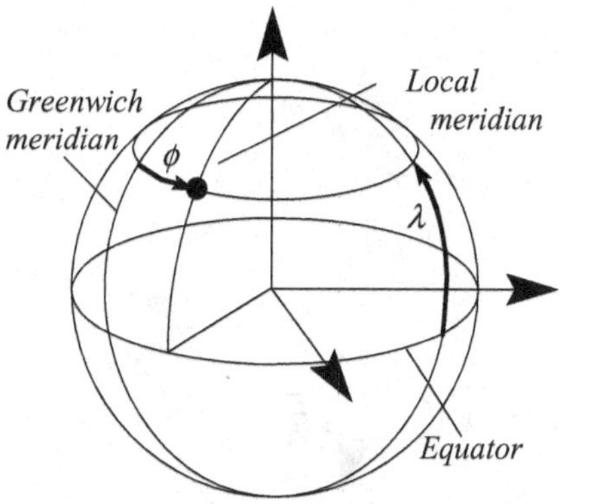

its origin coinciding with the center of the Earth. The x axis of this frame points to a fixed point in space known as the first point in Aries, which lies in the equatorial plane of the Earth as well as in the plane containing the Greenwich meridian.

The Earth-fixed reference frame is fixed in the Earth and also has its origin at the center of the Earth. The axis of the Earth's rotation is defined by the vector $\mathbf{K}_E$ (Figure 9.7). The position of a point in the vicinity of the Earth's surface is defined by the latitude λ, the longitude ϕ, and the altitude. The equatorial plane is normal to $\mathbf{K}_E$, and each meridian is the intersection of the surface of the sphere with a plane containing $\mathbf{K}_E$, whereas the equator is the intersection of the surface of the sphere with the equatorial plane. The reference direction $\mathbf{I}_E$, which passes through the equator, defines the zero-longitude meridian through Greenwich, England. Latitude (in degrees) is measured on both sides of the equator along a meridian. The locus of all points corresponding to the same latitude forms a latitude circle or parallel.

Thus the Earth-fixed reference frame is a rotating reference frame, rotating about the z axis of the Earth-centered inertial frame, denoted by $\mathbf{K}_I$, which is coincident with the Earth-fixed axis $\mathbf{K}_E$. The third axis is mutually perpendicular to the x and z axes of each of the two frames and points to the east. The two frames coincide once every year on the day of the vernal equinox (23rd June). On this day the x axes of the two frames pass through the Greenwich meridian. The angular velocity of the Earth-fixed frame with respect to the inertial frame is calculated from the fact that the Earth revolves about its axis of rotation 365.25 times in a year while revolving around the Sun once. Thus the total number of revolutions performed by the Earth in a year is 366.25, and the sidereal rate ω_s is given by

$$\omega_s = 2\pi \, (\text{rad/cycle}) \frac{(365.25 + 1) \, (\text{cycles})}{365.25 \times 24 \, (\text{h}) \times 3600 (\text{s/h})}$$

or

$$\omega_s = 2\pi \frac{1.0027379}{24 \times 3600} \, \text{rad/s},$$

which reduces to

$$\omega_s = 2\pi \frac{1.0027379}{86400} \text{rad/s} = 7.292115 \times 10^{-5} \, \text{rad/s}. \tag{9.33}$$

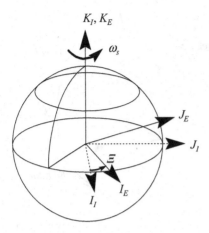

Figure 9.8. Earth and inertial reference frames.

The relationship between the Earth and the inertial reference frame is shown in Figure 9.8. In Figure 9.8 the hour angle of the vernal equinox Ξ is related to the sidereal rate by

$$\frac{d}{dt}\Xi = \omega_s. \tag{9.34}$$

It is standard practice to take this angle to be $\Xi_0 = (6\,h, 40\,min, 5.156\,s)$, on January 1, 1975, at zero hour Universal Time (UT; Greenwich mean solar time). The angle Ξ_0 is expressed in radians and is given in terms of D full days (4 years = 1461 days) plus S seconds from 0 hour UT on January 1, 1975:

$$\Xi_0 = 2\pi \left(0.0027379\,D + \frac{1.0027379\,(S + 24\,005.156)}{86\,400} \right). \tag{9.35}$$

The unit vectors $\mathbf{K}_I$ and $\mathbf{K}_E$ are equal and the unit vectors ($\mathbf{I}_E$, $\mathbf{J}_E$ and $\mathbf{I}_I$, $\mathbf{J}_I$) are related by

$$\mathbf{I}_E = \cos \Xi \mathbf{I}_I + \sin \Xi \mathbf{J}_I,$$

$$\mathbf{J}_E = -\sin \Xi \mathbf{I}_I + \cos \Xi \mathbf{J}_I.$$

These relations can expressed compactly in the matrix notation as

$$\begin{bmatrix} \mathbf{I}_E \\ \mathbf{J}_E \\ \mathbf{K}_E \end{bmatrix} = \begin{bmatrix} \cos \Xi & \sin \Xi & 0 \\ -\sin \Xi & \cos \Xi & 0 \\ 0 & 0 & 1 \end{bmatrix} \begin{bmatrix} \mathbf{I}_I \\ \mathbf{J}_I \\ \mathbf{K}_I \end{bmatrix}. \tag{9.36}$$

Earth-Fixed Geographic Frame of Reference (g_e frame). This frame has its axes pointing actual local north, east, and downward directions, respectively. The origin of the frame is fixed locally at a point on the Earth's surface defined by the local latitude and longitude. It is shown in Figure 9.9.

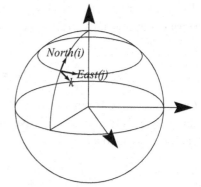

Figure 9.9. Geographic reference frame.

Unit vectors in the Earth-fixed geographic frame are related to those in the Earth-centered Earth-fixed frame by

$$
\begin{bmatrix} \mathbf{I}_G \\ \mathbf{J}_G \\ \mathbf{K}_G \end{bmatrix} = \begin{bmatrix} -\sin \lambda \cos \phi & -\sin \lambda \cos \phi & \cos \lambda \\ -\sin \phi & \cos \phi & 0 \\ -\cos \lambda \cos \phi & -\cos \lambda \sin \phi & -\sin \lambda \end{bmatrix} \begin{bmatrix} \mathbf{I}_E \\ \mathbf{J}_E \\ \mathbf{K}_E \end{bmatrix}.
\tag{9.37}
$$

Body-Centered Geographic Frame of Reference (g_b frame). This frame has its axes pointing actual local north, east, and downward directions, respectively. The origin of the frame lies on a vertical passing through the body center of gravity and translates with the body but is not fixed in the body.

Body-Centered, Body-Fixed System (b frame). This a reference frame centered at the body's center of gravity with the axes oriented along three mutually perpendicular reference directions fixed in the body.

Robot-Centered, Body-Fixed Frame (b_a frame). Consider a typical robot such as a mobile robot vehicle. Body-fixed robot coordinates are fixed to the robot and point forward along the axis of the vehicle, laterally outward along the right and toward the floor of the vehicle.

To understand and derive the governing kinematic equations of the position of a moving body, it is important to be able to transform coordinates expressed in one frame to another. If the directions of the coordinate axes differ, the coordinates have to be rotated. The direction cosine matrix (DCM) is used to transform coordinates between two frames when they are both centered at the same point. DCMs and their time-dependent behavior form the basis of the derivation of the equations governing the kinematics of position and orientation, as follows.

A transformation of coordinates $\mathbf{x}_t$ in the $\mathbf{t}$ frame to coordinates $\mathbf{x}_s$ in the $\mathbf{s}$ frame can be done as

$$
\mathbf{x}_s = \mathbf{T}_{st} \mathbf{x}_t,
\tag{9.38}
$$

where the DCM is

$$
\mathbf{T}_{st} = \begin{bmatrix} l_1 & m_1 & n_1 \\ l_2 & m_2 & n_2 \\ l_3 & m_3 & n_3 \end{bmatrix}.
\tag{9.39}
$$

The transformation matrix is an orthogonal matrix with both the $\mathbf{T}_{st}$ columns and rows being mutually orthogonal. This implies that $\mathbf{T}_{st}$ has the following important property:

$$\mathbf{T}_{ts} = (\mathbf{T}_{st})^{-1} = (\mathbf{T}_{st})^{\mathrm{T}}. \tag{9.40}$$

One important property that can be expressed in terms of rotation around the coordinate axes is the time derivative of a transition matrix, $\dot{\mathbf{T}}_{st}$. It may be shown that it can be expressed as

$$\dot{\mathbf{T}}_{st} = \mathbf{T}_{st}\boldsymbol{\Omega}_{st}. \tag{9.41}$$

Here, the quantity $\boldsymbol{\Omega}_{st}$ is a skew-symmetric matrix. The subscripts in $\boldsymbol{\Omega}_{st}$ imply that it is the rotation of the **t** frame relative to the **s** frame that is measured and that the rotation is measured in terms of coordinates in the **t** frame. As there are only three nonzero elements, the elements of the matrix $\boldsymbol{\Omega}_{st}$ may also be expressed as a vector,

$$\boldsymbol{\omega}_{st} = [\omega_1 \quad \omega_2 \quad \omega_3],$$

and the matrix $\boldsymbol{\Omega}_{st}$ may be defined as

$$\boldsymbol{\Omega}_{st} = [\times \boldsymbol{\omega}_{st}] = \begin{bmatrix} 0 & -\omega_3 & \omega_2 \\ \omega_3 & 0 & -\omega_1 \\ -\omega_2 & \omega_1 & 0 \end{bmatrix}. \tag{9.42}$$

There is a close relation between the vector cross product and the notation $[\times\boldsymbol{\omega}_{st}]$, which is sometimes represented by the matrix $\boldsymbol{\Omega}_{st}$. Time derivatives of small angles are given by angular rates and $\boldsymbol{\omega}_{st}$ describes rotation rates, and its components are similar in many respects to these of the angular rates.

The kinematic equations governing position and orientation are a set of differential equations describing how position, velocity, and attitude are changed depending on acceleration and angular rotations. In what follows, kinematic equations governing position and orientation will be expressed with coordinates in the $\mathbf{g}_e$ frame.

Because accelerometers are known to measure only the nongravitational component of the total acceleration, in the inertial frame (**i** frame) we have, after adding the gravity component,

$$\mathbf{A}^i = \mathbf{A}_{\mathrm{NG}}^i + \mathbf{G}^i. \tag{9.43}$$

A transformation of coordinates $\mathbf{x}_g$ in the $\mathbf{g}_e$ frame to coordinates $\mathbf{x}_i$ in the **i** frame can be done as

$$\mathbf{x}_i = \mathbf{T}_{ig}\mathbf{x}_g. \tag{9.44}$$

This generally is performed by two successive transformations: from the $\mathbf{g}_e$ frame to the **e** frame and from the **e** frame to the **i** frame. Hence,

$$\mathbf{x}_i = \mathbf{T}_{ie}\mathbf{T}_{eg}\mathbf{x}_g. \tag{9.45}$$

The time derivative of the transition matrix $\mathbf{T}_{ie}$ can be calculated if general equation (9.41) is applied to $\mathbf{T}_{ie}$:

$$\dot{\mathbf{T}}_{ie} = \mathbf{T}_{ie}\boldsymbol{\Omega}_{ie}, \qquad (9.46)$$

where $\boldsymbol{\Omega}_{ie}$ is a skew-symmetric matrix with elements given by the components of the angular rate vector $\boldsymbol{\omega}_{ie} = [\omega_1 \quad \omega_2 \quad \omega_3]$ between the **i** frame and the **e** frame:

$$\boldsymbol{\Omega}_{ie} = [\times\boldsymbol{\omega}_{ie}] = \begin{bmatrix} 0 & -\omega_3 & \omega_2 \\ \omega_3 & 0 & -\omega_1 \\ -\omega_2 & \omega_1 & 0 \end{bmatrix}. \qquad (9.47)$$

The only nonzero components in $\boldsymbol{\Omega}_{ie}$ is ω_3, because the **e** frame rotates about only the three-axis. This component is given by $\omega_3 = \omega_s$. We further need the second time derivative of $\mathbf{T}_{ie}$, which is

$$\ddot{\mathbf{T}}_{ie} = \dot{\mathbf{T}}_{ie}\boldsymbol{\Omega}_{ie} + \mathbf{T}_{ie}\dot{\boldsymbol{\Omega}}_{ie} = \mathbf{T}_{ie}\boldsymbol{\Omega}_{ie}\boldsymbol{\Omega}_{ie} + \mathbf{T}_{ie}\dot{\boldsymbol{\Omega}}_{ie}, \qquad (9.48)$$

where Equation (9.41) is applied twice in succession. It is possible to simplify this expression by noting that the sidereal rate ω_s is constant, and therefore $\dot{\omega}_s$ is zero, which gives

$$\ddot{\mathbf{T}}_{ie} = \mathbf{T}_{ie}\boldsymbol{\Omega}_{ie}\boldsymbol{\Omega}_{ie}. \qquad (9.49)$$

Now, differentiating Equation (9.45) and inserting the expressions for the first and second derivatives of $\mathbf{T}_{ie}$ from (9.46) and (9.49) yields

$$\ddot{\mathbf{x}}_i = \mathbf{T}_{ie}\left(\boldsymbol{\Omega}_{ie}\boldsymbol{\Omega}_{ie}\mathbf{T}_{eg}\mathbf{x}_g + 2\boldsymbol{\Omega}_{ie}\mathbf{T}_{eg}\dot{\mathbf{x}}_g + \mathbf{T}_{eg}\ddot{\mathbf{x}}_g\right).$$

Multiplying both sides with $(\mathbf{T}_{ig})^{-1} = \mathbf{T}_{ge}\mathbf{T}_{ei} = \mathbf{T}_{ge}(\mathbf{T}_{ie})^{-1}$ to resolve the inertial accelerations into components in the geographic frame, we obtain

$$\ddot{\mathbf{x}}_i^g = \mathbf{T}_{ge}(\mathbf{T}_{ie})^{-1}\ddot{\mathbf{x}}_i = \left(\mathbf{T}_{ge}\boldsymbol{\Omega}_{ie}\boldsymbol{\Omega}_{ie}\mathbf{T}_{eg}\mathbf{x}_g + 2\mathbf{T}_{ge}\boldsymbol{\Omega}_{ie}\mathbf{T}_{eg}\dot{\mathbf{x}}_g + \ddot{\mathbf{x}}_g\right).$$

Hence,

$$\ddot{\mathbf{x}}_g = -\left(\mathbf{T}_{ge}\boldsymbol{\Omega}_{ie}\boldsymbol{\Omega}_{ie}\mathbf{T}_{eg}\mathbf{x}_g + 2\mathbf{T}_{ge}\boldsymbol{\Omega}_{ie}\mathbf{T}_{eg}\dot{\mathbf{x}}_g\right) + \ddot{\mathbf{x}}_i^g.$$

The acceleration is sensed in the body frame and hence it follows that

$$\ddot{\mathbf{x}}_i^g = \mathbf{T}_{gb}\mathbf{A}_{NG}^b + \mathbf{G}^g.$$

Hence it follows that, in terms of the measured acceleration components,

$$\ddot{\mathbf{x}}_g = -\left(\mathbf{T}_{ge}\boldsymbol{\Omega}_{ie}\boldsymbol{\Omega}_{ie}\mathbf{T}_{eg}\mathbf{x}_g + 2\mathbf{T}_{ge}\boldsymbol{\Omega}_{ie}\mathbf{T}_{eg}\dot{\mathbf{x}}_g\right) + \mathbf{T}_{gb}\mathbf{A}_{NG}^b + \mathbf{G}^g. \qquad (9.50)$$

The matrix $\mathbf{T}_{gb}$ has direction cosine components relating to the body attitude relative to the Earth-fixed geographic frame. As the attitude of the body is time dependent so are the components of the $\mathbf{T}_{gb}$ matrix. But

$$\dot{\mathbf{T}}_{gb} = \mathbf{T}_{gb}\boldsymbol{\Omega}_{gb}, \qquad (9.51)$$

where $\boldsymbol{\Omega}_{gb}$ is a skew-symmetric matrix with elements given by the components of the angular rate vector $\boldsymbol{\omega}_{gb} = [\omega_1 \quad \omega_2 \quad \omega_3]$ between the **g** frame and the **b** frame:

$$\boldsymbol{\Omega}_{gb} = [\times \boldsymbol{\omega}_{gb}] = \begin{bmatrix} 0 & -\omega_3 & \omega_2 \\ \omega_3 & 0 & -\omega_1 \\ -\omega_2 & \omega_1 & 0 \end{bmatrix}, \tag{9.52}$$

where $\boldsymbol{\omega}_{gb}$ is the vector of the body components of the body angular rates relative to the geographic frame.

The gyros sense the body components of the body angular rates relative to an inertial frame. Hence, to obtain the body components of the body angular rates relative to the geographic frame, one must subtract the body components of the angular velocity of the geographic frame from the body components of the body angular rates relative to an inertial frame. Hence it follows that

$$\boldsymbol{\omega}_{gb} = \boldsymbol{\omega}_{ib} - \mathbf{T}_{bg}\boldsymbol{\omega}_{ig} = \boldsymbol{\omega}_{ib} - (\mathbf{T}_{gb})^{-1}\boldsymbol{\omega}_{ig}. \tag{9.53}$$

Three-dimensional second-order differential equation (9.50) can be transformed to a system with six first-order differential equations in terms of six state variables. Together with (9.51)–(9.53), a system of nine first-order differential equations that are sufficient for describing the position determination problem can be formed. The differential equations may be written as

$$\frac{d}{dt}\mathbf{x}_g = \dot{\mathbf{x}}_g, \quad \frac{d}{dt}\dot{\mathbf{x}}_g = -(\boldsymbol{\Omega}_{ig}\boldsymbol{\Omega}_{ig}\mathbf{x}_g + 2\boldsymbol{\Omega}_{ig}\dot{\mathbf{x}}_g) + \mathbf{A}_{\mathrm{NG}}^g + \mathbf{G}^g,$$
$$\dot{\mathbf{T}}_{gb} = \mathbf{T}_{gb}\boldsymbol{\Omega}_{gb}, \tag{9.54}$$

where $\mathbf{A}_{\mathrm{NG}}^g$ and $\boldsymbol{\Omega}_{gb}$ are the accelerometer and rate gyro outputs, respectively, in the geographic frame. These equations must be integrated numerically to determine the angular rates and the position coordinates of the body. Once the body angular rates are known, the quaternion attitude rate equations in Subsection 2.4.11 can be numerically integrated to determine the body orientation. Figure 9.6 is a diagrammatic representation of the position-determination methodology and is based on Equations (9.52)–(9.54).

There are indeed a number of methods for parameterizing the transformation matrix T_{gb} based on the representation of body attitude by *quaternions*, body displacements by *dual quaternions*, and *homogeneous coordinates*. However, no matter which method of parameterizing the body attitudes and displacements is adopted, the fundamental problem remains the estimation of the transformation T_{gb} or the body attitudes as well as the displacement vector in the Earth-fixed geographic frame.

9.4.1 Gyro-Free Inertial Measuring Units

Given any six accelerometers "arbitrarily" located and oriented, one can compute the linear and angular motion of a rigid body by using a simple algorithm, with the

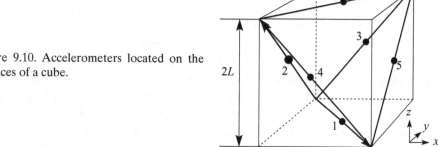

Figure 9.10. Accelerometers located on the six faces of a cube.

six accelerometer measurements as inputs to the algorithm. This is the basis of gyro-free position and orientation fixing. However, there are a number of advantages when the accelerometers are located on the six faces of a cube and mounted with their axis of sensitivity directed as shown in Figure 9.10.

In this case it can be shown that the time derivative of the angular rate and the body force per unit mass are respectively given in terms of the accelerometer outputs, $A_i, i = 1, 2, 3, \ldots, 6$, by

$$\dot{\boldsymbol{\omega}}_{ib}^b = \dot{\boldsymbol{\omega}}_{ib} = \frac{1}{2\sqrt{2}L} \begin{bmatrix} +A_1 - A_2 + A_5 - A_6 \\ -A_1 + A_3 - A_4 - A_6 \\ +A_2 - A_3 - A_4 + A_5 \end{bmatrix},$$

$$\boldsymbol{f}^b = \frac{1}{2\sqrt{2}} \begin{bmatrix} +A_1 + A_2 - A_5 - A_6 \\ -A_1 + A_3 - A_4 + A_6 \\ +A_2 + A_3 + A_4 + A_5 \end{bmatrix} + L \begin{bmatrix} \omega_2\omega_3 \\ \omega_1\omega_3 \\ \omega_1\omega_2 \end{bmatrix}, \qquad (9.55)$$

where $\boldsymbol{\omega}_{ib}^b = [\omega_1 \quad \omega_2 \quad \omega_3]^T$ is the angular rate vector of the body relative to inertial frame and f^b is the force per unit mass on the body in the body frame. Thus the body angular rates may be found by integration of the first of these sets of equations.

Once the body angular rates are estimated, the remaining computations of the position and orientations are then identical to those in the previous section.

9.4.2 Error Dynamics of Position and Orientation

Because the problem of estimating position is solved directly by double integration, and because integration is inherently an unstable process, small measurement errors when integrated have the effect of introducing drifts in the calculated position relative to the true position. Drift in an inertial navigation computation is particularly sensitive to errors in sensor biases. Higher-frequency noise is attenuated as the integrators act as low-pass filters. The situation is compensated for by complementing the system with position-measurement aids that correct the position error periodically. So an INS is expected to perform over only a finite time horizon. Thus various position- and velocity-measuring devices are used to complement an INS.

The performance of a strapped-down INS depends to a large extent on the accuracy with which it is initially aligned. The process of alignment refers to the process of estimating the initial errors in the position and attitude measurements and then using a feedback system to reduce the influence of these initial errors to a minimum. The accuracy of the initial residual attitudes and position variables governs the subsequent values of the errors in position and orientation. Thus it is extremely important that the alignment process be carried out rapidly and accurately in the initial phase before the INS is set in operation. Although many studies on and investigations into initial alignment have been carried out, the Kalman filter has emerged as a standard method for estimating the initial alignment errors. This filter is essentially constructed from a linear model for the propagation of errors to maximize its ability to reject the effects of errors in sensing as well as external disturbances. The design and construction of the Kalman filter, however, require an extremely good representative model of locally linear error dynamics. In the context of a strapped-down system, once estimated, a feedback system of some form is then used within the error-prediction scheme to reduce the influence of these initial errors to a minimum.

Although drift is one of the major shortcomings of an INS, in the determination of orientation it results mostly from gyro biases, defined as the output produced by a gyro at rest. On integration of the aforementioned equations, the fixed biases, if uncompensated, lead to a constant rate of drift. However, the start-up biases can usually be measured before takeoff and corrected. What matters, then, is bias stability. The typical drift performance of a ring-laser gyro (RLG) is about 0.001°/h, which is sufficient for an INS whose position indication needs to be accurate within about a mile for 1 h. Smaller and less costly are fiber-optic gyros (FOGs) and dynamically tuned gyros (DTGs), which can achieve drift rates in the 0.01–1°/h range, sufficient for short-duration tactical missile flights. The Coriolis vibratory gyroscopes (CVGs), including micromachined versions, have drift rates ranging from several degrees per hour to a degree per second.

Drift in the linear position determined by an INS arises from several sources. First, there are accelerometer instrument errors, such as bias stability, scale factor stability, nonlinearity, and misalignment. Inertial-grade accelerometers, such as those in the 1 mile/h INS just mentioned, must keep all these errors to a few micro g's where g is the local acceleration due to gravity. Considering that the maximum accelerations that these accelerometers must contend with are about 50g's the minimum and maximum accelerations represent a 10^6 dynamic range! Tactical-grade inertial systems require accelerometers with a resolution of the order of hundred micro g's or less. Because we obtain position by double integrating the acceleration, a fixed accelerometer bias error results in a position drift error, which grows quadratically in time. It is therefore especially critical to accurately estimate and eliminate any persistent bias errors. A second critical cause of error in position measurement is error in the orientation determined by the gyros. Because the INS interprets the direction of the measured acceleration according to the computed orientation of the platform, any error in this computed orientation will cause it to integrate the accelerometers in the wrong direction, thus deviating slightly from the true course of the vehicle. The addition of the gravity component is also performed imperfectly by the navigation computer, causing a horizontal acceleration of $1g \times \sin$ (error angle) to be erroneously added to the Earth-frame acceleration vector. Thus, to take proper advantage of hundred micro g-class accelerometers, the pitch-and-roll accuracy must be well below $0.005°$ for the duration of the flight, which implies a far more stringent requirement on the gyros than on the accelerometers.

Assuming small perturbations in the coefficients and variables of the navigation equations and using the fact that $\boldsymbol{\Omega}_{ig}$ is a constant, we may derive the navigation error equations. Thus,

$$\frac{d}{dt}\delta\dot{\mathbf{x}}_g = -\left(\boldsymbol{\Omega}_{ig}\boldsymbol{\Omega}_{ig}\delta\mathbf{x}_g + 2\boldsymbol{\Omega}_{ig}\delta\dot{\mathbf{x}}_g\right)$$

$$+ \delta\mathbf{T}_{gb}\mathbf{A}_{NG}^b + \mathbf{T}_{gb}\delta\mathbf{A}_{NG}^b + \delta\mathbf{G}^g + \frac{\partial\mathbf{G}^g}{\partial\mathbf{x}_g}\delta\mathbf{x}_g, \qquad (9.56)$$

$$\frac{d}{dt}\delta\mathbf{x}_g = \delta\dot{\mathbf{x}}_g, \quad \frac{d}{dt}\delta\mathbf{T}_{gb} = \delta\mathbf{T}_{gb}\boldsymbol{\Omega}_{gb} + \mathbf{T}_{gb}\delta\boldsymbol{\Omega}_{gb},$$

where $\delta\mathbf{A}_{NG}^b$ and $\delta\boldsymbol{\Omega}_{gb}$ are the accelerometer errors and rate gyro output errors, respectively, in the geographic frame. The error states are $\delta\mathbf{x}_g$, the position error vector, $\delta\dot{\mathbf{x}}_g$, the velocity error vector, and $\delta\mathbf{T}_{gb}$, the geographic to body transformation error matrix that could be parameterized in one of several ways. Finally, $\delta\mathbf{G}^g$ is the gravity model error and

$$\boldsymbol{\Gamma}_{ge} = \frac{\partial\mathbf{G}^g}{\partial\mathbf{x}_g}, \qquad (9.57)$$

is the gravity gradient error. Because we assume that the gravity model used is most accurate, the gravity error is assumed to be zero. The gravity gradient error may however be estimated from the approximate expression for the gravity vector given by

$$\mathbf{G}^g \approx \frac{-kM}{|\mathbf{x}_g|^3}\mathbf{x}_g = \frac{-kM}{|r|^3}\mathbf{x}_g, \qquad (9.58)$$

where kM is a constant. The gravity gradient is then given by

$$\boldsymbol{\Gamma}_{ge} = \frac{\partial\mathbf{G}^g}{\partial\mathbf{x}_g} = -kM\left[\frac{\mathbf{I}_{3\times3}}{|r|^3}\mathbf{x}_g - \frac{3}{|r|^5}\mathbf{x}_g\left(\mathbf{x}_g\right)^T\right]. \qquad (9.59)$$

Thus the navigation error equations are

$$\frac{d}{dt}\delta\mathbf{x}_g = \delta\dot{\mathbf{x}}_g,$$

$$\frac{d}{dt}\delta\dot{\mathbf{x}}_g = -\left([\boldsymbol{\Omega}_{ig}\boldsymbol{\Omega}_{ig} - \boldsymbol{\Gamma}_{ge}]\,\delta\mathbf{x}_g + 2\boldsymbol{\Omega}_{ig}\delta\dot{\mathbf{x}}_g\right) + \delta\mathbf{T}_{gb}\mathbf{A}_{NG}^b + \mathbf{T}_{gb}\delta\mathbf{A}_{NG}^b, \quad (9.60)$$

$$\frac{d}{dt}\delta\mathbf{T}_{gb} = \delta\mathbf{T}_{gb}\boldsymbol{\Omega}_{gb} + \mathbf{T}_{gb}\delta\boldsymbol{\Omega}_{gb}.$$

The navigation error equations must be integrated to estimate the drift rates as well as the major components of the drift. It is also possible to establish bounds on the maximum permitted biases in the accelerometers and rate gyros in order to limit the total drift over the time horizon to an acceptable level. Unlike the navigation equations, the navigation error equations are not integrated in real time. However, they provide valuable information into the nature and magnitude of the drift errors generated by the integration processes in an INS. A study of the navigation error equations usually precedes the design and implementation of an INS.

To obtain a better understanding of error dynamics, one may consider the special case of a vehicle following a great circular path at a speed v, which is not necessarily the orbital velocity. The vehicle is assumed to be at a true distance $\mathbf{x}_g = r_0 \mathbf{k}$ from the center of the Earth in the geographic frame, the local north-pointing rectangular reference frame that moves with the vehicle. To understand the dynamics of the navigation errors, we now assume that the true position is not given by $\mathbf{x}_g = r_0 \mathbf{k}$, as believed by the navigation system, but is really given by

$$\mathbf{x}_g + \delta\mathbf{x}_g = r_0\mathbf{k} + \delta x\mathbf{i} + \delta y\mathbf{j} + \delta z\mathbf{k}. \tag{9.61}$$

Then the time derivatives in the rotating frame are

$$\frac{d}{dt}\delta\mathbf{x}_g = \delta\dot{\mathbf{x}}_g = \delta\dot{x}\mathbf{i} + \delta\dot{y}\mathbf{j} + \delta\dot{z}\mathbf{k}, \quad \frac{d}{dt}\delta\dot{\mathbf{x}}_g = \delta\ddot{x}\mathbf{i} + \delta\ddot{y}\mathbf{j} + \delta\ddot{z}\mathbf{k}.$$

The gravitational component of acceleration in the directions of the reference axes is

$$\mathbf{G}^g + \delta\mathbf{G}^g = \frac{-kM}{|\mathbf{x}_g + \delta\mathbf{x}_g|^3}(\mathbf{x}_g + \delta\mathbf{x}_g) = \frac{-kM}{|r_0|^3}\left(\delta x\mathbf{i} + \delta y\mathbf{j} + r_0\mathbf{k} - 2\delta z\mathbf{k}\right).$$

Hence it follows that

$$\delta\mathbf{G}^g = \frac{-kM}{|r_0|^3}\left(\delta x\mathbf{i} + \delta y\mathbf{j} - 2\delta z\mathbf{k}\right). \tag{9.62}$$

The inertial angular velocity of the geographic frame is

$$\mathbf{\Omega}_{ig} = \frac{v_0}{r_0}\mathbf{j}, \tag{9.63}$$

where v_0 is the nominal steady speed of the vehicle.

Assuming that the outputs of the accelerometers are contaminated by errors in all three directions, and substituting in to the equation for position dynamics,

$$\frac{d}{dt}(\dot{\mathbf{x}}_g + \delta\dot{\mathbf{x}}_g) = -\left[\mathbf{\Omega}_{ig}\mathbf{\Omega}_{ig}(\mathbf{x}_g + \delta\mathbf{x}_g) + 2\mathbf{\Omega}_{ig}(\dot{\mathbf{x}}_g + \delta\dot{\mathbf{x}}_g)\right] \\ + \mathbf{A}_{NG}^g + \delta\mathbf{A}_{NG}^g + \mathbf{G}^g + \delta\mathbf{G}^g, \tag{9.64}$$

we obtain the component equations for the dynamics of position error given by

$$\delta\ddot{x} + \left(\frac{kM}{r_0^3} - \frac{v_0^2}{r_0^2}\right)\delta x = -2\frac{v_0}{r_0}\delta\dot{z} + \delta a_x, \quad \delta\ddot{y} + \frac{kM}{r_0^3}\delta y = \delta a_y,$$

$$\delta\ddot{z} - \left(\frac{2kM}{r_0^3} + \frac{v_0^2}{r_0^2}\right)\delta z = 2\frac{v_0}{r_0}\delta\dot{x} + \delta a_z. \tag{9.65}$$

Thus the three equations of (9.65) govern the stability of the position error in the INS. With the passage of time, a desirable situation would be that the errors decay to zero.

From the preceding set it may be observed that the lateral error equation in the y direction represents a forced harmonic oscillator with a fundamental frequency given by

$$\Omega_s = \sqrt{\frac{kM}{r_0^3}} \approx \sqrt{g\frac{R_0^2}{r_0^3}}, \tag{9.66}$$

where R_0 is the radius of the Earth and g is the acceleration that is due to gravity on the Earth's surface. It is referred to as the *Schuler frequency* and is the angular rate of a satellite in orbit just above the Earth's surface. From the in-track error equation in the x direction we may conclude that the in-track error would be stable provided that

$$v_0 < r_0\Omega_s, \tag{9.67}$$

which implies that the vehicle must move well below the circular orbital speed at r_0 for this error to be stable. Finally, the vertical error equation in the z direction indicates that it is always unstable. Thus the stability of the navigational position error can be guaranteed only provided the vertical position error is periodically corrected by independent sensing of the altitude of the vehicle. It is for this reason that most flight vehicles are provided with an independent altimeter that may then be used to correct the position error in the vertical channel at regular intervals of time.

Thus, in the general case, apart from the position error dynamics in the in-track or "north" and cross-track or "east" directions, one must also consider the errors in the three gyro measurements and the associated dynamics.

EXERCISES

9.1. Show that the basis matrix for a Hermite spline is given by

$$\mathbf{M} = \begin{bmatrix} 2 & -2 & 1 & 1 \\ -3 & 3 & -2 & -1 \\ 0 & 0 & 1 & 0 \\ 1 & 0 & 0 & 0 \end{bmatrix}.$$

9.2. Show that the basis matrix for a **B** spline is given by

$$\mathbf{M} = \frac{1}{6}\begin{bmatrix} -1 & 3 & -3 & 1 \\ 3 & -6 & 3 & 0 \\ -3 & 0 & 3 & 0 \\ 1 & 4 & 1 & 0 \end{bmatrix}.$$

9.3. It is well known in numerical analysis that it is often more economical to interpolate functions by use of rational approximations; that is, a function that is a ratio

of two polynomials. Given a quartic polynomial, find a rational approximation of
the form

$$\theta(t) = \frac{at^2 + bt + c}{dt^2 + et + 1}$$

with the same power-series expansion (Taylor's series) as that of the quartic polynomial. (The rational function approximation is known as a [2, 2] Padé approximant.)

9.4. The end-effector of a two-link planar manipulator with equal link lengths is
required to traverse a straight horizontal line at a height equal to one link length.
Determine a cubic spline interpolated trajectory in the joint space as the lower link
moves from $\theta_1 = 30°$ to $\theta_1 = 150°$.

9.5. A segment of a path, $p(t)$, is to be approximated by a quadratic polynomial
approximation, $p_a(t)$, where,

$$p_a(t) = a_0 + a_1 t + a_2 t^2, \quad 0 \le t \le T.$$

To obtain the best approximation, the least-squares data-fitting method is adopted
and an error function defined as

$$E = \int_0^T [p_a(t) - p(t)]^2 dt$$

is minimized. Further, it is required that the approximation match the path exactly
at the initial point $t = 0$ and the final point $t = T$.
  Show that the coefficients of the approximation $p_a(t)$ are given by

$$a_0 = p_a(0), \quad a_1 = \frac{30}{T^2} \int_0^T p(t) \left[\frac{t}{T} - \left(\frac{t}{T} \right)^2 \right] dt - \frac{[3p(T) + 7a_1]}{2T},$$

and

$$a_2 = \frac{(p(T) - a_0 - a_1 T)}{T^2}.$$

9.6. Reconsider Exercise 9.5 and assume that the polynomial approximation $p_a(t)$
is given by the cubic

$$p_a(t) = a_0 + a_1 t + a_2 t^2 + a_3 t^3, \quad 0 \le t \le T.$$

Obtain expressions for the coefficients of the approximation.

9.7. Suppose one is given coordinates $y_0, y_1, \ldots, y_N$ and one desires to pass a
smooth curve through the points $(x_0, y_0), (x_1, y_1), \ldots, (x_N, y_N)$. A spline $S(x)$ for
this purpose is defined by

$$\frac{S(x)}{6h} = M_{j-1}(x_j - x)^3 + M_j(x - x_{j-1})^3$$

$$+ (6y_{j-1} - M_{j-1}h^2)(x_j - x) + (6y_j - M_j h^2)(x - x_{j-1}), \quad x_{j-1} \le x \le x_j,$$

where the coefficients $M_j, j = 0, 1, \ldots, N$ are to be determined.

(a) Given that $S'(x_0) = y_0'$ and $S'(x_N) = y_N'$, show that

$$
\frac{1}{6}
\begin{bmatrix}
2 & 1 & 0 & 0 & \cdots & 0 & 0 & 0 \\
1 & 4 & 1 & 0 & \cdots & 0 & 0 & 0 \\
0 & 1 & 4 & 1 & \cdots & 0 & 0 & 0 \\
\vdots & \vdots & \vdots & \vdots & \cdots & \vdots & \vdots & \vdots \\
0 & 0 & 0 & 0 & \cdots & 1 & 4 & 1 \\
0 & 0 & 0 & 0 & \cdots & 0 & 1 & 2
\end{bmatrix}
\begin{bmatrix}
M_0 \\ M_1 \\ M_2 \\ \vdots \\ M_{N-1} \\ M_N
\end{bmatrix}
= \frac{1}{h^2}
\begin{bmatrix}
6(y_1 - y_0 - hy_0') \\
y_2 - 2y_1 + y_0 \\
y_3 - 2y_2 + y_1 \\
\vdots \\
y_N - 2y_{N-1} + y_{N-2} \\
6(y_{N-1} - y_N + hy_N')
\end{bmatrix}.
$$

(b) Given that $S''(x_0) = 0$ and $S''(x_N) = 0$, show that

$$
\frac{1}{6}
\begin{bmatrix}
6 & 0 & 0 & 0 & \cdots & 0 & 0 & 0 \\
1 & 4 & 1 & 0 & \cdots & 0 & 0 & 0 \\
0 & 1 & 4 & 1 & \cdots & 0 & 0 & 0 \\
\vdots & \vdots & \vdots & \vdots & \cdots & \vdots & \vdots & \vdots \\
0 & 0 & 0 & 0 & \cdots & 1 & 4 & 1 \\
0 & 0 & 0 & 0 & \cdots & 0 & 0 & 6
\end{bmatrix}
\begin{bmatrix}
M_0 \\ M_1 \\ M_2 \\ \vdots \\ M_{N-1} \\ M_N
\end{bmatrix}
= \frac{1}{h^2}
\begin{bmatrix}
0 \\
y_2 - 2y_1 + y_0 \\
y_3 - 2y_2 + y_1 \\
\vdots \\
y_N - 2y_{N-1} + y_{N-2} \\
0
\end{bmatrix}.
$$

9.8. Reconsider the previous exercise and assume that $S(x)$ is periodic. It follows that $M_0 = M_N$. Show that

$$
\frac{1}{6}
\begin{bmatrix}
4 & 1 & 0 & \cdots & 0 & 0 & 1 \\
1 & 4 & 1 & \cdots & 0 & 0 & 0 \\
\vdots & \vdots & \vdots & \cdots & \vdots & \vdots & \vdots \\
0 & 0 & 0 & \cdots & 1 & 4 & 1 \\
1 & 0 & 0 & \cdots & 0 & 1 & 4
\end{bmatrix}
\begin{bmatrix}
M_1 \\ M_2 \\ \vdots \\ M_{N-1} \\ M_N
\end{bmatrix}
= \frac{1}{h^2}
\begin{bmatrix}
y_2 - 2y_1 + y_0 \\
y_3 - 2y_2 + y_1 \\
\vdots \\
y_N - 2y_{N-1} + y_{N-2} \\
y_1 - 2y_N + y_{N-1}
\end{bmatrix}.
$$

9.9. Consider the gyro-free IMU illustrated in Figure 9.10 and show that the time derivative of the angular rate and the body force per unit mass are respectively given in terms of the accelerometer outputs $A_i, i = 1, 2, 3, \ldots 6$, by

$$
\dot{\boldsymbol{\omega}}_{ib}^b = \dot{\boldsymbol{\omega}}_{ib} = \frac{1}{2\sqrt{2}L}
\begin{bmatrix}
+A_1 - A_2 + A_5 - A_6 \\
-A_1 + A_3 - A_4 - A_6 \\
+A_2 - A_3 - A_4 + A_5
\end{bmatrix},
$$

$$
\boldsymbol{f}^b = \frac{1}{2\sqrt{2}}
\begin{bmatrix}
+A_1 + A_2 - A_5 - A_6 \\
-A_1 + A_3 - A_4 + A_6 \\
+A_2 + A_3 + A_4 + A_5
\end{bmatrix}
+ L
\begin{bmatrix}
\omega_2 \omega_3 \\
\omega_1 \omega_3 \\
\omega_1 \omega_2
\end{bmatrix},
$$

where $\boldsymbol{\omega}_{ib}^b = [\omega_1 \;\; \omega_2 \;\; \omega_3]^T$ is the angular rate vector of the body relative to inertial frame and $\boldsymbol{f}^b$ is the force per unit mass on the body in the body frame.

9.10. In a gimballed platform mechanization of an INS designed to operate over the Earth's surface, the outputs of three attitude gyros mounted rigidly on the platform with their axes of sensitivity directed in three mutual perpendicular directions, one of which is normal to the platform, are used to compute feedback torquing commands to servomotors driving the platform gimbals so the platform is always oriented to the local geographic axes.

Three accelerometers mounted rigidly on the platform with their axes of sensitivity directed in three mutual perpendicular directions, one of which is normal to the platform, are used to measure the nongravitational component of the acceleration in the platform fixed-axes system, which is ideally coincident with the local geographic frame. The outputs of the accelerometers are then integrated to compute the velocities of the mobile platform and integrated once again to compute the position coordinates.

For purposes of deriving the navigational equations, the Earth is assumed to be an oblate spheroid (a spheroid flattened in the polar direction). The radius of curvature in a meridian at any latitude is

$$R_M = R_P \frac{(1 - e_E^2)}{(1 - e_E^2 \sin^2 \lambda)},$$

where

$$R_P = \frac{a_E}{\sqrt{(1 - e_E^2 \sin^2 \lambda)}},$$

a_E is the equatorial radius, and e_E is the eccentricity of the equatorial ellipse. However, because $e_E^2 \approx 0.0067$, the Earth may be considered to be sphere of radius $a_E = 6378.2$ km.

(a) Show that the latitude rate $\dot{\lambda}$ is related to the velocity component along a meridian V_N and the altitude h by

$$\dot{\lambda} = \frac{V_N}{R_M + h}.$$

(b) Show that the longitude rate $\dot{\phi}$ is related to the velocity component along a latitude V_E and the altitude h by

$$\dot{\phi} = \frac{V_E \sec \lambda}{R_P + h}.$$

(c) Show that angular velocity of the geographic frame is given by

$$\omega_G = \begin{bmatrix} \cos \lambda \\ 0 \\ -\sin \lambda \end{bmatrix} \omega_s + \begin{bmatrix} 1 \\ 0 \\ -\tan \lambda \end{bmatrix} \frac{V_E}{R_P + h} + \begin{bmatrix} 0 \\ -1 \\ 0 \end{bmatrix} \frac{V_N}{R_M + h}.$$

(d) Assume a spherical Earth model, i.e., assume that

$$R_M + h = R_P + h = R$$

and that the acceleration vector acts normal to the platform in the downward direction and show that

$$\frac{d}{dt} \begin{bmatrix} V_N \\ V_E \\ V_V \end{bmatrix} = \begin{bmatrix} A_N \\ A_E \\ A_V \end{bmatrix} + \begin{bmatrix} 0 \\ 0 \\ g \end{bmatrix} + 2\omega_s \begin{bmatrix} -V_E \sin \lambda \\ V_N \sin \lambda + V_V \cos \lambda \\ -V_E \cos \lambda \end{bmatrix}$$

$$+ \frac{1}{R} \begin{bmatrix} V_N V_V - V_E^2 \\ V_E (V_V + V_N \tan \lambda) \\ -(V_N^2 + V_E^2) \end{bmatrix}.$$

9.11. A simplified and approximate model for the error dynamics of an INS is given by

$$\frac{d}{dt}\begin{bmatrix} \Delta x \\ \Delta v \\ \Delta \psi \end{bmatrix} = \begin{bmatrix} \Delta v \\ -g\Delta\psi + E_A \\ (\Delta v/R) + E_G \end{bmatrix},$$

where Δx is the position error, Δv is the error in the velocity, $\Delta \psi$ is the platform tilt error, g is the acceleration that is due to gravity, R is the radius of the aircraft's flight, and E_A and E_G are the accelerometer and gyro errors, respectively. Assuming a constant gyro bias, show that

$$\Delta x(t) = -(g/\Omega^3)(\Omega t - \sin \Omega t)E_G, \quad \Omega = \sqrt{g/R} = 0.001235\,\text{rad/s}.$$

9.12. (a) Derive the dynamical equations for the errors in a gimballed navigation system and show that the characteristic equation has at least two roots on the imaginary axis corresponding to a radian frequency of $\Omega = 0.001235\,\text{rad/s}$ with a time period equal to about 84.4 min (the Schuler period).

(b) Show that the time period of a simple pendulum of length equal to the radius of the Earth is approximately equal to about 84.4 min (the Schuler period).

(c) Comment on the significance of the results in (a) and (b).

10

Hamiltonian Systems and Feedback Linearization

10.1 Dynamical Systems of the Liouville Type

William Rowan Hamilton was a remarkable dynamicist who was considered to be the greatest mathematician, after Sir Isaac Newton, of the English-speaking world. Born in Dublin in 1805 and named after an outlawed Irish patriot, Archibald Hamilton Rowan, William R. Hamilton died in 1865 as the Astronomer Royal of Ireland. In the span of 60 years he was responsible for a remarkable number of new concepts, including the calculus of the quaternion, a generalization of a complex number to three-dimensional space. Hamilton's discovery not only led to Arthur Cayley's application, in 1854, of the quaternion to the representation of spatial rotations but also led to the development of new algebras, including the theory of matrices and the algebra associated with biquaternions, defined in 1873 by William Kingdon Clifford. In 1834, when he was just 29 years of age, Hamilton wrote to his uncle: "It is my hope and purpose to remodel the whole of dynamics, in the most extensive sense of the word, by the idea of my characteristic function." He proceeded to apply this principle to the motion of systems of bodies, and in the following year expressed these equations of motion in a form that established a duality between the components of momentum of a system of bodies and the coordinates of their positions. He largely achieved his objectives and, in the process, set out on a quest for simplicity that yielded remarkable dividends.

Before Hamilton's general equations of motion are introduced, it is instructive to consider a restricted class of dynamical systems associated with name of Liouville. Liouville, in 1849, showed that by a proper choice of coordinates the kinetic and potential energies could be defined in such a way that the Lagrangian equations can be integrated. To recognize this feature, consider the Lagrangian equations of a system in the form

$$\frac{d}{dt}\frac{\partial L}{\partial \dot{q}_i} - \frac{\partial L}{\partial q_i} = 0, \tag{10.1}$$

where $L = L(q_i, \dot{q}_i)$ is the Lagrangian, which is defined as

$$L = T(q_i, \dot{q}_i) - V(q_i). \tag{10.2}$$

198

Here $T(q_i, \dot{q}_i)$ denotes the total kinetic energy of the dynamical system, $V(q_i)$ is the total potential energy, and q_i are the generalized coordinates. Assuming that the kinetic and potential energies are not explicit functions of time,

$$\frac{d}{dt} L = \sum_i \frac{\partial L}{\partial \dot{q}_i} \frac{d\dot{q}_i}{dt} + \frac{\partial L}{\partial q_i} \frac{dq_i}{dt} = \sum_i \frac{\partial L}{\partial \dot{q}_i} \frac{d\dot{q}_i}{dt} + \frac{d}{dt} \frac{\partial L}{\partial \dot{q}_i} \frac{dq_i}{dt}. \tag{10.3}$$

Hence it follows that

$$\frac{d}{dt} L = \frac{d}{dt} \left(\sum_i \frac{dq_i}{dt} \times \frac{\partial L}{\partial \dot{q}_i} \right) \tag{10.4}$$

and that

$$L = \sum_i \frac{dq_i}{dt} \times \frac{\partial L}{\partial \dot{q}_i} - C, \tag{10.5}$$

where C is a constant of integration. Further, because

$$L = T(q_i, \dot{q}_i) - V(q_i) \tag{10.6}$$

and V is not an explicit function of $\dot{q}_i$, we obtain

$$C = \sum_i \frac{dq_i}{dt} \times \frac{\partial L}{\partial \dot{q}_i} - L = \sum_i \frac{dq_i}{dt} \times \frac{\partial T}{\partial \dot{q}_i} - T + V. \tag{10.7}$$

When T is assumed to a homogeneous quadratic function of $\dot{q}_i$,

$$\sum_i \frac{dq_i}{dt} \times \frac{\partial T}{\partial \dot{q}_i} = 2T, \tag{10.8}$$

and the constant of integration C may be expressed as

$$C = 2T - T + V = T + V. \tag{10.9}$$

Thus it is possible in principle to solve the integral of the Lagrangian for $\dot{q}_i$, which may be again integrated to obtain q_i implicitly. Moreover, the relation for the constant of integration is the sum of the kinetic and potential energies. This indicates that the total energy is always a constant and that it is conserved in situations in which there are no other external forces.

10.1.1 Hamilton's Equations of Motion

The implications of Liouville's result are of greater significance than is apparent and could be applied to a wider class of systems than the conservative systems previously considered. To see this, we formulate the equations of motion in a form first introduced by Hamilton in 1834. The simplest and probably the most direct way of formulating these equations is to reduce the Lagrangian equations to a system of first-order differential equations. To this end, we introduce an additional set of generalized coordinates, the generalized momentum vector **p**, by means of

$$\mathbf{p} = \left[\frac{\partial L}{\partial \dot{q}_i} \right]^T = \nabla_{\dot{q}}^T L, \tag{10.10}$$

where the right-hand side is a column vector of partial derivatives of L. These relations allow us to regard the generalized momentum vector $\mathbf{p}$ as functions of $\dot{q}_i$ and vice versa. The Lagrangian equations may be expressed as

$$\frac{d}{dt}\mathbf{p} = \left[\frac{\partial L}{\partial q_i}\right]^T + \mathbf{Q} = \nabla_q^T L + \mathbf{Q}, \qquad (10.11)$$

where $\mathbf{Q}$ is the vector of generalized forces not included in the potential energy function. The Lagrangian is generally a function of the generalized coordinates and time:

$$L = L(q_i, \dot{q}_i, t) = L(\mathbf{q}, \mathbf{p}, t). \qquad (10.12)$$

We define a new function, the *Hamiltonian H*, by

$$H(\mathbf{q}, \mathbf{p}, t) = \mathbf{p}^T \dot{\mathbf{q}} - L(\mathbf{q}, \mathbf{p}, t). \qquad (10.13)$$

Hence,

$$\nabla_q^T H(\mathbf{q}, \mathbf{p}, t) = -\nabla_q^T L(\mathbf{q}, \mathbf{p}, t) = -\frac{d}{dt}\mathbf{p} + \mathbf{Q}. \qquad (10.14)$$

Furthermore,

$$\nabla_p^T H(\mathbf{q}, \mathbf{p}, t) = \dot{\mathbf{q}} + \nabla_{\dot{q}}^T \dot{\mathbf{q}}^T \times \mathbf{p} - \nabla_{\dot{q}}^T \dot{\mathbf{q}}^T \times \nabla_{\dot{q}}^T L(\mathbf{q}, \mathbf{p}, t) = \dot{\mathbf{q}}. \qquad (10.15)$$

Thus we have the two Hamiltonian equations, which are also known as the *canonical equations*, given by

$$\frac{d}{dt}\mathbf{p} = -\nabla_q^T H(\mathbf{q}, \mathbf{p}, t) + \mathbf{Q}, \quad \frac{d}{dt}\mathbf{q} = \nabla_p^T H(\mathbf{q}, \mathbf{p}, t). \qquad (10.16)$$

The two equations in (10.16) may be expressed together in a compact form as a single set of matrix equations,

$$\boldsymbol{J}\dot{\mathbf{x}} = \frac{dH^T}{d\mathbf{x}} - \begin{bmatrix} \mathbf{Q} \\ \mathbf{0} \end{bmatrix}, \qquad (10.17)$$

where

$$\mathbf{x} = \begin{bmatrix} \mathbf{q} \\ \mathbf{p} \end{bmatrix}, \quad \boldsymbol{J} = \begin{bmatrix} 0 & -\mathbf{I} \\ \mathbf{I} & 0 \end{bmatrix}, \quad \boldsymbol{J}^T = \boldsymbol{J}^{-1} = -\boldsymbol{J}\boldsymbol{J}^2 = -\mathbf{I},$$

and the row vector of derivatives is

$$\frac{dH}{d\mathbf{x}} = \nabla_x H.$$

The matrix $\boldsymbol{J}$, because of its nature, is said to be symplectic and is the matrix version of $j = \sqrt{-1}$, $\mathbf{Q}$ is the vector of generalized forces not included in the potential energy function, and H is the Hamiltonian function in terms of the generalized displacements $\mathbf{q}$ and the generalized momenta $\mathbf{p}$, defined in terms of the Lagrangian by

$$H = \mathbf{p}^T \dot{\mathbf{q}} - L, \quad \mathbf{p} = \frac{\partial L^T}{\partial \dot{\mathbf{q}}}, \quad L = T - V,$$

where T and V are the kinetic and potential energies.

In the absence of the generalized force vector $\mathbf{Q}$, the Hamiltonian equations are

$$\mathbf{J}\dot{\mathbf{x}} = \frac{dH^T}{d\mathbf{x}}. \tag{10.18}$$

Considering the time rate of change of the Hamiltonian, we have

$$\frac{d}{dt}H(\mathbf{q}, \mathbf{p}, t) = \frac{\partial H}{\partial t} + \frac{\partial H}{\partial \mathbf{q}}\frac{d\mathbf{q}}{dt} + \frac{\partial H}{\partial \mathbf{p}}\frac{d\mathbf{p}}{dt}. \tag{10.19}$$

Hence,

$$\frac{d}{dt}H(\mathbf{q}, \mathbf{p}, t) = \frac{\partial H}{\partial t} - \frac{d\mathbf{p}^T}{dt}\frac{d\mathbf{q}}{dt} + \frac{d\mathbf{q}^T}{dt}\frac{d\mathbf{p}}{dt} = \frac{\partial H}{\partial t}. \tag{10.20}$$

Thus if the Hamitonian that is not an explicit function of time is a constant, it follows that H is conserved. Furthermore,

$$\frac{d}{dt}H(\mathbf{q}, \mathbf{p}, t) = \frac{\partial H}{\partial t} + \frac{\partial H}{\partial \mathbf{x}}\frac{d\mathbf{x}}{dt} = \frac{\partial H}{\partial t} + \frac{d\mathbf{x}^T}{dt}\nabla_x^T H = \frac{\partial H}{\partial t} + \nabla_x H \mathbf{J} \nabla_x^T H.$$

Thus,

$$\frac{d}{dt}H(\mathbf{q}, \mathbf{p}, t) = \frac{\partial H}{\partial t} + \nabla_x H \mathbf{J} \nabla_x^T H = \frac{\partial H}{\partial t}, \tag{10.21}$$

and it follows that, as a result of the Hamiltonian structure of the equations of motion,

$$\nabla_x H \mathbf{J} \nabla_x^T H = 0. \tag{10.22}$$

In the case of a conservative mechanical system, it can be shown that the Hamiltonian is equal to the total energy of the system and hence the total energy is conserved. It must be recognized that, even in mechanical dynamical systems involving rotating coordinates or gyroscopic forces, one often comes across Hamiltonian systems where H is as previously defined in terms of the Lagrangian, but does not represent the total energy. A Hamiltonian system is said to be conservative only if the Hamiltonian is the total energy of the system and when the total energy is conserved. In general this may not be the case. If the kinetic energy T is given by

$$T = \frac{1}{2}[\dot{\mathbf{q}}^T \mathbf{T}_2 \dot{\mathbf{q}} + \mathbf{T}_1(\mathbf{q})\dot{\mathbf{q}} + \mathbf{T}_0(\mathbf{q})], \tag{10.23}$$

then

$$\sum_i \frac{\partial L}{\partial \dot{q}_i}\dot{q}_i = \dot{\mathbf{q}}^T \mathbf{T}_2 \dot{\mathbf{q}} + \mathbf{T}_1(\mathbf{q})\dot{\mathbf{q}} = 2T - \mathbf{T}_1(\mathbf{q})\dot{\mathbf{q}} - 2\mathbf{T}_0(\mathbf{q}),$$

$$H = 2T - \mathbf{T}_1(\mathbf{q})\dot{\mathbf{q}} - 2\mathbf{T}_0(\mathbf{q}) - T + V = T + V - \mathbf{T}_1(\mathbf{q})\dot{\mathbf{q}} - 2\mathbf{T}_0(\mathbf{q}).$$

Hence $H = T + V$ only if $\mathbf{T}_1(\mathbf{q})\dot{\mathbf{q}} + 2\mathbf{T}_0(\mathbf{q}) \equiv 0$; i.e., when

$$T = \frac{1}{2}\dot{\mathbf{q}}^T \mathbf{T}_2 \dot{\mathbf{q}}. \tag{10.24}$$

Further, if H is not an explicit function of time, it is a constant and H is conserved. A mechanical system is energy conservative only if $H = T + V = C$, where C is a constant and equal to the total energy.

In general, a linear system

$$\dot{\mathbf{x}} = \mathbf{Ax} + \mathbf{Bu}, \ \ \mathbf{y} = \mathbf{Cx} + \mathbf{Du} \tag{10.25}$$

is a Hamiltonian system if

$$\boldsymbol{J}\mathbf{A} = (\boldsymbol{J}\mathbf{A})^T, \ \ \mathbf{B}^T\boldsymbol{J} = \mathbf{C}, \mathbf{D} = \mathbf{D}^T, \tag{10.26}$$

where $\boldsymbol{J}$ is the symplectic form defined earlier. For purposes of comparison, the linear system is a gradient system provided there exits a transformation $\mathbf{T}$ that is a symmetric nonsingular matrix and

$$\mathbf{T}\mathbf{A} = (\mathbf{T}\mathbf{A})^T, \ \ \mathbf{B}^T\mathbf{T} = \mathbf{C}, \ \ \mathbf{D} = \mathbf{D}^T. \tag{10.27}$$

Thus a gradient system cannot also be a Hamiltonian system. Again, for purposes of comparison, it may be observed that a reversible system satisfies the conditions

$$\mathbf{T}\mathbf{A} = -\mathbf{A}\mathbf{T}, \ \ \mathbf{T}\mathbf{B} = -\mathbf{B}, \ \ \mathbf{C}\mathbf{T} = \mathbf{C}, \ \ \mathbf{T} = \mathbf{T}^{-1}, \tag{10.28}$$

as the system must be invariant under both a transformation of the independent variable t as well as a linear transformation $\mathbf{T}$ of the states.

10.1.2 Passivity of Hamiltonian Dynamics

There is another important consequence of the Hamiltonian nature of the dynamics of engineering systems that is applicable to a large class of electromechanical systems. This is the property of *passivity*. When the generalized force vector $\mathbf{Q}$ is the control input, the property has several important implications that facilitate the design of globally stable control systems. The Hamiltonian nature of the dynamics implies the passivity of the generalized velocity vector $\dot{\mathbf{q}}$ with respect to the generalized force vector $\mathbf{Q}$, i.e.,

$$\int_0^t \dot{\mathbf{q}}^T(\tau)\,\mathbf{Q}(\tau)\,d\tau \geq \gamma, \tag{10.29}$$

for any $t \geq 0$ and a fixed constant γ depending on only the initial state. Because

$$\mathbf{Q} = \frac{d}{dt}\mathbf{p} + \nabla_q^T H(\mathbf{q}, \mathbf{p}, t), \ \ \frac{d}{dt}\mathbf{q} = \nabla_p^T H(\mathbf{q}, \mathbf{p}, t), \tag{10.30}$$

it follows that

$$\int_0^t \dot{\mathbf{q}}^T(\tau)\,\mathbf{Q}(\tau)\,d\tau = \int_0^t \frac{\partial H}{\partial \mathbf{x}}\frac{d\mathbf{x}}{d\tau}d\tau. \tag{10.31}$$

Thus, when H is not an explicit function of time,

$$\int_0^t \dot{\mathbf{q}}^T(\tau)\,\mathbf{Q}(\tau)\,d\tau = \int_0^t \frac{\partial H}{\partial \mathbf{x}}\frac{d\mathbf{x}}{d\tau}d\tau = H(t) - H(0) \geq -H(0) = \gamma. \tag{10.32}$$

This relationship could also be shown to be true for the case when, in addition to the control inputs, there are other generalized forces that are dissipative in nature.

The consequence of the property of *passivity* is that, in a system governed by a dynamics of a Hamiltonian nature, energy cannot be created and can be either conserved or dissipated. Thus passive systems are inherently not unstable, which is an important property in the context of the design of robot-control systems. Moreover the principle of conservation of energy that applies to systems governed by Hamiltonian dynamics is but a special case of the first law of thermodynamics, which states that the increase in energy of a system during any transformation is precisely equal to the energy the system receives from its environment.

10.1.3 Hamilton's Principle

The most important aspect of Hamilton's equations of motion of a mechanical system is not so much the derivation of the equations but the physical interpretations and the basis for the derivation. Hamilton himself called it the *principle of least action*, following the corpuscular theory of light, as the path traced by a ray of light minimizes a quantity that he referred to as the "action." The equations of motion of a dynamic system were derived on the basis that the motion minimizes the action integral. An extended version of the principle is related as the principle of virtual work.

Hamilton's principle of least action for a conservative system may be stated as

$$\delta S = \delta \int_0^t L(\mathbf{q}, \mathbf{p}, t) \, dt = \delta \int_0^t [T(\mathbf{q}, \mathbf{p}, t) - V(\mathbf{q}, \mathbf{p}, t)] \, dt = 0. \quad (10.33)$$

Physically, it implies that, as a body executes a dynamic motion, the increase in the kinetic energy or the virtual work done during the process must be exactly equal to the decrease in the potential energy or the internal energy stored in the system. Thus the principle is simply a restatement of the principle of energy conservation in the context of a conservative system. The paths of motion followed by the dynamic system of particles are such that the action integral has a stationary value; i.e., the integral along the given path has the same value to within first-order infinitesimals as that along all neighboring paths. The difference between two paths for a given t is called the *variation* of q, δq. Thus the variation of the integral of the Lagrangian over a time interval is zero when the integral has a stationary value.

Hamilton's principle may be considered to be an "integral principle," as it considers the entire motion of a system between times $t_1 = 0$ and $t_2 = t$. The instantaneous configuration of the system is described by the n generalized coordinates $q_1, q_2, q_3, \ldots, q_N$, and corresponds to a particular point in Cartesian hyperspace where the q–s form the n coordinate axes. This n-dimensional space is the configuration space. As time evolves, the point representing the current configuration in the configuration space moves and traces a curve. This curve describes the path of motion of the system. The configuration space is not to be confused with the physical three-dimensional space, for which only three coordinates are needed to describe a position at any give time. For example, a system that is being described both by the spatial coordinates and the velocities would have a six-dimensional configuration space at any given point in time. It is these paths of motion in the configuration space that are such that the integral of the Lagrangian over a time interval has a stationary value.

10.2 Contact Transformation

Hamilton's equations of motion are fundamentally nonlinear in general. It is therefore quite natural to seek to transform these equations to simpler forms such as

$$\dot{q} = 0 \quad \text{or} \quad \dot{q} = -\varpi_n^2 q,$$

and

$$\dot{p} = 0 \quad \text{or} \quad \dot{p} = -\varpi_m^2 p,$$

where ϖ_i^2 are constants, while preserving the Hamitonian structure of the equations of motion. Although the study of Hamiltonian systems triggered the study of symplectic systems, the development of contact transformations arose from a need for simplicity. It is indeed ironic that the adjective symplectic in mathematics was coined by replacing the Latin roots in "com-plex" with their Greek equivalents "sym-plectic."

A general transformation of the coordinates

$$\mathbf{x} = [\mathbf{q} \quad \mathbf{p}]^T \tag{10.34}$$

to the coordinates

$$x = [q \quad p]^T \tag{10.35}$$

takes the form

$$x = x(\mathbf{x}), \tag{10.36}$$

and the corresponding Jacobian relationship is given by

$$dx = \mathbf{J} d\mathbf{x}. \tag{10.37}$$

We observe that the quantity $d\mathbf{x}^T \nabla_x^T H(\mathbf{x})$ is a scalar and require that this quantity be invariant under a transformation in order to preserve the Hamiltonian structure of the equations. Thus, considering the difference,

$$d\mathbf{x}^T \nabla_x^T H(\mathbf{x}) - d\mathbf{x}^T \nabla_x^T \tilde{H}(x) = d\mathbf{x}^T \left[\nabla_x^T H(\mathbf{x}) - \mathbf{J}^T \nabla_x^T \tilde{H}(x) \right],$$

where $\tilde{H}(x) \equiv \tilde{H}[x(\mathbf{x})] = H(\mathbf{x})$. However,

$$d\mathbf{x}^T \left[\nabla_x^T H(\mathbf{x}) - \mathbf{J}^T \nabla_x^T \tilde{H}(x) \right] = d\mathbf{x}^T \left(\mathbf{J}\dot{\mathbf{x}} - \mathbf{J}^T \mathbf{J}\dot{x} \right) = d\mathbf{x}^T \left(\mathbf{J} - \mathbf{J}^T \mathbf{J}\mathbf{J} \right) \dot{\mathbf{x}}.$$

Hence it follows that the Jacobian of the transformation, $\mathbf{J}$, must satisfy the constraint

$$\mathbf{J}^T \mathbf{J} \mathbf{J} = \mathbf{J}, \mathbf{J} = \begin{bmatrix} 0 & -\mathbf{I} \\ \mathbf{I} & 0 \end{bmatrix}, \tag{10.38}$$

in order that the Hamiltonian structure of the equations of motion be preserved; i.e., the transformed equations of motion are

$$\mathbf{J}\dot{\mathbf{x}} = \frac{d\tilde{H}(\mathbf{x})}{d\mathbf{x}}^T. \tag{10.39}$$

Transformations that satisfy the preceding constraint are said to be *contact* or *canonical* transformations. The constraint equation itself is often associated with name of Poisson and is therefore known as the *Poisson constraint equation*; the elements of this matrix equation are known as *Poisson brackets*. Contact or canonical transformations are said to be extended point transformations when $\boldsymbol{q} = \boldsymbol{q}(\mathbf{q})$ and $\boldsymbol{p} = \boldsymbol{p}(\mathbf{p})$. In general, contact or canonical transformations are not extended point transformations.

An important feature of the canonical or contact transformations is that, under the operation of matrix multiplication, they form a *group* as they satisfy the four primary requirements that must be met to form one; i.e., the product of two transformations is again a contact transformation (*closure*); the multiplication is *associative*, implying that if A, B, and C are contact transformations, $(AB)C = A(BC)$; an *inverse* transformation exists as does an *identity* transformation, which leaves the variables unaltered. Although it is not essential that two transformations commute, the identity transformation commutes with every member of the group and every member of the group commutes with its own inverse transformation.

10.2.1 Hamilton–Jacobi Theory

One method of constructing a contact or canonical transformation is based on the so-called *generating function*, $S(\boldsymbol{p}, \mathbf{q}, t)$, which implicitly defines a contact transformation $(\mathbf{p}, \mathbf{q}) \rightarrow (\boldsymbol{p}, \boldsymbol{q})$ by

$$\mathbf{p} = \frac{\partial}{\partial \mathbf{q}} S(\boldsymbol{p}, \mathbf{q}, t), \boldsymbol{q} = \frac{\partial}{\partial \boldsymbol{p}} S(\boldsymbol{p}, \mathbf{q}, t), \tag{10.40}$$

and the Hamiltonian is concomitantly transformed by

$$\tilde{H}(\boldsymbol{p}, \boldsymbol{q}, t) = H(\mathbf{p}, \mathbf{q}, t) + \frac{\partial}{\partial t} S(\boldsymbol{p}, \mathbf{q}, t). \tag{10.41}$$

A by-product of this approach is the Hamilton–Jacobi equation, which is a partial differential equation encapsulating the solutions of the Hamiltonian equations in the form of a special contact transformation.

To prove the preceding relation we observe that the variation of the integral

$$\delta \int_0^t [L(\mathbf{q}, \mathbf{p}, t) - \tilde{L}(\boldsymbol{q}, \boldsymbol{p}, t)] \, dt = \delta(S - \boldsymbol{p}^T \boldsymbol{q})|_0^t, \tag{10.42}$$

where $\tilde{L}(\boldsymbol{q}, \boldsymbol{p}, t) = \boldsymbol{p}^T \dot{\boldsymbol{q}} - \tilde{H}(\boldsymbol{q}, \boldsymbol{p}, t)$. Thus when the variations satisfy the appropriate initial final boundary conditions and when

$$\delta \int_0^t L(\mathbf{q}, \mathbf{p}, t)\, dt = 0, \tag{10.43}$$

it follows that

$$\delta \int_0^t \tilde{L}(\boldsymbol{q}, \boldsymbol{p}, t)\, dt = 0. \tag{10.44}$$

Hence the equations of motion in the transformed variables satisfy Hamilton's principle and the associated first-order equations in these variables constitute a canonical set of equations.

10.2.2 Significance of the Hamiltonian Representations

Hamiltonian mechanics was actually discovered first by Lagrange in relation to his work on celestial mechanics. Lagrange discovered that the equations expressing the perturbation of elliptical planetary motion that is due to interactions could be written as a simple system of first-order differential equations (known today as Lagrange's planetary equations). However, it was undoubtedly Hamilton who realized, some 24 years later, the theoretical importance of Lagrange's planetary equations and exploited it to a fuller extent.

Hamilton's equations form a system of first-order coupled differential equations, and the ordinary theory of existence and uniqueness of solutions may be applied to these equations. In practice, Hamilton's equations are almost impossible to solve exactly, except in a few cases. Two of these exceptions are (1) the time-independent Hamiltonians with quadratic potentials (they lead to Hamilton equations, which are linear, and can thus be explicitly solved); (2) the Kepler problem in spherical polar coordinates and, more generally, all "integrable" Hamiltonian systems (they can be solved by successive quadratures).

On the other hand, one can obtain the solutions for the evolution of the generalized position and momenta in time and space by solving the Hamilton–Jacobi equation without solving Hamilton's equations of motion. The Hamilton–Jacobi formulation bears a close relationship to the technique of dynamic programming used for purposes of synthesizing feedback-control laws. Dynamic programming provides a unifying principle for the analysis and solution of optimal control problems. When applied to continuous systems, such as robots seeking to optimally emulate models of human and animal behavior, dynamic programming leads to the Hamilton–Jacobi–Bellman equation, a generalization of the Hamilton–Jacobi equation. A solution to the Hamilton–Jacobi–Bellman equation for a particular control problem provides an optimal controller in feedback form.

The primary limitation of the Hamilton–Jacobi-based methods is that the cost of the associated numerical algorithms, in both space and time, rises exponentially with the number of dimensions of the *phase space* of the model. The phase space is the space formed by the generalized position and velocity coordinates of the model, and therefore has twice the number of dimensions as the model has DOFs. This exponential rise in computational cost is sometimes called the "curse of dimensionality."

Yet in many cases the Hamilton–Jacobi-type formulation, applied in conjunction with Lyapunov's second method based on the Lyapunov function, permits the design of suboptimal but acceptable feedback-control laws for a host of robot-control problems.

Further, Hamiltonian systems are closely associated with energy-conserving systems. In fact, they belong to wider class of systems known as *passive systems*, which do not generate energy but could conserve or dissipate it. The Hamilton–Jacobi theory has been extended and applied to such systems, and it is actually possible to construct robust feedback-control laws that ensure that the property of passivity is preserved. A detailed presentation of these control synthesis techniques as well as the concept of *passivity* is well beyond the scope of this chapter. However, the Hamilton–Jacobi theory forms the backbone of this entire class of techniques.

10.3 Canonical Representations of the Dynamics

Dynamics formulations usually have two main areas of applications: They are used in control and in design analysis and operations. In control systems design, the designer is often concerned about only the equations of dynamics of the system, in which constraint forces are eliminated from the formulation. In these types of systems, it is customary to adopt a minimal set of coordinates to describe the dynamics of the system. The derivation of the governing system dynamics in terms of a minimal set of coordinates is generally difficult because of the additional complications arising from the need to eliminate constraint equations that may or may not be integrable. In the former case, the constraints are said to be *holonomic*, whereas in the latter they are said to be *nonholonomic*. A *nonholonomic* constraint can be viewed as a restriction on the allowable velocities of the object. Whenever it is possible to eliminate the constraint equations from the formulation of the system dynamics, it is possible to express the system dynamics in terms of a minimal set of coordinates. The benefit would be a faster solution of the governing equations of motion, although no information may yet be available about the constraint variables. An alternative is to use a nonminimal or redundant set of coordinates. Descriptions of the system dynamics result in a set of governing equations that is usually slower to be solved, although they are more easily derived, simpler in form, and provide information about the constraint variables. Generally, in the most common class of dynamical problems, one is most interested in the motion variables and their evolution with time. There is little or no interest in the constraint forces and their behavior over a period of time. However, in design and operations of manipulator system, the constraint forces are of key importance. Constraint forces can also be of relevance in the design of advanced controllers. The methods subsequently described can be advantageous in various problems of serial and parallel manipulator dynamics. Rather than solving the Hamiltonian equations, it is often desirable that these equations be transformed to *canonical forms* or *canonical representations* that may then be used for synthesizing feedback-control laws or in the determination and control of constraint forces or moments.

Canonical forms facilitate the synthesis of nonlinear feedback controllers with relative ease. However, the study of the transformation of the nonlinear equations to canonical representations requires the introduction of new mathematical objects such as *Lie algebras* and *Lie groups*, as well as continuous geometrical and topological objects known as *manifolds*. For our purposes, we restrict our attention to the application of these concepts to the transformation of Hamiltonian systems of equations. Only a brief description of Lie derivatives and Lie brackets and their relationship to Lie algebras is presented in the next subsection. Lie brackets are particularly useful in representing nonholonomic motion constraints.

10.3.1 Lie Algebras

To begin, we observe that every vector function of $\mathbf{x}$, say $\mathbf{f}(\mathbf{x})$, corresponds to field of vectors in n-dimensional space with the vector $\mathbf{f}(\mathbf{x})$ emanating from the point $\mathbf{x}$. Thus, following the terminology used in differential geometry, a vector function $\mathbf{f}(\mathbf{x}) : \mathbf{R}^n \to \mathbf{R}^n$ is a *vector field* in $\mathbf{R}^n$. The *Lie derivatives* of $h(\mathbf{x})$ in the direction of a vector field $\mathbf{f}(\mathbf{x})$ are denoted as $L_f h(\mathbf{x})$. Explicitly, the Lie derivative is the directional derivative of $h(\mathbf{x})$ in the "directions" $f_i(\mathbf{x})$:

$$L_f h(\mathbf{x}) = \sum_{i=1}^{n} \frac{\partial h}{\partial x_i} f_i(\mathbf{x}) = [\nabla_x h(\mathbf{x})] \mathbf{f}(\mathbf{x}). \tag{10.45}$$

The adjoint operators or *Lie brackets* are defined as

$$\mathrm{ad}_f^0 \mathbf{s}(\mathbf{x}) = \mathbf{s}, \, \mathrm{ad}_f^1 \mathbf{s}(\mathbf{x}) = [\mathbf{f}, \mathbf{s}] = (\nabla_x \mathbf{s}) \mathbf{f} - (\nabla_x \mathbf{f}) \mathbf{s}, \tag{10.46}$$

and the higher operators are defined recursively as

$$\mathrm{ad}_f^{i+1} \mathbf{s}(\mathbf{x}) = \left[\mathbf{f}, \mathrm{ad}_f^i \mathbf{s}(\mathbf{x}) \right]. \tag{10.47}$$

Two vector fields $\mathbf{g}_1$ and $\mathbf{g}_2$ are said to be involutive if and only if there exist scalar functions $\alpha(\mathbf{x})$ and $\beta(\mathbf{x})$ such that the Lie bracket

$$[\mathbf{g}_1, \mathbf{g}_2] = \alpha(\mathbf{x})\mathbf{g}_1 + \beta(\mathbf{x})\mathbf{g}_2. \tag{10.48}$$

Lie algebra L, named in honor of Sophus Lie (pronounced "lee"), a Norwegian mathematician who pioneered the study of these mathematical objects, is a vector space over some field together with a bilinear multiplication called the bracket. A nonassociative algebra obeyed by objects such as the Lie bracket and Poisson bracket is a Lie algebra. Elements f, g, and h of a Lie algebra satisfy $[f, f] = 0$, $[f + g, h] = [f, h] + [g, h]$, and the *Jacobi identity* defined by

$$[f, [g,h]] + [g, [h, f]] + [h, [f,g]] = 0. \tag{10.49}$$

The relation $[f, f] = 0$ implies that $[f, g] = -[g, f]$. The binary operation of a Lie algebra is the bracket

$$[fg, h] = f[g, h] + [f, h]g. \tag{10.50}$$

Lie algebras permit the recursive formulation of rigid body dynamics for a general class of mechanisms. In particular, the Lie algebra formulation allows one to easily derive the analytic gradient of the recursive dynamics transformation. They may then be used to develop optimal solutions for a class of mechanisms. The Lie algebra should be thought of as a set of special vector fields, and the brackets associated with it play a key role in its application.

The main use of a Lie algebra is the study a special kind of group known as the Lie group. These relate to groups of continuous transformations of the type considered in the Appendix and Chapter 3, particularly those that satisfy differential equations. A Lie algebra may be associated with a Lie group, in which case it reflects the local structure of the Lie group. Briefly, whenever a Lie group G has a group representation on V, its tangent space at the identity, which is a Lie algebra, has a Lie algebra representation on V given by the differential at the identity. Conversely, if a connected Lie group G corresponds to a Lie algebra that has a Lie algebra representation on V, then G has a group representation on V given by the matrix exponential. Two typical examples illustrate this aspect.

The special orthogonal group $SO(3)$ (see the Appendix) is the set of all 3×3 real orthogonal matrices with a unit determinant. It is a subgroup of the general linear group $GL(3)$:

$$SO(3) = \{\boldsymbol{R} \in GL(3) : \boldsymbol{RR}^T = \mathbf{I}, |\boldsymbol{R}| = 1\}. \tag{10.51}$$

$SO(3)$ is a Lie group because it has the structure of a group (under matrix multiplication) and it is a differentiable manifold.

The set of all homogeneous transformation matrices (Chapter 3) is called the $SE(3)$, the special Euclidean group of rigid body transformations in three dimensions, represented by

$$SE(3) = \left\{ \begin{bmatrix} \mathbf{R} & \mathbf{d} \\ \mathbf{0} & 1 \end{bmatrix}, \mathbf{R} \in R^{3 \times 3}, \mathbf{d} \in R^3, \mathbf{RR}^T = \mathbf{I}, |\mathbf{R}| = 1 \right\}. \tag{10.52}$$

On a Lie group the tangent space at the group identity defines a Lie algebra. The Lie algebra of $SO(3)$ is given by $so(3)$, and the Lie algebra of $SE(3)$ is given by $se(3)$ (see Subsection 3.3.3), where

$$so(3) = \left\{ \hat{\mathbf{w}} \in R^{3 \times 3}, \hat{\mathbf{w}}^T = -\hat{\mathbf{w}} \right\}, \tag{10.53a}$$

$$se(3) = \left\{ \begin{bmatrix} \hat{\mathbf{w}} & \mathbf{u} \\ \mathbf{0} & 0 \end{bmatrix}, \hat{\mathbf{w}} \in R^{3 \times 3}, \mathbf{u} \in R^3, \hat{\mathbf{w}}^T = -\hat{\mathbf{w}} \right\}, \tag{10.53b}$$

$$\hat{\mathbf{w}} = \begin{bmatrix} 0 & -w_z & w_y \\ w_z & 0 & -w_x \\ -w_y & w_x & 0 \end{bmatrix}, \tag{10.53c}$$

and $\hat{\mathbf{w}}$ is a skew-symmetric matrix uniquely identified by $\mathbf{w} \in R^3$.

10.3.2 Feedback Linearization

Consider a controlled n-dimensional dynamical system, in which the Hamiltonian equations of motion are expressed in the form

$$\dot{\mathbf{x}} = \mathbf{f}(\mathbf{x}) + \mathbf{g}(\mathbf{x})u, \tag{10.54}$$

where $\mathbf{f}[\mathbf{x}(t)]$ and $\mathbf{g}[\mathbf{x}(t)]$ are vector fields of the vector $\mathbf{x}(t)$. We restrict our attention to single-input (u) to single-output, control-affine systems, characterized by an n-dimensional state response

$$\mathbf{x}(t) = \vec{\chi}\,[t, \mathbf{x}_0, t_0, u(t)]. \tag{10.55}$$

The output function of time is assumed to be given by

$$y(t) = h\,[\mathbf{x}(t)]. \tag{10.56}$$

Before considering the general case, we assume that the n-dimensional dynamical system, in which the Hamiltonian equations of motion are expressed in the so-called pure feedback form, is given by

$$\begin{aligned} \dot{x}_1 &= f_1(x_1,x_2), \\ \dot{x}_2 &= f_2(x_1,x_2,x_3) \\ &\vdots \\ \dot{x}_{n-1} &= f_{n-1}(x_1,x_2,x_3,\ldots,x_n), \\ \dot{x}_n &= f_n(x_1,x_2,x_3,\ldots,x_n) + g_n(x_1,x_2,x_3,\ldots,x_n)u. \end{aligned} \tag{10.57}$$

We define a new set of coordinates and a new control input variable by the transformations

$$\begin{aligned} \xi_1 &= x_1, \\ \xi_2 &= \dot{\xi}_1 = \frac{\partial \xi_1}{\partial x_1} f_1 + \frac{\partial \xi_1}{\partial x_2} f_2 = L_f \xi_1, \\ \xi_3 &= \dot{\xi}_2 = \frac{\partial \xi_2}{\partial x_1} f_1 + \frac{\partial \xi_2}{\partial x_2} f_2 + \frac{\partial \xi_2}{\partial x_3} f_3 = L_f \xi_2 = L_f^2 \xi_1, \\ &\vdots \\ \xi_n &= \dot{\xi}_{n-1} = L_f \xi_{n-1} = L_f^{n-1} \xi_1, \end{aligned} \tag{10.58}$$

to obtain the controlled linear system

$$\dot{\xi}_1 = \xi_2,\ \dot{\xi}_2 = \xi_3, \ldots, \dot{\xi}_{n-1} = \xi_n,\ \dot{\xi}_n = \bar{u}, \tag{10.59}$$

when the feedback-augmented input variable $\bar{u}$ is given by

$$\bar{u} = \dot{\xi}_n = L_f \xi_n + u L_g \xi_n = L_f^n \xi_1 + u L_g L_f^{n-1} \xi_1. \tag{10.60}$$

The system is said to be locally feedback linearizable because it is now a controllable linear system and the coordinate transformations may be defined locally. The property of controllability shows whether the states of the system are sufficiently coupled so that they are influenced, to some extent, by the control input. Needless to say, we have restricted ourselves to a special form of equations of motion, and we would like, in general, to identify the class of systems that can be reduced in this way to a controllable linear system.

Consider the single-input (u) to single-output (y) open-loop system:

$$\sum_\infty = \begin{cases} \dot{\mathbf{x}} = \mathbf{f}(\mathbf{x}) + \mathbf{g}(\mathbf{x})\, u \\ y = h(\mathbf{x}) \end{cases}, \tag{10.61}$$

where h is called the output function, u is the control input, and $\dim(\mathbf{x}) = n$. This system is said to have *relative degree r* at a point $\mathbf{x}_0$ if

1. $L_g L_f^k h(\mathbf{x}) = 0$ for all $\mathbf{x}$ in a neighborhood of $\mathbf{x}_0$ for $k = 0, 1, \ldots, (r-2)$, and
2. $L_g L_f^{r-1} h(\mathbf{x}_0) \neq 0$ for all $\mathbf{x}$,

where $L_g h(\mathbf{x})$ is the Lie derivative of the function h with respect to the vector field $\mathbf{g}$.

By virtue of the *Jacobi identity* satisfied by elements of a Lie algebra, the conditions

$$L_g h(\mathbf{x}) = L_g L_f h(\mathbf{x}) = L_g L_f^2 h(\mathbf{x}) = \cdots L_g L_f^{r-2} h(\mathbf{x}) = 0 \tag{10.62a}$$

are equivalent to the conditions

$$L_g h(\mathbf{x}) = L_{\mathrm{ad}_f g} h(\mathbf{x}) = L_{\mathrm{ad}_f^2 g} h(\mathbf{x}) = \cdots L_{\mathrm{ad}_f^{r-2} g} h(\mathbf{x}) = 0, \tag{10.62b}$$

where the adjoint operator $\mathrm{ad}_f^k()$ is defined in Subsection 10.3.1.

If the relative degree of the control system is equal to its dimension, that is, $r = n$ at all points $\mathbf{x}$, then the system is said to be full state-feedback linearizable and we can use the following change of coordinates:

$$\xi_1 = h,$$

$$\xi_2 = \dot{\xi}_1 = \sum_{i=1}^n \frac{\partial \xi_1}{\partial x_i} f_i + u \frac{\partial \xi_1}{\partial x_1} g_i = L_f h(\mathbf{x}) + u L_g h(\mathbf{x}) = L_f h(\mathbf{x}),$$

$$\xi_3 = \dot{\xi}_2 = \sum_{i=1}^n \frac{\partial \xi_2}{\partial x_i} f_i + u \frac{\partial \xi_2}{\partial x_1} g_i = (L_f + u L_g) L_f h(\mathbf{x}) = L_f^2 h(\mathbf{x}) \tag{10.63}$$

$$\vdots$$

$$\xi_n = \dot{\xi}_{n-1} = L_f^{n-1} h(\mathbf{x}),$$

$$\dot{\xi}_n = L_f^n h(\mathbf{x}) + L_g L_f^{n-1} h(\mathbf{x})\, u,$$

and because $L_g L_f^{n-1} h(\mathbf{x}) \neq 0\, \forall \mathbf{x}$, we can define

$$u = \frac{1}{L_g L_f^{n-1} h(\mathbf{x})} \left[\bar{u} - L_f^n h(\mathbf{x}) \right] \tag{10.64}$$

so that $\dot{\xi}_n = \bar{u}$ and $y = \xi_1$. Furthermore, we can verify that the first state satisfies the conditions

$$\nabla \xi_1 \, \text{ad}^i_f \mathbf{g} = 0, i = 0, 1, 2, \ldots, r - 2, \nabla \xi_1 \, \text{ad}^{r-1}_f \mathbf{g} \neq 0. \qquad (10.65)$$

In this manner, the open-loop system $\sum_\infty$ has been transformed into a linear system (a chain of integrators), which is in a controllable canonic form:

$$\dot{\xi}_1 = \xi_2, \dot{\xi}_2 = \xi_3, \ldots, \dot{\xi}_{n-1} = \xi_n, \dot{\xi}_n = \bar{u}, \qquad (10.66)$$

We may then use standard pole-placement techniques in control theory to select an appropriate control law to define $\bar{u}$ to satisfy the dynamic control requirements. Because this system is in a linear and controllable form it is also stabilizable. We can also specify an arbitrary trajectory by specifying a value of h along it. If $h_d [\mathbf{x}(t)]$ is the desired "trajectory," we can choose

$$\bar{u} = h_d^{(n)} + a_{n-1} \left(h_d^{(n-1)} - \xi_n \right) + \cdots + a_0 (h_d - \xi_1), \qquad (10.67)$$

where the a_i are chosen so that $s^n + a_{n-1}s^{n-1} + \cdots + a_1 s + a_0$ is a stable Hurwitz polynomial in the sense that all the roots of this polynomial lie in the left half of the complex plane, a condition that guarantees the asymptotic stability of the response of the closed-loop system. Arbitrary trajectories may or may not be possible, depending on the form of the output function $h [\mathbf{x}(t)]$ because the feasible trajectories must be compatible with that function.

The conditions for reducing the system to an equivalent linear system could be relaxed to allow for linear systems with "zeros." A zero or *transmission zero* is essentially a situation in which the input to the system does not appear at the output. Thus a single-input–single-output nonlinear system is state-space equivalent to a controllable linear system with a linear output if and only if there exist constants $c_i, i = 0, 1, 2, \ldots, 2n - 1$, such that

$$L_g L^i_f h(\mathbf{x}) = c_i. \qquad (10.68)$$

The earlier result may be recovered provided $c_i = 0$ for all i given by $i = 0, 1, 2, \ldots, 2n - 2$ and $c_{2n-1} \neq 0$. In this case it can be shown that we obtain a linear system with no zeros. However, by relaxing the requirement that there exist constants $c_i, i = 0, 1, 2, \ldots, 2n - 1$, we still obtain a linear system; it is now no longer free of the presence of transmission zeros. Although a detailed treatment and discussion of transmission zeros is beyond the scope of this chapter, the generalization of the requirements for feedback linearization permits the concept to be extended relatively easily to systems with multiple inputs and outputs.

The extension of this result to multiple-input–multiple-output systems is straightforward. A multi-input–multi-output nonlinear system with m linear inputs given as

$$\dot{\mathbf{x}} = \mathbf{f}(\mathbf{x}) + \sum_{k=1}^{m} \mathbf{g}_k(\mathbf{x})u_k, \qquad (10.69)$$

$$\mathbf{y}(t) = \mathbf{h}[\mathbf{x}(t)] = \{h_1 [\mathbf{x}(t)], h_2 [\mathbf{x}(t)], h_3 [\mathbf{x}(t)], \ldots, h_p [\mathbf{x}(t)]\}, \qquad (10.70)$$

is state-space equivalent to a controllable linear system with p linear outputs if and only if there exist constants $c_{i,j,k}, i = 0, 1, 2, \ldots, 2n - 1$, such that

$$L_{g_k} L_f^i h_j (\mathbf{x}) = c_{i,j,k}. \tag{10.71}$$

The ability to linearize multi-input–multi-output systems is particularly useful in the context of the dynamic control of robot arms and robot-manipulator systems, although the subject of feedback linearization and its application to the field of robotics is still in its infancy.

10.3.3 Partial State–Feedback Linearization

Reconsidering the single-input–single-output case, if the relative degree of the control system is not equal to its dimension, that is, $r < n$ at all points $\mathbf{x}$, then the system is said to be partially state-feedback linearizable, and we can use the following change of coordinates:

$$\begin{aligned}
&\xi_1 = h, \xi_2 = \dot{\xi}_1 = L_f h (\mathbf{x}), \; \xi_3 = \dot{\xi}_2 = L_f^2 h (\mathbf{x}), \ldots, \\
&\xi_r = \dot{\xi}_{r-1} = L_f^{r-1} h (\mathbf{x}), \; \xi_{r+1} = x_{r+1}, \ldots, \xi_n = x_n,
\end{aligned} \tag{10.72}$$

and it follows that

$$\begin{aligned}
&\dot{\xi}_1 = \xi_2, \dot{\xi}_2 = \xi_3, \ldots, \dot{\xi}_{r-1} = \xi_r, \dot{\xi}_r = f_r (\xi) + g_r (\xi) \bar{u}, \\
&\dot{\xi}_{r+1} = f_{r+1} (\xi) + g_{r+1} (\xi) \bar{u}, \ldots, \dot{\xi}_n = f_n (\xi) + g_n (\xi) \bar{u}.
\end{aligned} \tag{10.73}$$

The preceding transformed dynamics can be shown to represent the partially feedback linearizable dynamics. Furthermore, when

$$\bar{u} = -[f_r (\xi) + a_{r-1} \xi_r + \cdots + a_0 \xi_1] / g_r (\xi), \tag{10.74}$$

where the a_i are chosen so that $s^r + a_{r-1} s^{r-1} + \cdots + a_1 s + a_0$ is a stable Hurwitz polynomial in the sense that all the roots of this polynomial lie in the left half of the complex plane. This condition guarantees the asymptotic stability of the response of the first r states, and the first r states may be assumed to tend to zero after some finite time. Therefore, by setting these states identically to zero in the equations of motion, we have

$$\begin{aligned}
&\xi_1 = 0, \dot{\xi}_1 = \xi_2 = 0, \dot{\xi}_2 = \xi_3 = 0, \ldots, \dot{\xi}_{r-1} = \xi_r = 0, \dot{\xi}_r = 0, \ldots, \\
&\dot{\xi}_{r+1} = f_{r+1} (\xi) + g_{r+1} (\xi) \hat{u}, \dot{\xi}_n = f_n (\xi) + g_n (\xi) \hat{u},
\end{aligned} \tag{10.75}$$

where

$$\hat{u} = -f_r (\xi) / g_r (\xi), y = \xi_1 = 0. \tag{10.76}$$

The output now being zero, the associated dynamics defines the *zero dynamics* of the system. In particular, when the zero dynamics is asymptotically stable in the sense that all states associated with it tend to zero asymptotically, partial feedback linearization is still extremely useful for constructing feasible controllers to meet dynamic control requirements. Although a complete discussion of the control engineering aspects of the feedback linearization is beyond the scope of this chapter, the technique has several other useful and practical applications.

The methodology of partial feedback linearization could also be extended to the case of multi-input–multi-output systems without much difficulty. Although the details are not presented here, the exercises should assist the interested reader to apply the technique to robotic systems.

10.3.4 Involutive Transformations

The more difficult problem is determining an output function $h[\mathbf{x}(t)]$ that satisfies the requirements of the partial differential equations appearing in the definition of relative degree. The critical issue is the ability to find an artificial output function $h[\mathbf{x}(t)]$ with the maximal relative degree over a finite and known region in the state space. In particular, full state-feedback linearization requires that the relative degree be equal to n. The ability to find such an artificial input hinges on two conditions being met. The first of these is the nonlinear system equivalent of the condition for *full state controllability* in linear systems. The second is the condition of *involutivity*, which is unique to nonlinear systems as it is generically satisfied by linear systems.

In Subsection 10.3.2 it was stated that the transformed first state satisfies the conditions

$$\nabla \xi_1 \, \mathrm{ad}_f^i \mathbf{g} = 0, i = 0, 1, 2, \ldots, r - 2, \nabla \xi_1 \, \mathrm{ad}_f^{r-1} \mathbf{g} \neq 0. \tag{10.77}$$

These equations are a total of r linear equations that define the first state ξ_1, and the remaining states may then be constructed as illustrated in Subsection 10.3.2. Based on a well-known *theorem* in differential geometry, the *Frobenius theorem*, it is known that a solution to this problem exists only when each and every pair of the vector fields,

$$\mathrm{ad}_f^i \mathbf{g} = 0, i = 0, 1, 2, \ldots, r - 2, \tag{10.78}$$

is *involutive*; i.e., the Lie bracket of each and every pair can be expressed as a weighted linear combination of the $r - 2$ vector fields. Furthermore, it is also required that the vector fields,

$$\mathrm{ad}_f^i \mathbf{g} = 0, i = 0, 1, 2, \ldots, r - 1, \tag{10.79}$$

be linearly independent. This latter condition is the nonlinear equivalent of the condition for full state controllability when $r = n$. Thus coordinate transformations that satisfy both conditions with $r = n$ are said to be involutive transformations. The existence of an involutive transformation guarantees full state-feedback linearization. The transformation itself could be constructed by the solutions of

$$\nabla \xi_1 \, \mathrm{ad}_f^i \mathbf{g} = 0, i = 0, 1, 2, \ldots, r - 2, \nabla \xi_1 \, \mathrm{ad}_f^{r-1} \mathbf{g} = 1. \tag{10.80}$$

10.4 Applications of Feedback Linearization

We consider a number of robotics-related examples to illustrate the application of feedback linerization to real, practical systems and their component subsystems. Although relatively simple, these examples illustrate the application of the transformation techniques associated with involutive transformations.

Example 10.1. Third-Order Noninvolutive System

We consider the following third-order noninvolutive system,

$$\dot{\mathbf{x}} = \mathbf{f}(\mathbf{x}) + \mathbf{g}(\mathbf{x})\,\mathbf{u}, \tag{10.81}$$

where

$$\dot{x}_1 = x_2 + x_3^2, \quad \dot{x}_2 = x_3 + x_2^2, \quad \dot{x}_3 = u + x_1^2.$$

The vectors $\mathbf{f(x)}$ and $\mathbf{g(x)}$ may be identified as

$$\mathbf{f(x)} = \begin{bmatrix} x_2 + x_3^2 & x_3 + x_2^2 & x_1^2 \end{bmatrix}^T, \quad \mathbf{g(x)} = \begin{bmatrix} 0 & 0 & 1 \end{bmatrix}^T. \tag{10.82}$$

The Jacobian is given by

$$J(\mathbf{f}) = \begin{bmatrix} 0 & 1 & 2x_3 \\ 0 & 2x_2 & 1 \\ 2x_1 & 0 & 0 \end{bmatrix}. \tag{10.83}$$

The vectors $\mathbf{g}_1(\mathbf{x}) = [\mathbf{f}, \mathbf{g}]$ and $\mathbf{g}_2(\mathbf{x}) = [[\mathbf{f}, \mathbf{g}], \mathbf{g}]$ are obtained from the Lie bracket equation

$$[\mathbf{f}, \mathbf{g}] = J(\mathbf{g})\mathbf{f} - J(\mathbf{f})\mathbf{g}, \tag{10.84}$$

where $J()$ is the Jacobian of the argument.
Hence

$$\mathbf{g}_1(\mathbf{x}) = -[2x_3 \quad 1 \quad 0]^T, \quad \mathbf{g}_2(\mathbf{x}) = \begin{bmatrix} 1 - 2x_1^2 & 2x_2 & 4x_1x_2 \end{bmatrix}^T.$$

To check if the system is feedback linearizable, the system must be involutive. For the system to be involutive, it is necessary to check if the Lie bracket $[\mathbf{g}_1, \mathbf{g}]$ is a weighted linear combination of $\mathbf{g}, \mathbf{g}_1, \mathbf{g}_2$. In fact, the Lie bracket $[\mathbf{g}_1, \mathbf{g}]$ is

$$[\mathbf{g}_1, \mathbf{g}] = -[2 \quad 0 \quad 0]^T,$$

and the system is therefore not involutive and not feedback linearizable.
It is easy to see that the "offending" term is the component vector

$$\mathbf{f}_1(\mathbf{x}) = [x_3^2 \quad 0 \quad 0]^T$$

in the vector $\mathbf{f(x)}$. Thus, if this component is ignored or approximated by a linear term (such as a describing function), the system is feedback linearizable. Hence the system is only partially feedback linearizable. Partial feedback linearization is an important step in the transformation of a system to near linear form.

Example 10.2. Third-Order Locally Feedback Linearizable System

We consider the following third-order system that is locally feedback linearizable, i.e., not globally feedback linearizable,

$$\dot{\mathbf{x}} = \mathbf{f}(\mathbf{x}) + \mathbf{g}(\mathbf{x})\,\mathbf{u}, \tag{10.85}$$

where

$$\dot{x}_1 = x_2 + x_2^2 + x_3^2 + u, \ \ \dot{x}_2 = x_3 + \sin(x_1 - x_3), \ \ \dot{x}_3 = u + x_3^2.$$

By examining the linearized equations, we first introduce the transformation

$$\begin{bmatrix} x_1 \\ x_2 \\ x_3 \end{bmatrix} = \begin{bmatrix} 1 & 0 & 1 \\ 0 & 1 & 0 \\ 0 & 0 & 1 \end{bmatrix} \begin{bmatrix} y_1 \\ y_2 \\ y_3 \end{bmatrix}$$

and reduce the equations to

$$\dot{y}_1 = y_2 + y_2^2, \ \dot{y}_2 = y_3 + \sin(y_1), \ \dot{y}_3 = u + y_3^2.$$

The vectors, $\mathbf{f}, \mathbf{g}, \mathbf{g}_1, \mathbf{g}_2$ are

$$\mathbf{f}(\mathbf{y}) = \begin{bmatrix} y_2 + y_2^2 \\ y_3 + \sin(y_1) \\ y_3^2 \end{bmatrix}, \ \mathbf{g}(\mathbf{x}) = \begin{bmatrix} 0 \\ 0 \\ 1 \end{bmatrix}, \ \mathbf{g}_1(\mathbf{x}) = \begin{bmatrix} 0 \\ 1 \\ 2y_3 \end{bmatrix}, \ \mathbf{g}_2(\mathbf{x}) = \begin{bmatrix} 2y_2 + 1 \\ 2y_3 \\ 2y_3^2 \end{bmatrix}.$$

The Lie bracket $[\mathbf{g}_1, \mathbf{g}]$ is

$$[\mathbf{g}_1, \mathbf{g}] = -[0 \ \ 0 \ \ 2]^T = -2\mathbf{g},$$

and the system is feedback linearizable. However, the vectors $\mathbf{g}, \mathbf{g}_1, \mathbf{g}_2$ are linearly independent only if $y_2 \neq 1/2$, and therefore the system is only locally feedback linearizable. To feedback linearize the system, we let

$$z_1 = y_1, z_2 = L_f z_1 = y_2 + y_2^2, z_3 = L_f z_2 = (1 + 2y_2)[y_3 + \sin(y_1)],$$
$$v = \{y_3 + \sin(y_1) + 2[y_3 + \sin(y_1)]^2\} + (1 + 2y_2)\left[u + y_3^2 + \cos(y_1)(y_2 + y_2^2)\right],$$

and the system reduces to

$$\dot{z}_1 = z_2, \dot{z}_2 = z_3, \dot{z}_3 = v.$$

Example 10.3. Third-Order Feedback Linearizable System:
The Rössler System

We consider the following third-order system,

$$\dot{\mathbf{x}} = \mathbf{f}(\mathbf{x}) + \mathbf{g}(\mathbf{x})\,\mathbf{u}, \tag{10.86}$$

where

$$\dot{x}_1 = ax_1 + x_2, \dot{x}_2 = -x_1 - x_3, \dot{x}_3 = u + b + x_3(x_1 - c).$$

This is indeed one of the simplest of autonomous (time-independent coefficients) systems that is also quite easily feedback linearizable while exhibiting a complex behavior. In fact, transforming the states by the nonlinear transformations

$$z_1 = x_1, \; z_2 = x_2 + ax_1, \; z_3 = (a^2 - 1)x_1 + ax_2 - x_3$$

reduces the equations to

$$\dot{z}_1 = z_2, \; \dot{z}_2 = z_3, \; \dot{z}_3 = a z_3 - z_2 - (a z_2 - z_1 - z_3)(z_1 - c) - u = v.$$

Example 10.4. Inverted Pendulum Connected by a Flexible Joint

This is a two-DOF pendulum described by the equations of motion

$$I\ddot{x}_1 + Mgl \sin(x_1) + k(x_1 - x_3) = 0,$$
$$J\ddot{x}_3 + k(x_3 - x_1) = u.$$

It is a fourth-order system that is linearized by the sequential application of the transformation of coordinates. The state equations are

$$\dot{x}_1 = x_2, \; \dot{x}_2 = -\frac{Mgl}{I}\sin(x_1) - \frac{k}{I}x_1 + \frac{k}{I}x_3,$$

$$\dot{x}_3 = x_4, \; \dot{x}_4 = \frac{k}{J}x_1 - \frac{k}{J}x_3 + \frac{1}{J}u.$$

The controllability matrix for the preceding nonlinear system is obtained by use of the Lie bracket formula and is given by

$$\mathbf{C} = \frac{1}{IJ^2} \begin{bmatrix} kJ & 0 & 0 & 0 \\ 0 & kJ & 0 & 0 \\ -kI & 0 & IJ & 0 \\ 0 & -kI & 0 & IJ \end{bmatrix}.$$

Because the preceding matrix is a linear transformation (all elements are constant) we first apply the transformation

$$\mathbf{y} = \mathbf{Cx},$$

where $\mathbf{y} = [y_1 \; y_2 \; y_3 \; y_4]^{\mathrm{T}}$ and $\mathbf{x} = [x_1 \; x_2 \; x_3 \; x_4]^{\mathrm{T}}$. The nonlinear equations reduce to

$$\dot{y}_1 = y_2, \; \dot{y}_2 = y_3 - \frac{Mgl}{I}\sin\left(\frac{k}{IJ}y_1\right) - \frac{k}{I}y_1 + \frac{k}{J}y_1,$$

$$\dot{y}_3 = y_4, \; \dot{y}_4 = Mgl \sin\left(\frac{k}{IJ}y_1\right) + u.$$

It is interesting to note that, although the equations are feedback linearizable, even the linear part is not in the control canonic form. It is known from linear control theory that a second transformation is essential to reduce even the linear part to the control canonic form. It is often convenient to first transform the linear part to control canonic form and then apply the feedback linearization algorithm. Of course, this is possible only if the system is controllable. Feedback linearization, after a linear transformation is performed, requires the nonlinear controllability matrix to be involutive as well. Sometimes this condition may not be satisfied.

Example 10.5: Sequential Transformation of the Model of a Nonlinear Fourth-Order System

We now apply a linear transformation to a nonlinear system so that the linear part is in the control canonic form. The example considered here is similar to the two-mass, three-spring system and is illustrated in Figure 10.1. The springs $k_1(x)$ and $k_2(x)$ are of the form

$$k_i(x) = k_{i0}[1 + k_{in}(x)], i = 1, 2,$$

where $k_{in}(x)$ represents the nonlinear part of the spring stiffness.

The first-order equations of motion are

$$\dot{x}_1 = y_1, \dot{y}_1 = -\frac{k_1 + k_c}{m_1}x_1 + \frac{k_c}{m_1}x_2,$$

$$\dot{x}_2 = y_2, \dot{y}_2 = \frac{k_c}{m_2}x_1 - \frac{k_c + k_2}{m_2}x_2 + \frac{1}{m_2}u. \tag{10.87}$$

To transform the linear part of the equations to the control canonic form we must apply the transformation

$$z_1 = x_1$$

and obtain

$$z_2 = L_f z_1,$$

where $L_f z = \Delta z \cdot f(x)$ is the Lie derivative in the direction of $f(x)$. Hence,

$$z_2 = y_1.$$

Furthermore,

$$z_3 = L_f z_2, z_3 = -\frac{k_{10} + k_c}{m_1}x_1 + \frac{k_c}{m_1}x_2,$$

$$z_4 = L_f z_3, z_4 = -\frac{k_{10} + k_c}{m_1}y_1 + \frac{k_c}{m_1}y_2.$$

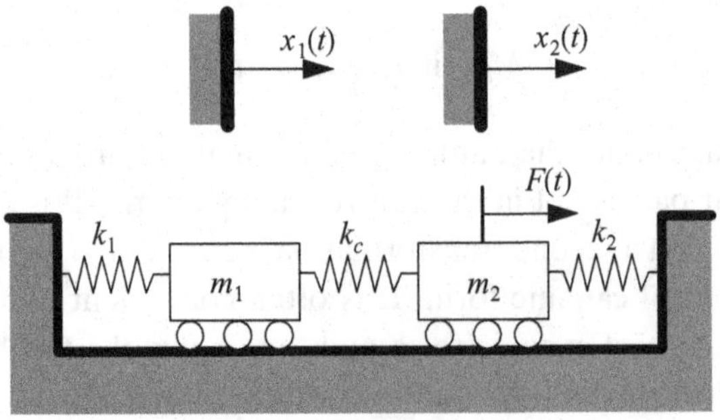

Figure 10.1. Nonlinear vibrating system.

Therefore it follows that

$$x_1 = z_1, y_1 = z_2,$$

$$y_2 = \frac{m_1}{k_c}\left[z_4 + \frac{(k_{10} + k_c)}{m_1}z_2\right], \quad x_2 = \frac{m_1}{k_c}\left[z_3 + \frac{(k_{10} + k_c)}{m_1}z_1\right].$$

The equations of motion for the linear part transform to

$$\dot{z}_1 = z_2, \dot{z}_2 = z_3, \dot{z}_3 = z_4$$

and

$$\dot{z}_4 = \frac{k_{20}}{m_1 m_2}u - \left[\frac{k_{10}k_{20} + k_c(k_{10} + k_{20})}{m_1 m_2}\right]z_1 - \left(\frac{k_{10} + k_c}{m_1} + \frac{k_{20} + k_c}{m_2}\right)z_3.$$

Applying the same transformation to the nonlinear equations, we obtain

$$\dot{z}_1 = z_2, \dot{z}_2 = z_3 - \frac{k_{10}}{m_1}k_{1n}(z_1)z_1, \dot{z}_3 = z_4,$$

$$\dot{z}_4 = \frac{k_c}{m_1}\frac{1}{m_2}u - \frac{k_{10} + k_c}{m_1}\left[z_3 - \frac{k_{10}}{m_1}k_{1n}(z_1)z_1\right] + \frac{k_c}{m_1}\frac{k_c}{m_2}z_1$$

$$- \frac{k_c + k_2\left\{\frac{m_1}{k_c}\left[z_3 + \frac{(k_{10}+k_c)}{m_1}z_1\right]\right\}}{m_2}\left[z_3 + \frac{(k_{10} + k_c)}{m_1}z_1\right].$$

Although the equations are not in the control canonic form, one may now apply feedback linearization and relate the linear coordinates to the corresponding nonlinear ones. Assuming the nonlinear controllability matrix is involutive, the final equations obtained are identical to those obtained by direct feedback linearization of the original nonlinear equations.

Example 10.6. Feedback Linearization of a Two-Input–Two-Output Nonlinear System

Consider the following two-input (u_1, u_2) and two-output (**y**) system:

$$\dot{x}_1 = x_2 + 2x_2(x_3 + u_1 + u_2), \dot{x}_2 = x_3 + u_1 + u_2, \dot{x}_3 = x_4 - x_3^2 - x_5^2 + u_1,$$

$$\dot{x}_4 = x_5 + 2x_3\left(x_4 - x_3^2 - x_5^2 + u_1\right) + 2x_4 u_2, \dot{x}_5 = 0, \tag{10.88}$$

$$\mathbf{y} = [h_1(\mathbf{x}), h_2(\mathbf{x})] = \left[x_1 - x_2^2 \quad x_3\right].$$

We set $\xi_1 = x_1 - x_2^2$, $\xi_2 = x_2$, $\xi_3 = x_3$, $\xi_4 = x_4 - x_3^2 - x_5^2$, and $\xi_5 = x_5$,

It can be easily verified that the resulting system is linear and is given by

$$\dot{\xi}_1 = \xi_2, \dot{\xi}_2 = \xi_3 + u_1 + u_2, \dot{\xi}_3 = \xi_4 + u_1, \dot{\xi}_4 = \xi_5, \dot{\xi}_5 = u_2.$$

The outputs are given as

$$\mathbf{y} = [\xi_1, \xi_3].$$

Example 10.7. Computed Torque Control of a Two-Link Robot Manipulator

Consider the ideal case of a two-link manipulator with an end-effector carrying a typical payload.

The end-effector and its payload are modeled as a lumped mass and assumed to be located at the tip of the outer link. The two-link manipulator with the tip mass is illustrated in Figure 10.2. The general equations of motion of a two-link manipulator may be expressed as

$$\dot{\theta}_1 = \omega_1, \dot{\theta}_2 = \omega_2 - \omega_1,$$

$$\begin{bmatrix} I_{11} & I_{12} \\ I_{21} & I_{22} \end{bmatrix} \begin{bmatrix} \dot{\omega}_1 \\ \dot{\omega}_2 \end{bmatrix} + (m_2 L_{2cg} + M L_2) L_1 \sin(\theta_2) \begin{bmatrix} \omega_1^2 - \omega_2^2 \\ \omega_1^2 \end{bmatrix} \qquad (10.89)$$

$$= [T_1 \quad T_2]^T - g[\Gamma_1 \quad \Gamma_2]^T,$$

where m_i, L_i, L_{icg}, and k_{icg} are the mass, length, the position of the cg with reference to the ith joint, and radius of gyration about the cg of the ith link:

$$I_{11} = m_1 \left(L_{1cg}^2 + k_{1cg}^2 \right) + (m_2 + M) L_1^2 + (m_2 L_{2cg} + M L_2) L_1 \cos(\theta_2),$$
$$I_{12} = m_2 \left(L_{2cg}^2 + k_{2cg}^2 \right) + M L_2^2 + (m_2 L_{2cg} + M L_2) L_1 \cos(\theta_2),$$
$$I_{21} = (m_2 L_{2cg} + M L_2) L_1 \cos(\theta_2), \ I_{22} = m_2 \left(L_{2cg}^2 + k_{2cg}^2 \right) + M L_2^2,$$

and

$$\Gamma_1 = (m_1 L_{1cg} + m_2 L_1 + M L_1) \cos(\theta_1) + (m_2 L_{2cg} + M L_2) \cos(\theta_1 + \theta_2),$$
$$\Gamma_2 = (m_2 L_{2cg} + M L_2) \cos(\theta_1 + \theta_2).$$

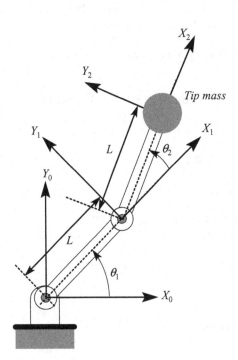

Figure 10.2. Two-link planar anthropomorphic manipulator; the Z axes are all aligned normal to the plane of the paper.

In the special case in which the two links are uniform and identical to each other, assuming that gravity torques are small, the equations may be written in standard state-space form as

$$\dot{\theta}_1 = \omega_1, \dot{\theta}_2 = \omega_2 - \omega_1,$$

$$\begin{bmatrix} \mu + \dfrac{4}{3} + \left(\mu + \dfrac{1}{2}\right)\cos\theta_2 & \left(\mu + \dfrac{1}{3}\right) + \left(\mu + \dfrac{1}{2}\right)\cos\theta_2 \\ \left(\mu + \dfrac{1}{2}\right)\cos\theta_2 & \mu + \dfrac{1}{3} \end{bmatrix}\begin{bmatrix} \dot{\omega}_1 \\ \dot{\omega}_2 \end{bmatrix}$$

$$= -\left(\mu + \dfrac{1}{2}\right)\sin\theta_2 \begin{bmatrix} \omega_1^2 - \omega_2^2 \\ \omega_1^2 \end{bmatrix} + \begin{bmatrix} T_1/mL^2 \\ T_2/mL^2 \end{bmatrix},$$

where $\mu = m/M$ is the ratio of the link mass m to the tip mass M, L is the link length, and T_1 and T_2 are the externally applied torques at the two joints.

The last pair of equations may also be expressed as

$$\Delta(\mu)\begin{bmatrix} \dot{\omega}_1 \\ \dot{\omega}_2 \end{bmatrix} = \begin{bmatrix} \mu + \dfrac{1}{3} & -\left(\mu + \dfrac{1}{3}\right) - \left(\mu + \dfrac{1}{2}\right)\cos\theta_2 \\ -\left(\mu + \dfrac{1}{2}\right)\cos\theta_2 & \mu + \dfrac{4}{3} + \left(\mu + \dfrac{1}{2}\right)\cos\theta_2 \end{bmatrix}$$

$$\times \left[-\left(\mu + \dfrac{1}{2}\right)\sin\theta_2 \begin{bmatrix} \omega_1^2 - \omega_2^2 \\ \omega_1^2 \end{bmatrix} + \begin{bmatrix} T_1/mL^2 \\ T_2/mL^2 \end{bmatrix} \right],$$

where

$$\Delta(\mu) = \left[\dfrac{7}{36} + \dfrac{2}{3}\mu + \left(\mu + \dfrac{1}{2}\right)^2 \sin^2\theta_2\right].$$

In the preceding form the governing equations of motion are expressed in the standard state-space form, thus facilitating their numerical integration to obtain the complete time history of the state vector:

$$\mathbf{x} = \begin{bmatrix} \theta_1 & \theta_2 & \omega_1 & \omega_2 \end{bmatrix}^T.$$

The two servocontrol torque motors at the joints may be assumed to be identical, and the feedback law may be assumed to be a computed torque controller, as it may be estimated from the known time history of the state vector. In the general case, the computed torque controller inputs are

$$\begin{bmatrix} T_1 \\ T_2 \end{bmatrix} = \begin{bmatrix} T_{1c} \\ T_{2c} \end{bmatrix} \equiv \begin{bmatrix} I_{11} + I_{12} & I_{12} \\ I_{21} + I_{22} & I_{22} \end{bmatrix}\begin{bmatrix} v_1 \\ v_2 \end{bmatrix}$$

$$+ (m_2 L_{2cg} + ML_2) L_1 \sin(\theta_2)\begin{bmatrix} \omega_1^2 - \omega_2^2 \\ \omega_1^2 \end{bmatrix} + g\begin{bmatrix} \Gamma_1 \\ \Gamma_2 \end{bmatrix},$$

where the auxiliary control inputs are defined as

$$\begin{bmatrix} v_1 \\ v_2 \end{bmatrix} = \begin{bmatrix} \ddot{\theta}_{1d} \\ \ddot{\theta}_{2d} \end{bmatrix} + 2\omega_n\begin{bmatrix} \dot{\theta}_{1d} \\ \dot{\theta}_{2d} \end{bmatrix} + \omega_n^2\begin{bmatrix} \theta_{1d} \\ \theta_{2d} \end{bmatrix} - 2\omega_n\begin{bmatrix} \dot{\theta}_1 \\ \dot{\theta}_2 \end{bmatrix} - \omega_n^2\begin{bmatrix} \theta_1 \\ \theta_2 \end{bmatrix}$$

and $[\theta_{1d} \quad \theta_{2d}]^T$ are the demanded angular position commands to the joint servo-motors representing the desired angular position states. The last four terms in the equation defining the auxiliary control input are based on the so-called proportional-derivative (PD) controller concept. A PD control scheme emulates the effect of a linear spring and damper acting between the current state and the desired state.

The tracking error between the desired angular position states and the current angular position states is given by

$$\begin{bmatrix} e_1 \\ e_2 \end{bmatrix} = \begin{bmatrix} \theta_{1d} \\ \theta_{2d} \end{bmatrix} - \begin{bmatrix} \theta_1 \\ \theta_2 \end{bmatrix}$$

and satisfies the equation

$$\begin{bmatrix} \ddot{e}_1 \\ \ddot{e}_2 \end{bmatrix} + 2 \begin{bmatrix} \omega_{n1} & 0 \\ 0 & \omega_{n2} \end{bmatrix} \begin{bmatrix} \dot{e}_1 \\ \dot{e}_2 \end{bmatrix} + \begin{bmatrix} \omega_{n1}^2 & 0 \\ 0 & \omega_{n2}^2 \end{bmatrix} \begin{bmatrix} e_1 \\ e_2 \end{bmatrix} = \begin{bmatrix} 0 \\ 0 \end{bmatrix}$$

and is exponentially stable as the response emulates a pair of critically damped and independently vibrating masses, each of which is supported on a spring and damper.

Consider the general case of a two-link manipulator and assume that the computed control torques are the baseline torques and express the external joint torques, in general, as

$$\begin{bmatrix} T_1 \\ T_2 \end{bmatrix} = \begin{bmatrix} T_{1c} \\ T_{2c} \end{bmatrix} + \begin{bmatrix} T_1 \\ T_2 \end{bmatrix} - \begin{bmatrix} T_{1c} \\ T_{2c} \end{bmatrix} \equiv \begin{bmatrix} T_{1c} \\ T_{2c} \end{bmatrix} + \begin{bmatrix} \Delta T_1 \\ \Delta T_2 \end{bmatrix}.$$

Hence observe that the tracking error satisfies

$$\begin{bmatrix} \ddot{e}_1 \\ \ddot{e}_2 \end{bmatrix} + 2 \begin{bmatrix} \omega_{n1} & 0 \\ 0 & \omega_{n2} \end{bmatrix} \begin{bmatrix} \dot{e}_1 \\ \dot{e}_2 \end{bmatrix} + \begin{bmatrix} \omega_{n1}^2 & 0 \\ 0 & \omega_{n2}^2 \end{bmatrix} \begin{bmatrix} e_1 \\ e_2 \end{bmatrix} = \begin{bmatrix} I_{11} + I_{12} & I_{12} \\ I_{21} + I_{22} & I_{22} \end{bmatrix}^{-1} \begin{bmatrix} \Delta T_1 \\ \Delta T_2 \end{bmatrix},$$

where

$$\begin{bmatrix} \Delta T_1 \\ \Delta T_2 \end{bmatrix} = \begin{bmatrix} T_1 \\ T_2 \end{bmatrix} - \begin{bmatrix} T_{1c} \\ T_{2c} \end{bmatrix}$$

are the additional residual torques acting on the joints. This indicates that any additional residual torques in the system tend to drive the tracking error. One of the objectives of a robot controller is then to ensure that the tracking error is exponentially stable even in the presence of residual torques. When the additional residual torques are predictable, a number of classical control schemes may be used to modify the simple PD scheme.

When the additional residual torques are not predictable, one of the many adaptive control schemes must be applied rather than the simple PD scheme used in this example. These controllers not only have the ability to continuously adapt to the dynamic variations in the additional residual torques but are also able to adequately compensate for their dynamical characteristics so the controlled robot arm has the desired stability properties. Thus this last example illustrates the need and role of a typical robot-control system.

10.5 Optimal Control of Hamiltonian and Near-Hamiltonian Systems

An interesting feature of most uncontrolled robotic mechanical systems is that they are usually Hamiltonian in nature or could be idealized as Hamiltonian systems by ignoring the dissipative forces that may be present. We discuss the simplest problem of obtaining the optimal control of a Hamiltonian and a near-Hamiltonian system. Given the system modeled by the set of n governing system of equations,

$$J\dot{\mathbf{x}} = \frac{dH^T}{d\mathbf{x}} - \begin{bmatrix} \mathbf{F} \\ \mathbf{0} \end{bmatrix} - \begin{bmatrix} \mathbf{B} \\ \mathbf{0} \end{bmatrix} \mathbf{u}, \quad \mathbf{x} = \begin{bmatrix} \mathbf{q} \\ \mathbf{p} \end{bmatrix}, \quad J = \begin{bmatrix} 0 & -\mathbf{I} \\ \mathbf{I} & 0 \end{bmatrix}, \quad J^T = J^{-1} = -J,$$

(10.90)

where J is given by the symplectic form, H is the Hamiltonian function in terms of the generalized displacements $\mathbf{q}$ and the generalized momenta $\mathbf{p}$, defined in terms of the Lagrangian by the equations

$$H = \mathbf{p}^T \dot{\mathbf{q}} - L, \quad \mathbf{p} = \frac{\partial L^T}{\partial \dot{\mathbf{q}}}, \quad L = T - V,$$

(10.91)

where T and V are the kinetic and potential energies, respectively, $\mathbf{F}$ is the vector of normalized external forces that cannot be derived from the potential function V, and $\mathbf{Bu}$ are the normalized control forces.

It is often desirable to choose the control $\mathbf{u} = \mathbf{u}^*$ so as to minimize the performance or cost functional:

$$J = \frac{1}{2} \int \left(\mathbf{x}^T \mathbf{Q} \mathbf{x} + \mathbf{u}^T \mathbf{R} \mathbf{u} \right) dt = \int \mathrm{L}(\mathbf{x}, \mathbf{u}) \, dt.$$

(10.92)

The matrices $\mathbf{Q}$ and $\mathbf{R}$ are assumed to be symmetric and $\mathbf{R}$ is assumed to be invertible. To find $\mathbf{u}$ we augment the cost functional and minimize

$$J_{\mathrm{augmented}} = \int \left[\mathrm{L}(\mathbf{x}, \mathbf{u}) + \lambda^T \left(\frac{dH^T}{d\mathbf{x}} - \begin{bmatrix} \mathbf{F} \\ \mathbf{0} \end{bmatrix} - \begin{bmatrix} \mathbf{B} \\ \mathbf{0} \end{bmatrix} \mathbf{u} - J\dot{\mathbf{x}} \right) \right] dt.$$

Minimizing the augmented cost functional with respect to the n-dimensional multiplier λ^T, which is also known as the costate vector, results in the given system model equations. An approach to solving this augmented minimization problem defines another Hamiltonian H_{OC}, where H_{OC} is defined by

$$H_{\mathrm{OC}} = \int \mathrm{H}(\mathbf{x}, \lambda, \mathbf{u}^*) \, dt,$$

(10.93)

where

$$\mathrm{H}(\mathbf{x}, \lambda, \mathbf{u}^*) = \min_u \left[\mathrm{L}(\mathbf{x}, \mathbf{u}) + \lambda^T \left(\frac{dH^T}{d\mathbf{x}} - \begin{bmatrix} \mathbf{F} \\ \mathbf{0} \end{bmatrix} - \begin{bmatrix} \mathbf{B} \\ \mathbf{0} \end{bmatrix} \mathbf{u} \right) \right].$$

(10.94)

In defining H($\mathbf{x}$, λ, $\mathbf{u}^*$), an *n*-dimensional multiplier or costate vector λ^T has been introduced. The multiplier raises the dimensionality of the problem to $2n$ but allows the conditions for the optimal $\mathbf{u}^*$ to be obtained. Integrating the last term in the integral defining the augmented cost functional by parts and minimizing the augmented cost functional with respect to $\mathbf{u}$ and $\mathbf{x}$, we obtain the conditions for the optimal $\mathbf{u}^*$ as

$$\partial H(\mathbf{x}, \lambda, \mathbf{u})/\partial \mathbf{u} = 0, \Rightarrow \mathbf{u}^* = -\mathbf{R}^{-1}[\mathbf{B}^T 0]\lambda,$$

and

$$\dot{\lambda}^T J = -\frac{\partial H^T(\mathbf{x}, \lambda, \mathbf{u})}{\partial \mathbf{x}} = -\lambda^T \left[\frac{d^2 H}{d\mathbf{x}^2} - \begin{bmatrix} \frac{d\mathbf{F}}{d\mathbf{x}} \\ 0 \end{bmatrix} \right] - \mathbf{x}^T \mathbf{Q}. \tag{10.95}$$

It is customary to introduce a further transformation $\lambda = \mathbf{P}x$, so

$$\mathbf{u}^* = -\mathbf{R}^{-1}[\mathbf{B}^T 0]\mathbf{P}\mathbf{x}. \tag{10.96}$$

If we let $\mathbf{H} = d^2 H/d\mathbf{x}^2$, then, from Equation (10.95),

$$J\dot{\lambda} = \left[\mathbf{H} - \left[\frac{d\mathbf{F}^T}{d\mathbf{x}} \, 0 \right] \right] \lambda + \mathbf{Q}\mathbf{x}. \tag{10.97}$$

When $\mathbf{F} = \mathbf{0}$ and $\mathbf{Q} = \mathbf{0}$, $J\dot{\lambda} = \mathbf{H}\lambda$. If we assume that $\lambda = \dot{\mathbf{x}}$,

$$\dot{\lambda} = \ddot{\mathbf{x}} = \frac{d}{dt}J^{-1}\frac{dH^T}{d\mathbf{x}} = J^{-1}\frac{d^2 H}{d\mathbf{x}^2}\dot{\mathbf{x}},$$

and this results in

$$J\dot{\lambda} = \frac{d^2 H}{d\mathbf{x}^2}\dot{\mathbf{x}} = \frac{d^2 H}{d\mathbf{x}^2}\lambda = \mathbf{H}\lambda \tag{10.98a}$$

and

$$\mathbf{u}^* = -\mathbf{R}^{-1}\left[\mathbf{B}^T 0\right]\dot{\mathbf{x}} = -\mathbf{R}^{-1}\mathbf{B}^T\dot{\mathbf{q}}. \tag{10.98b}$$

This result is quite physically reasonable as it implies that the application of optimal control theory to a strictly Hamiltonian system results in a controller that requires a feedback directly proportional to the velocity vector. The feedback is a special case of the control law [Equation (10.96)] that increases the damping in the system that is already not unstable.

Further, when $\mathbf{R} = \mathbf{0}$ and $\mathbf{u} = \mathbf{0}$,

$$H = \frac{dH}{d\mathbf{x}}\dot{\mathbf{x}} = \frac{dH}{dt}.$$

Hence it follows that

$$H_{OC} \equiv H. \tag{10.99}$$

It follows therefore that the two Hamiltonians in this case are identical. This is an important conclusion because the Hamiltonian H_{OC} may viewed as a generalization of the classical Hamiltonian H. It must be noted, however, that although H_{OC}

is a constant, it does not represent the total energy, and hence the system under consideration is not a conservative system. Further, the general equation for the multiplier vector λ,

$$J\dot{\lambda} = \left[\mathbf{H} - \left[\frac{d\mathbf{F}^T}{d\mathbf{x}} \; 0\right]\right]\lambda + \mathbf{Qx} = \partial \mathrm{H}^T(\mathbf{x}, \lambda, \mathbf{u})/\partial \mathbf{x}, \mathbf{u}^* = -\mathbf{R}^{-1}\left[\mathbf{B}^T 0\right]\lambda,$$

is a linear equation in the costate vector λ. Together with the governing system of equations, the pair of equation sets

$$J\dot{\mathbf{x}} = \frac{dH^T}{d\mathbf{x}} - \begin{bmatrix}\mathbf{F}\\\mathbf{0}\end{bmatrix} - \begin{bmatrix}\mathbf{B}\\\mathbf{0}\end{bmatrix}\mathbf{u}, J\dot{\lambda} = \left[\mathbf{H} - \left[\frac{d\mathbf{F}^T}{d\mathbf{x}} \; 0\right]\right]\lambda + \mathbf{Qx}, \quad (10.100)$$

constitute a Hamiltonian system of equations. When the Hamiltonian H is a homogeneous quadratic function, $\mathbf{u}$ and $\mathbf{P}$ [Equation (10.96)] are obtained by a method involving eigenvalue–eigenvector decomposition.

The limits of the integral are omitted in the definition of the Hamiltonian H_{OC}, and these limits are generally fixed; i.e., time is considered an independent variable. It is important to point out that there is another variational principle in mechanics, *the principle of least action*, in which the time variable is no longer treated as an independent variable but as a dependent variable. The *action* integral is defined as a definite time integral of the total kinetic energy. General transformations of coordinates must therefore involve not only the states but also the time variable. For purposes of linearization it may be necessary to transform the time variable as well. For feedback-control synthesis there may not be any benefit in transforming the time variable, although it may be possible to transform the Hamiltonian H_{OC} to a quadratic form.

10.6 Dynamics of Nonholonomic Systems

The occurrence of nonholonomic constraints in robot dynamics was mentioned in passing in a previous section. These systems occur quite commonly in robotics, and it is well nigh impossible to ignore them. Generally, in uncontrolled systems, nonholonomic constraints arise from the presence of one or more rolling contacts between rigid bodies or from conservation laws (usually angular momentum conservation in space robot manipulators). In controlled systems, they arise from the nature of controls that can possibly be applied to the open-loop system. Not only are the dynamics of such systems complex, but they also require careful motion planning.

Motion planning with nonholonomic constraints is considerably more difficult than motion planning in the presence of only position constraints. The difficulties arise, for instance, in planning problems involving (1) wheeled robots navigating in a cluttered environment; (2) multifingered robot hands rolling on the surface of grasped objects; (3) space robots that are capable of freely orbiting a planet; and (4) hopping robots. Typical examples of nonholonomic systems are the unicycle and bicycle. The dynamics of these archetypal systems provide a rich set of problems as they are generally "nonminimum" phase systems subject to nonholonomic contact constraints associated with the rolling constraints on the wheels. The nonminimum phase implies that the inverse models are generally not stable. Like the rolling disk, they are, when considered to traverse a flat ground surface, systems subject to symmetries; their Lagrangians and constraints are invariant with respect to translations and rotations in the ground plane. Yet there is a fundamental difficulty in deriving the equations of motion of such systems, and for this reason we consider the derivation of the equations of motion of a typical nonholonomic system by means of the Lagrangian formalism.

Reconsider Hamilton's principle of least action for an unconstrained conservative system:

$$\delta S = \delta \int_0^t L(\mathbf{q}, \mathbf{p}, t)\, dt = 0. \tag{10.101}$$

Performing the variations and integrating by parts gives

$$\delta S = \int_0^t \left(\delta \mathbf{q}^T \frac{\delta L}{\delta \mathbf{q}^T} + \delta \dot{\mathbf{q}}^T \frac{\delta L}{\delta \dot{\mathbf{q}}^T} \right) dt = \int_0^t \delta \mathbf{q}^T \left(\frac{\delta L}{\delta \mathbf{q}^T} - \frac{d}{dt} \frac{\delta L}{\delta \dot{\mathbf{q}}^T} \right) dt = 0,$$

and the Euler–Lagrange equations are

$$\frac{\delta L}{\delta \mathbf{q}^T} - \frac{d}{dt} \frac{\delta L}{\delta \dot{\mathbf{q}}^T} = 0. \tag{10.102}$$

In the nonconservative case we may include the additional nonpotential forces by considering the additional virtual work done by these forces on the virtual displacements, which gives

$$\frac{\delta L}{\delta \mathbf{q}^T} - \frac{d}{dt} \frac{\delta L}{\delta \dot{\mathbf{q}}^T} + \mathbf{Q}^{\mathrm{NP}} = 0. \tag{10.103}$$

We may now include the effect of constraints. Considering in the first instance holonomic constraints, we may assume the constraints in their most general form to be

$$f_k(\mathbf{q}, t) = 0, \ k = 1, 2, 3, \ldots, K. \tag{10.104}$$

We may generalize Hamilton's principle to yield the corresponding equations of motion simply by replacing the Lagrangian $L(\mathbf{q}, \dot{\mathbf{q}}, t)$ with a generalized Lagrangian defined by

$$L^*(\mathbf{q}, \dot{\mathbf{q}}, t) = L(\mathbf{q}, \dot{\mathbf{q}}, t) + \sum_{k=1}^{K} \lambda_k f_k(\mathbf{q}, t), \tag{10.105}$$

where the λ_k are known as the Lagrange multipliers.

The Euler–Lagrange equations corresponding to the generalized Lagrangian may be obtained from

$$\frac{\delta L^*}{\delta \mathbf{q}^T} - \frac{d}{dt} \frac{\delta L^*}{\delta \dot{\mathbf{q}}^T} + \mathbf{Q}^{\mathrm{NP}} = \frac{\delta L}{\delta \mathbf{q}^T} + \sum_{k=1}^{K} \lambda_k \frac{\delta f_k(\mathbf{q}, t)}{\delta \mathbf{q}^T} - \frac{d}{dt} \frac{\delta L}{\delta \dot{\mathbf{q}}^T} + \mathbf{Q}^{\mathrm{NP}} = 0,$$

and we may recover the constraints themselves by considering virtual increments in the Lagrange multipliers λ_k and performing the variations with respect to these virtual increments. Hence the modified forces that are due to the presence of the constraints may be identified as

$$\frac{d}{dt}\frac{\delta L}{\delta \dot{\mathbf{q}}^T} - \frac{\delta L}{\delta \mathbf{q}^T} = \mathbf{Q}^{\text{NP}} + \sum_{k=1}^{K}\lambda_k \frac{\delta f_k(\mathbf{q},t)}{\delta \mathbf{q}^T} = \mathbf{Q}^{*\text{NP}}. \tag{10.106}$$

Considering the nonholonomic constraints, we may assume them in their most general form to be

$$g_k(\mathbf{q}, \dot{\mathbf{q}}, t) = 0, k = 1, 2, 3, \dots, K. \tag{10.107}$$

However, to generalize Hamilton's equations by application of the additional virtual displacement field and the principle of virtual work, it is essential that the constraints satisfied by the true states $\mathbf{q}$ and $\dot{\mathbf{q}}$ are also satisfied by the virtually displaced states. Thus it is an essential requirement that

$$g_k(\mathbf{q} + \delta\mathbf{q}, \dot{\mathbf{q}} + \delta\dot{\mathbf{q}}, t) = 0, k = 1, 2, 3, \dots, K. \tag{10.108}$$

However, this implies that the virtual displacements are applied instantaneously with $\delta t = 0$. In general, it is quite difficult for the virtually displaced paths to also satisfy the nonholonomic constraints unless these are linear in the generalized velocities. If they are not already linear in the generalized velocities, there must exist a nonlinear canonical transformation such that in the transformed domain the nonholonomic constraints are linear in the transformed generalized velocities. In this case the constraints may be expressed as

$$\mathbf{G}_k(\mathbf{q}, t)\dot{\mathbf{q}} = 0, k = 1, 2, 3, \dots, K. \tag{10.109a}$$

In variational form these constraints may then be written as

$$\mathbf{G}_k(\mathbf{q}, t)\delta\mathbf{q} = 0, k = 1, 2, 3, \dots, K. \tag{10.109b}$$

It is now possible to include these constraints in the generalized Lagrangian by including the additional contribution to the virtual work done by external forces, and it follows that

$$\frac{d}{dt}\frac{\delta L}{\delta \dot{\mathbf{q}}^T} - \frac{\delta L}{\delta \mathbf{q}^T} = \mathbf{Q}^{\text{NP}} + \sum_{k=1}^{K}\lambda_k \mathbf{G}_k(\mathbf{q}, t). \tag{10.110}$$

The classic examples of systems involving holonomic and nonholonomic constraints are centered around the case of a this disk rolling down an inclined plane.

Consider, for example, the case of a thin disk of radius R rolling down a planar inclined plane, as illustrated in Figure 10.3(a). For pure rolling we require the translational velocity of the disk's center of mass down the incline to be equal to the velocity of point on the outer periphery of the disk.

Hence,

$$\dot{x} = R\dot{\theta}. \tag{10.111}$$

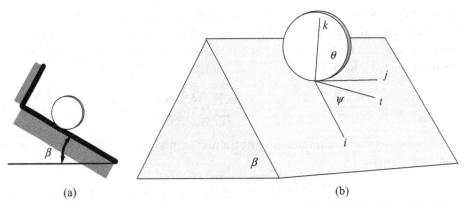

Figure 10.3. (a) Thin disk rolling down a planar inclined plane, (b) thin disk rolling by an angle θ down a plane inclined to the horizontal at an angle β. The disk rolls in a direction at an angle ψ to the direction i, representing the direction of maximum slope.

This is a typical example of a holonomic constraint as it is integrable. Now consider the case of this disk rolling down a three-dimensional inclined surface, as illustrated in Figure 10.3(b).

In this case, at a particular instant, the disk is rolling down in a direction at an angle ψ to the x axis. For this reason, in the direction of translation we have

$$\dot{x}\cos\psi + \dot{y}\sin\psi = R\dot{\theta}, \; \dot{x}\sin\psi - \dot{y}\cos\psi = 0, \tag{10.112}$$

as there is no displacement normal to this direction.

Although these constraints are linear in the velocities $\dot{x}$, $\dot{y}$, and $\dot{\theta}$, they are essentially nonholonomic as they are not integrable.

10.6.1 The Bicycle

Although not a biomimetic vehicle, the bicycle is probably one of the most ingenious vehicles ever invented primarily for the purpose of human locomotion. To illustrate the application of the preceding concepts to practical problems of robot dynamics, we consider the archetypal example of an autonomous wheeled mobile robot, the autonomously driven bicycle illustrated in Figure 10.4. In this model all the inertial properties are assumed to be concentrated in the body labeled as the "payload" with a mass m, and the wheels, the motors, and frame are assumed to be light. Thus its mass could be assumed to be negligible. The wheels are assumed to roll on level ground without slipping.

Figure 10.5 illustrates a top view of the bicycle model and a ground fixed inertial reference frame with the x and y axes in the ground plane and the z axis perpendicular to the ground plane and pointing upward. The gravity vector is assumed to act in the negative z direction. The line joining the two points of contact of each of the two wheels is the *line of contact*. The line of contact may be considered to be the *ground contact vector*, directed from the rear-wheel contact point to the front-wheel contact point, and the angle it makes with respect to the x axis defines the yaw angle, ψ.

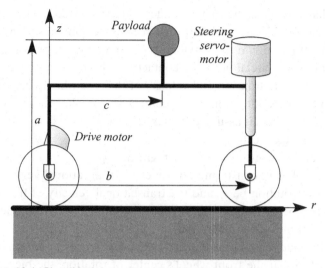

Figure 10.4. Simplified dynamic model of a bicycle driven autonomously.

The angle that nominally the upright plane of symmetry of the bicycle makes with the vertical plane defines the roll angle ϕ.

The steering angle that the steering column makes with the reference direction, the line of contact, is denoted by α. A direct drive, speed-controlled motor, driving the rear wheel, provides the force to propel it forward along the contact vector. The bicycle may be steered, and the steering commands are provided to the steering

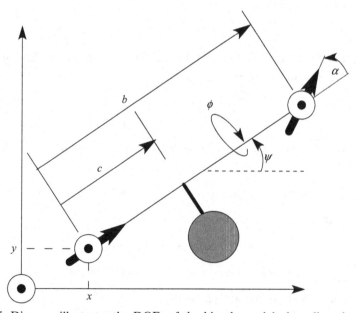

Figure 10.5. Diagram illustrates the DOFs of the bicycle model, the roll angle ϕ, the yaw angle ψ, the position coordinates of the rear-wheel contact point, and the control variable (the steering angle α).

servomotor that drives the steering column about a fixed steering axis. The steering servomotor thus provides the steering angle set-point commands to the steering column. The radii of gyration of the payload about the body axes passing through the center of mass in the direction of the contact vector and an axis parallel to the z axis are k_x and k_z. The translational velocity of the rear wheel's center of rotation is assumed to be in the direction of the contact vector whereas the corresponding vector for the front wheel is assumed to make the same angle as the steering column makes with the contact vector.

The generalized coordinates are then given as ϕ, ψ, x, and y. Our first task is to establish the relevant constraints. If we let the translational velocity of the rear wheel's center of rotation be v_r and the translational velocity of the front wheel's center of rotation be v_f, it then follows that the component of v_f in the direction of the contact vector must be equal to v_r.

Because the component of v_f normal to the contact vector determines the bicycle's lateral velocity at the front wheel's center of rotation, the bicycle's yaw rate about a vertical axis passing through the rear wheel's center of rotation is given by

$$\dot{\psi} = v_f \sin \alpha / b = v_r \tan \alpha / b, \ v_f = v_r / \cos \alpha.$$

Hence,

$$v_r = \dot{\psi} b / \tan \alpha, \ v_f = \dot{\psi} b / \sin \alpha. \tag{10.113}$$

However, we may also express v_r in terms of component velocities in the ground fixed inertial reference frame. Furthermore, as v_r is directed in the direction of the contact vector, the transverse velocity component v_t normal to the contact vector is zero. Hence it follows that

$$v_r = \dot{x} \cos \psi + \dot{y} \sin \psi, \ v_t = \dot{y} \cos \psi - \dot{x} \sin \psi = 0. \tag{10.114}$$

In fact, the pair of equations in (10.114) defines a coordinate transformation to natural velocity coordinates given by

$$\begin{bmatrix} v_r \\ v_t \end{bmatrix} = \begin{bmatrix} \cos \psi & \sin \psi \\ -\sin \psi & \cos \psi \end{bmatrix} \begin{bmatrix} \dot{x} \\ \dot{y} \end{bmatrix}. \tag{10.115}$$

Multiplying the equation for v_r by $\sin \alpha$ and v_t by $-\cos \alpha$ in (10.115) and adding gives

$$v_r \sin \alpha = \dot{x} \sin (\alpha + \psi) - \dot{y} \cos (\alpha + \psi) = \dot{\psi} b \times \cos \alpha. \tag{10.116}$$

The two constraints that must be imposed follow:

$$v_t = \dot{y} \cos \psi - \dot{x} \sin \psi = 0,$$

$$\dot{\psi} b \cos \alpha - \dot{x} \sin (\alpha + \psi) + \dot{y} \cos (\alpha + \psi) = 0, \tag{10.117}$$

which could be expressed in matrix form as

$$\mathbf{G}\dot{\mathbf{q}} = \begin{bmatrix} \mathbf{G}_1 \\ \mathbf{G}_2 \end{bmatrix} \dot{\mathbf{q}} = \begin{bmatrix} 0 \\ 0 \end{bmatrix} = \mathbf{0}, \tag{10.118}$$

where

$$\begin{bmatrix} \mathbf{G}_1 \\ \mathbf{G}_2 \end{bmatrix} \dot{\mathbf{q}} \equiv \begin{bmatrix} 0 & b\cos\alpha & -\sin(\alpha+\psi) & \cos(\alpha+\psi) \\ 0 & 0 & -\sin\psi & \cos\psi \end{bmatrix} \begin{bmatrix} \dot\phi \\ \dot\psi \\ \dot{x} \\ \dot{y} \end{bmatrix}.$$

The generalized velocity vector is defined as

$$\dot{\mathbf{q}} = [\dot\phi \quad \dot\psi \quad \dot{x} \quad \dot{y}]^T, \tag{10.119}$$

and the constraint matrix **G** is defined as

$$\mathbf{G} = \begin{bmatrix} 0 & b\cos\alpha & -\sin(\alpha+\psi) & \cos(\alpha+\psi) \\ 0 & 0 & -\sin\psi & \cos\psi \end{bmatrix}. \tag{10.120}$$

An auxiliary variable is introduced for convenience and is defined by

$$s = \tan\alpha/b. \tag{10.121}$$

Hence the relation between the bicycle's yaw rate and v_r is

$$\dot\psi = v_r s. \tag{10.122}$$

As a consequence of the assumptions made, the mass of the bicycle may be assumed to be m. The center of mass of the bicycle is collocated with the center of mass of the payload. The rolling moment of inertia of the bicycle about the contact vector and the yawing moment of inertia about an axis parallel to the z axis passing through the rear wheel's center of rotation are

$$I_\phi = m\left(k_x^2 + a^2\right) \equiv m \times j_\phi, \; I_\psi = m\left(k_z^2 + c^2\right) \equiv m \times j_\psi. \tag{10.123}$$

Furthermore, it is assumed that the payload's pitching moment of inertia about an axis mutually perpendicular to the roll and yaw axes is identical to the yawing moment of inertia.

The translational velocity components of the bicycle's center of mass in the direction of the contact vector, transverse and normal to it, are

$$\begin{bmatrix} u \\ v \\ w \end{bmatrix} = \begin{bmatrix} v_r \\ v_t \\ 0 \end{bmatrix} + \begin{bmatrix} \dot\phi \\ 0 \\ \dot\psi \end{bmatrix} \times \begin{bmatrix} c \\ -a\sin\phi \\ a\cos\phi \end{bmatrix} = \begin{bmatrix} v_r + \dot\psi a \sin\phi \\ v_t + c\dot\psi - a\dot\phi\cos\phi \\ -\dot\phi a \sin\phi \end{bmatrix}. \tag{10.124}$$

Hence, from (10.124),

$$\begin{bmatrix} u \\ v \\ w \end{bmatrix} = \begin{bmatrix} \dot{x}\cos\psi + \dot{y}\sin\psi + \dot\psi a \sin\phi \\ \dot{y}\cos\psi - \dot{x}\sin\psi + c\dot\psi - a\dot\phi\cos\phi \\ -\dot\phi a \sin\phi \end{bmatrix}. \tag{10.125}$$

The bicycle's body axes' angular rates are

$$[\omega_1 \quad \omega_2 \quad \omega_3]^T = [\dot\phi \quad \dot\psi\sin\phi \quad \dot\psi\cos\phi]^T. \tag{10.126}$$

The kinetic and gravitational potential energies of the bicycle may then be expressed as

$$T = \frac{1}{2}m\left(u^2 + v^2 + w^2 + k_x^2\dot{\phi}^2 + k_z^2\dot{\psi}^2\right),$$

$$V = -mga\left(1 - \cos\phi\right). \tag{10.127}$$

The kinetic energy may be expressed as

$$T = T_1 + T_2 + T_3, \tag{10.128}$$

where

$$T_1 = \frac{1}{2}m\left[a^2\dot{\phi}^2 + \left(c^2 + a^2 s_\phi^2\right)\dot{\psi}^2 - 2acc_\phi\dot{\phi}\dot{\psi} + 2as_\phi\left(\dot{x}c_\psi + \dot{y}s_\psi\right)\dot{\psi}\right],$$

$$T_2 = \frac{1}{2}m(\dot{x}^2 + \dot{y}^2),\; T_3 = \frac{1}{2}m\left(k_x^2\dot{\phi}^2 + k_z^2\dot{\psi}^2\right),$$

and

$$c_\phi = \cos\phi,\, s_\phi = \sin\phi,\, c_\psi = \cos\psi \text{ and } s_\psi = \sin\psi.$$

For convenience we partition the generalized coordinate vector as

$$\mathbf{q} = [\,\phi \quad \psi \quad x \quad y\,]^T = [\,\mathbf{q}_1 \quad \mathbf{q}_1\,]^T, \tag{10.129}$$

with

$$\mathbf{q}_1 = [\,\phi \quad \psi\,]^T, \quad \mathbf{q}_2 = [\,x \quad y\,]^T.$$

The Euler–Lagrange equations may then be expressed as

$$\mathbf{M}\left(\mathbf{q}_1\right)\ddot{\mathbf{q}} = \mathbf{C}\left(\mathbf{q}_1, \dot{\mathbf{q}}\right) + \mathbf{B}\left(\mathbf{q}_1\right)u_2 + \lambda\mathbf{G}^T\left(\tilde{u}_1, \mathbf{q}_1\right),\, \mathbf{G}\left(\tilde{u}_1, \mathbf{q}_1\right)\dot{\mathbf{q}} = \mathbf{0}, \tag{10.130}$$

where

$$\mathbf{M}\left(\mathbf{q}_1\right) = \begin{bmatrix} j_\phi & -acc_\phi & ac_\phi s_\psi & -ac_\phi c_\psi \\ -acc_\phi & j_\psi + a^2 s_\phi^2 & ac_\psi s_\phi - cs_\psi & as_\phi s_\psi + cc_\psi \\ ac_\phi s_\psi & ac_\psi s_\phi - cs_\psi & 1 & 0 \\ -ac_\phi c_\psi & as_\phi s_\psi + cc_\psi & 0 & 1 \end{bmatrix},$$

$$\mathbf{C}\left(\mathbf{q}_1, \dot{\mathbf{q}}\right) = \begin{bmatrix} a^2 s_\phi c_\phi \dot{\psi}^2 \\ -cas_\phi\dot{\phi}^2 - 2a^2 s_\phi c_\phi\dot{\phi}\dot{\psi} \\ as_\phi s_\psi\dot{\phi}^2 - 2ac_\phi c_\psi\dot{\phi}\dot{\psi} + \left(cc_\psi + as_\phi s_\psi\right)\dot{\psi}^2 \\ -as_\phi c_\psi\dot{\phi}^2 - 2ac_\phi s_\psi\dot{\phi}\dot{\psi} + \left(cs_\psi - as_\phi c_\psi\right)\dot{\psi}^2 \end{bmatrix} + g\begin{bmatrix} as_\phi \\ 0 \\ 0 \\ 0 \end{bmatrix},$$

$$\mathbf{B}\left(\mathbf{q}_1\right) = \frac{1}{m}\begin{bmatrix} 0 \\ 0 \\ c_\psi \\ s_\psi \end{bmatrix},\, \mathbf{G}\left(\tilde{u}_1, \mathbf{q}_1\right) = \begin{bmatrix} 0 & b\cos\tilde{u}_1 & -\sin\left(\tilde{u}_1 + \psi\right) & \cos\left(\tilde{u}_1 + \psi\right) \\ 0 & 0 & -\sin\psi & \cos\psi \end{bmatrix},$$

$\tilde{u}_1$ is the steering angle α, and u_2 is the propulsive force driving the bicycle forward. When the relation

$$s = \tan \alpha / b \qquad (10.131)$$

is used, it is sometimes convenient to use s as the control input rather than the steering angle. The Lagrange multiplier λ is usually determined by a combination of empirical physical laws and the relevant initial conditions.

To effect the reduction of the governing equations of motion to a simpler set and eliminate the nonholonomic constraints, consider a transformation of the variables as desirable. In particular, consider the velocity transformation

$$\dot{\mathbf{q}} \equiv \begin{bmatrix} \dot{\phi} \\ \dot{\psi} \\ \dot{x} \\ \dot{y} \end{bmatrix} = \begin{bmatrix} 1 & 0 & 0 & 0 \\ 0 & 0 & 1 & 0 \\ 0 & c_\psi & 0 & -s_\psi \\ 0 & s_\psi & 0 & c_\psi \end{bmatrix} \begin{bmatrix} \dot{\phi} \\ v_r \\ \dot{\psi} \\ v_t \end{bmatrix} \equiv \begin{bmatrix} 1 & 0 & 0 & 0 \\ 0 & 0 & 1 & 0 \\ 0 & c_\psi & 0 & -s_\psi \\ 0 & s_\psi & 0 & c_\psi \end{bmatrix} \dot{\tilde{\mathbf{q}}}, \qquad (10.132)$$

which reduces the nonholonomic constraint equations, (10.118), to

$$\mathbf{G}\dot{\mathbf{q}} \equiv \begin{bmatrix} 0 & b\cos\alpha & -\sin(\alpha+\psi) & \cos(\alpha+\psi) \\ 0 & 0 & -\sin\psi & \cos\psi \end{bmatrix} \begin{bmatrix} 1 & 0 & 0 & 0 \\ 0 & 0 & 1 & 0 \\ 0 & c_\psi & 0 & -s_\psi \\ 0 & s_\psi & 0 & c_\psi \end{bmatrix} \begin{bmatrix} \dot{\phi} \\ v_r \\ \dot{\psi} \\ v_t \end{bmatrix} = \begin{bmatrix} 0 \\ 0 \end{bmatrix},$$

which simplify to

$$\mathbf{G}\dot{\tilde{\mathbf{q}}} \equiv \begin{bmatrix} 0 & -\sin\alpha & b\cos\alpha & \cos\alpha \\ 0 & 0 & 0 & 1 \end{bmatrix} \begin{bmatrix} \dot{\phi} \\ v_r \\ \dot{\psi} \\ v_t \end{bmatrix} = \begin{bmatrix} 0 \\ 0 \end{bmatrix}. \qquad (10.133)$$

Dividing the first of the constraint equations by $b\cos\alpha$ and using the second constraint $v_t = 0$ results in the transformed constraint equations, we obtain

$$\mathbf{G}\dot{\tilde{\mathbf{q}}} \equiv \begin{bmatrix} 0 & -s & 1 & 0 \\ 0 & 0 & 0 & 1 \end{bmatrix} \dot{\tilde{\mathbf{q}}} = \begin{bmatrix} 0 \\ 0 \end{bmatrix}. \qquad (10.134)$$

Assuming that the second of the constraints is satisfied, we may express the transformed state velocity vector $\dot{\tilde{\mathbf{q}}}$ in terms of the first two components of itself as a projection, and it follows that

$$\dot{\tilde{\mathbf{q}}} = \begin{bmatrix} 1 & 0 \\ 0 & 1 \\ 0 & s \\ 0 & 0 \end{bmatrix} \begin{bmatrix} \dot{\phi} \\ v_r \end{bmatrix} \equiv \begin{bmatrix} 1 & 0 \\ 0 & 1 \\ 0 & s \\ 0 & 0 \end{bmatrix} \dot{\tilde{\mathbf{q}}}_1. \qquad (10.135)$$

Substituting the assumed projection into the constraint equations, we observe that they are satisfied. Substituting for $\dot{\tilde{\mathbf{q}}}$ into the velocity transformation results in

$$
\dot{\mathbf{q}} =
\begin{bmatrix}
1 & 0 & 0 & 0 \\
0 & 0 & 1 & 0 \\
0 & c_\psi & 0 & -s_\psi \\
0 & s_\psi & 0 & c_\psi
\end{bmatrix}
\dot{\tilde{\mathbf{q}}} =
\begin{bmatrix}
1 & 0 & 0 & 0 \\
0 & 0 & 1 & 0 \\
0 & c_\psi & 0 & -s_\psi \\
0 & s_\psi & 0 & c_\psi
\end{bmatrix}
\begin{bmatrix}
1 & 0 \\
0 & 1 \\
0 & s \\
0 & 0
\end{bmatrix}
\dot{\tilde{\mathbf{q}}}_1 =
\begin{bmatrix}
1 & 0 \\
0 & s \\
0 & c_\psi \\
0 & s_\psi
\end{bmatrix}
\dot{\tilde{\mathbf{q}}}_1 ,
$$

which could be expressed as

$$
\dot{\mathbf{q}} = \tilde{\mathbf{T}} \dot{\tilde{\mathbf{q}}}_1 , \tag{10.136}
$$

where

$$
\tilde{\mathbf{T}} =
\begin{bmatrix}
1 & 0 & 0 & 0 \\
0 & s & c_\psi & s_\psi
\end{bmatrix}^T .
$$

Introducing the auxiliary input variable u_1 such that

$$
\dot{s} = u_1 , \tag{10.137}
$$

we may express the time derivative of the transformation $\tilde{\mathbf{T}}$ as

$$
\dot{\tilde{\mathbf{T}}} =
\begin{bmatrix}
0 & 0 \\
0 & u_1 \\
0 & -v_r s \times s_\psi \\
0 & v_r s \times c_\psi
\end{bmatrix}, \tag{10.138}
$$

and it follows that

$$
\ddot{\mathbf{q}} = \tilde{\mathbf{T}} \ddot{\tilde{\mathbf{q}}}_1 + \dot{\tilde{\mathbf{T}}} \dot{\tilde{\mathbf{q}}}_1 . \tag{10.139}
$$

Hence the Euler–Lagrange equations may be expressed in terms of the transformed velocities and accelerations as

$$
\tilde{\mathbf{T}}^T \mathbf{M}(\mathbf{q}_1) \tilde{\mathbf{T}} \ddot{\tilde{\mathbf{q}}}_1 = \tilde{\mathbf{T}}^T \mathbf{C}(\mathbf{q}_1, \dot{\mathbf{q}}) - \tilde{\mathbf{T}}^T \mathbf{M}(\mathbf{q}_1) \dot{\tilde{\mathbf{T}}} \dot{\tilde{\mathbf{q}}}_1 + \tilde{\mathbf{T}}^T \mathbf{B}(\mathbf{q}_1) u_2 .
$$

The governing equations of motion may therefore be expressed as

$$
\tilde{\mathbf{M}}(\tilde{\mathbf{q}}_1) \ddot{\tilde{\mathbf{q}}}_1 = \tilde{\mathbf{C}}(\tilde{\mathbf{q}}_1, \dot{\tilde{\mathbf{q}}}_1) + \tilde{\mathbf{B}}(\tilde{\mathbf{q}}_1)
\begin{bmatrix}
u_1 \\
u_2
\end{bmatrix}, \tag{10.140}
$$

where

$$
\tilde{\mathbf{M}}(\tilde{\mathbf{q}}_1) =
\begin{bmatrix}
j_\phi & -acc_\phi s \\
-acc_\phi s & 1 + \left(j_\psi + a^2 s_\phi^2\right) s^2 + 2sas_\phi
\end{bmatrix},
$$

$$
\tilde{\mathbf{C}}(\tilde{\mathbf{q}}_1, \dot{\tilde{\mathbf{q}}}_1) =
\begin{bmatrix}
ac_\phi v_r^2 s + a^2 s_\phi c_\phi v_r^2 s^2 \\
-acss_\phi \dot{\phi}^2 - 2asc_\phi v_r \dot{\phi} - 2a^2 s^2 s_\phi c_\phi v_r \dot{\phi}
\end{bmatrix}
+ g
\begin{bmatrix}
as_\phi \\
0
\end{bmatrix},
$$

$$
\tilde{\mathbf{B}}(\tilde{\mathbf{q}}_1) =
\begin{bmatrix}
acc_\phi v_r & 0 \\
-\left[\left(j_\psi + a^2 s_\phi^2\right) s + as_\phi\right] v_r & 1/m
\end{bmatrix}.
$$

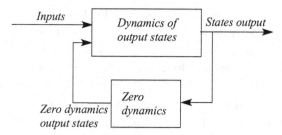

Figure 10.6. Partitioning of plant model to reveal the zero dynamics.

Because $\tilde{\mathbf{B}}$ is square nonsingular, we choose an auxiliary control input vector defined by

$$\begin{bmatrix} u_1 \\ u_2 \end{bmatrix} = \tilde{\mathbf{B}}^{-1}(\tilde{\mathbf{q}}_1)\,[\tilde{\mathbf{M}}(\tilde{\mathbf{q}}_1)\mathbf{Q}_{1d} - \tilde{\mathbf{C}}(\tilde{\mathbf{q}}_1, \dot{\tilde{\mathbf{q}}}_1)], \tag{10.141}$$

and the governing dynamical equations reduce to

$$\ddot{\tilde{\mathbf{q}}}_1 = \mathbf{Q}_{1d}. \tag{10.142}$$

If we choose to define the auxiliary control input vector as

$$\mathbf{Q}_{1d} = \ddot{\tilde{\mathbf{q}}}_{1d} - 2\begin{bmatrix} \omega_{n1} & 0 \\ 0 & \omega_{n2} \end{bmatrix}\dot{\mathbf{e}} - \begin{bmatrix} \omega_{n1}^2 & 0 \\ 0 & \omega_{n2}^2 \end{bmatrix}\mathbf{e}, \tag{10.143}$$

in terms of the tracking error between the current and desired states,

$$\mathbf{e} = \tilde{\mathbf{q}}_1 - \tilde{\mathbf{q}}_{1d}, \tag{10.144}$$

the tracking error satisfies

$$\ddot{\mathbf{e}} + 2\begin{bmatrix} \omega_{n1} & 0 \\ 0 & \omega_{n2} \end{bmatrix}\dot{\mathbf{e}} + \begin{bmatrix} \omega_{n1}^2 & 0 \\ 0 & \omega_{n2}^2 \end{bmatrix}\mathbf{e} = \mathbf{0}. \tag{10.145}$$

The tracking error is then exponentially stable as the response emulates a pair of critically damped and independently vibrating masses, each of which is supported on a spring and damper. One could select different natural frequencies for each of the tracking error modes to avoid any possibilities of coupling in the presence of unmodeled dynamics. Thus, in principle, the bicycle may be driven in a stabilized mode, although the practical implementation of the controller requires a few other additional considerations. The main issues arise because the zero dynamics of the system are unstable. The zero dynamics represents the internal dynamics of the system, which may be represented in block-diagram form as shown in Figure 10.6.

When the bicycle is moving forward with a uniform forward velocity along a straight line on a level surface and in a plane, the zero dynamics is unstable. Thus it is essential that the zero dynamics, representing an inverted pendulum, must be completely stabilized before the bicycle can be considered stable. This is completely in keeping with the human experience of riding a bicycle.

The static stability of a bicycle is primarily due to a distance known as the *trail*, produced by the geometry of the front forks. The point of contact of the front wheel with the ground must be located behind the point where the steering axis intersects the ground. The in-plane distance from the contact point to the steering axis defines the trail. With a positive trail, a tilt of the bicycle results in a gravitational steering moment in the direction of the tilt. The forward momentum of a rolling bicycle resists the resultant change in heading, swinging the bicycle upright. The larger the trail distance, the greater this reaction. It is stabilizing in the sense that it reduces the effective height of the inverted pendulum and has the tendency to restore the tilt of the bicycle such that the contact area is back under the center of gravity of the bicycle and rider. The effect of the positive trail is that it provides internal feedback to make the zero dynamics relatively less unstable. In a *unicycle* the trail is zero, and consequently the steering requires constant adjustment to ensure stability. In the model considered in Figure 10.4 the trail has been deliberately chosen to be zero. The situation is therefore quite similar to the case of a unicycle.

The dynamic stability that a bicycle rider experiences is due to the gyroscopic restoring moment. It is generated by the wheel's rotation and has a stabilizing influence on a moving bicycle, although it is small in magnitude. The angular momentum of the wheels and the gravitational torque applied to them because of tilt generate a precession. The precession tends to steer the bike into the direction of whichever side toward which the bicycle is tilted. In the absence of the trail, the gyroscopic moment is the only stabilizing effect. Like the trail, the gyroscopic moment also has the effect of stabilizing the zero dynamics, although it is dynamic in nature. Consequently the bicycle is stable above and beyond a certain critical forward velocity. The detailed design of a dynamic stabilizer for the zero dynamics is not an easy task and requires additional considerations, beyond the scope of this chapter.

EXERCISES

10.1. The bob of a simple pendulum is replaced with a heavy slender uniform rod, which is attached to the cord at one end, as shown in Figure 10.7. The lengths of the cord and that of the rod are each equal to L.

Figure 10.7. Unifilar suspension.

Obtain the Lagrangian and the corresponding Hamiltonian for this system. Hence determine the Hamiltonian equations of motion.

10.2. Consider the two-link manipulator in Figure 10.2. The two links, of lengths L_i, $i = 1, 2$, are assumed to be uniform and homogeneous with the mass per unit length equal to m_i, the corresponding angular displacements being θ_i. The tip mass is assumed to be M.

Obtain the Lagrangian and the corresponding Hamiltonian for this system. Hence determine the Hamiltonian equations of motion.

10.3. Consider the two-link manipulator in Figure 4.3. The two links, of lengths a_i, $i = 1, 2$, are assumed to be uniform and homogeneous with the mass per unit length equal to m_i, the corresponding angular and linear displacements being $q_1 = \theta_1$ and $q_2 = d_2$, respectively.

Obtain the Lagrangian and the corresponding Hamiltonian for this system. Hence determine the Hamiltonian equations of motion.

10.4. Consider the cable-stayed square planar platform $ABCD$, of side a and of mass m, as illustrated in Figure 5.6. Assume the weight to be negligible. Reconsider exercise 8.10 and obtain the Hamiltonian and the Hamiltonian equations of motion of the platform.

10.5. Reconsider Exercise 8.11 and consider just the first three DOFs; find the Hamiltonian equations of a SCARA robot.

10.6. The governing equations of motion of an oscillating eccentric rotor are non-linear, and partial feedback linearization could be applied to reduce the equations to pure feedback form,

$$\dot{x}_1 = x_2, \dot{x}_2 = -x_1 + \varepsilon \sin(y_1), \dot{y}_1 = y_2, \dot{y}_2 = u.$$

Show that the equations could be reduced to the control canonic form by local feedback linearization for all values of the states except near the points where $y_1 = \pm(2n + 1)\pi/2, n = 0, 1, 2, 3, \ldots$, when the system is not controllable.

10.7. Consider a pair of Hénon–Heiles-type coupled oscillators. This is a Hamiltonian system for which the Hamiltonian is given by

$$H = \frac{1}{2}\left(x_1^2 + x_2^2 + y_1^2 + y_2^2\right) + a y_2 \left(y_1^2 - \frac{1}{3}y_2^2\right) - u x_2.$$

The system represents a pair of coupled oscillators in the form

$$\ddot{x}_1 + x_1 = -2a x_1 y_1, \ddot{y}_1 + y_1 = u - a\left(x_1^2 - y_1^2\right).$$

(a) Show that the Hamiltonian equations of motion are

$$\dot{x}_1 = x_2, \dot{x}_2 = -x_1 - 2a\, x_1\, y_1, \dot{y}_1 = y_2, \dot{y}_2 = u - y_1 - a\left(x_1^2 - y_1^2\right).$$

(b) Show that the equations could be reduced to the control canonic form by local feedback linearization.

10.8. A spherical pendulum is adopted as a model of the load on a boom crane. The slowly varying vertical coordinate of the boom tip is assumed to be the control variable.

We assume that the swinging load of mass m is influenced by, but does not influence, the inertial coordinates of the boom tip, x_a, y_a, and z_a, which are assumed to be known functions of time. The coordinates are assumed to be x, y, and z. The distance between the boom and the load is a constant:

$$(x - x_a)^2 + (y - y_a)^2 + (z - z_a)^2 = l^2.$$

The Lagrangian L is

$$L = \frac{1}{2}m\left(\dot{x}^2 + \dot{y}^2 + \dot{z}^2\right) - mg\left(l - z + z_a\right).$$

Determine whether the equations of motion are feedback linearizable.

10.9. The inverted pendulum on a moving cart is a classic model of a loaded kinematic double joint. The motion of the system is defined by the horizontal displacement of the cart y, from some reference point, and the angle that the pendulum rod makes with respect to the vertical, θ. The kinetic and potential energies of the system are

$$T = \frac{1}{2}M\dot{y}^2 + \frac{1}{2}m\left(\dot{y}^2 + 2l\dot{y}\dot{\theta}\cos\theta + l^2\dot{\theta}^2\right), V = mgl\cos\theta.$$

Neglecting all friction, the only generalized force is the one acting on the cart and in the direction of the y coordinate, which is the control input, u. Determine whether the equations of motion are feedback linearizable.

10.10. Consider the system defined by the equations of motion,

$$\dot{\mathbf{x}} = \mathbf{f}(\mathbf{x}) + \mathbf{g}(\mathbf{x})\mathbf{u}, \ y = h\left(\mathbf{x}\right),$$

where

$$\dot{x}_1 = x_2 + x_4^2 + u, \ \dot{x}_2 = x_4^2 + u, \ \dot{x}_3 = -\left(x_3 - x_2\right)^2 + x_4^2 + x_4^3 + u,$$

$$\dot{x}_4 = \left(x_2 - x_3\right)x_4^2 - x_4, \ y = x_1 - x_2.$$

Show that the system is partially feedback linearizable with index $r = 2$, and hence find the transformation.

10.11. A commonly adopted model of a kinematic joint is the ball and beam model. A ball of mass m, considered to be a particle, is in a tube of infinite length and of moment of inertia J, at the midpoint of which a torque τ is exerted. All friction is neglected, and equations of motion in terms of the position of the ball with respect to the origin r and the angle φ, by which the tube is rotated about the horizontal, are

$$\dot{r} = v, \dot{v} = -g\sin\varphi + r\omega^2, \dot{\varphi} = \omega,$$

$$(J + mr^2)\dot{\omega} = -2mrv\omega - mgr\cos\varphi + \tau.$$

Show that the system is not feedback linearizable about any neighborhood of the origin.

10.12. Consider the autonomously driven model of bicycle and suppose that we may choose the initial conditions and appropriate inputs such that the bicycle is moving forward with a uniform forward velocity along a straight line on a level surface and in a plane. Then the resulting restricted zero dynamics may be found by setting $\ddot{x}$, $\ddot{y}$, $\ddot{\psi}$, and $\dot{\psi}$ to zero.

(a) Show that the restricted zero dynamics represents an equivalent inverted pendulum. Comment on the consequences of this property.

(b) Consider the influence of the gyroscopic torques that is due to the rotation of the front and rear wheels and discuss the consequences insofar as the zero dynamics is concerned. Hence derive the governing equations of motion of the bicycle and obtain the modified zero dynamics when the gyroscopic effects are included in the analysis. Identify any additional design parameters that may arise.

(c) Recognizing that the gyroscopic restoring torque is a function of the forward speed of the bicycle, determine the minimum critical speed beyond which the modified zero dynamics is not unstable.

10.13. Figure 10.8 illustrates a three-link manipulator. For dynamical purposes it is assumed that links 1 and 3 are uniform and homogeneous with a mass per unit length equal to m_i, $i = 1$ or 3.

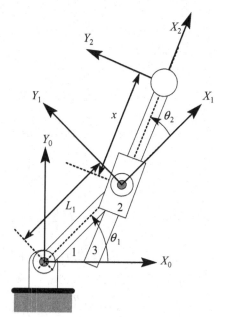

Figure 10.8. Two-link nonanthropomorphic manipulator.

It is also assumed that link 2 is a coupler pinned to link 1 at its center of mass and prismatically coupled to link 3. The moment of inertia of link 2 is assumed to be I_i, $i = 2$. In addition, the wrist is assumed to be a lumped mass attached to link 3 at the top end and a balancing mass of the same weight is attached at the other. Link 1 is assumed to be length L_1 and link 3 is assumed to be of length $2L_3$. A control force Q_3 is applied to link 3 to position it along the prismatic joint and a control torque Q_2 is applied by link 1 to the coupler.

Obtain the Hamiltonian equations of motion and determine whether the system is partially feedback linearizable. Hence find the relative degree and the corresponding transformation.

10.14. Figure 4.8 illustrates a three-link planar manipulator with three parallel rotational joints. The dynamic equations may be expressed as

$$\mathbf{I}(\mathbf{q})\,\ddot{\mathbf{q}} + \mathbf{C}(\mathbf{q}, \dot{\mathbf{q}}) - g\Gamma_g(\mathbf{q}) = \tau,$$

where τ is the vector of joint torques and $\mathbf{q} = [q_1 \ q_2 \ q_3]^T$ is the vector of joint angles, measured relative to the inertial vertical for link 1 and with respect to the $(i-1)$ link axis for link i, $i = 1, 2$. The centers of mass of the links are assumed to at the midpoints of the link lengths. The masses, lengths, and centroidal moments of inertia about an axis perpendicular to the plane of rotation of the three links are m_i, l_i, and I_{iz}, respectively. Let $c_i = \cos\theta_i$, $s_i = \sin\theta_i$, $c_{ij} = \cos(\theta_i + \theta_j)$, $s_{ij} = \sin(\theta_i + \theta_j)$, $c_{ijk} = \cos(\theta_i + \theta_j + \theta_k)$, and $s_{ijk} = \sin(\theta_i + \theta_j + \theta_k)$.

(a) Show that

$$\mathbf{I}(\mathbf{q}) = \begin{bmatrix} I_{11} & I_{12} & I_{13} \\ I_{12} & I_{22} & I_{23} \\ I_{13} & I_{23} & I_{33} \end{bmatrix}, \quad \mathbf{C}(\mathbf{q}, \dot{\mathbf{q}}) = \begin{bmatrix} c_1 \\ c_2 \\ c_3 \end{bmatrix}, \quad \Gamma_g(\mathbf{q}) = \begin{bmatrix} \Gamma_1 \\ \Gamma_2 \\ \Gamma_3 \end{bmatrix},$$

where

$$I_{11} = \left(\frac{m_1}{4} + m_2 + m_3\right)l_1^2 + \left(\frac{m_2}{4} + m_3\right)l_2^2 + \frac{m_3}{4}l_3^2 + \sum_{i=1}^{3} I_{iz}$$

$$+ (m_2 + 2m_3)\,l_1 l_2 c_2 + m_3 l_3\,(l_1 c_{23} + l_2 c_3),$$

$$I_{12} = \left(\frac{m_2}{4} + m_3\right)l_2^2 + \frac{m_3}{4}l_3^2 + \sum_{i=2}^{3} I_{iz}$$

$$+ \left(\frac{m_2}{2} + m_3\right)l_1 l_2 c_2 + \frac{m_3}{2}l_3\,(l_1 c_{23} + 2l_2 c_3),$$

$$I_{13} = \frac{m_3}{4}l_3^2 + I_{3z} + \frac{m_3}{2}l_3\,(l_1 c_{23} + l_2 c_3),$$

$$I_{22} = \left(\frac{m_2}{4} + m_3\right)l_2^2 + \frac{m_3}{4}l_3^2 + \sum_{i=2}^{3} I_{iz} + m_3 l_2 l_3 c_3,$$

$$I_{23} = \frac{m_3}{4}l_3^2 + I_{3z} + \frac{m_3}{2}l_2 l_3 c_3, \quad I_{33} = \frac{m_3}{2}l_3^2 + I_{3z},$$

$$c_1 = -\left(\frac{m_2}{2} + m_3\right)l_1 l_2 s_2 \dot{q}_2\,(2\dot{q}_1 + \dot{q}_2)$$

$$- \frac{m_3}{2}l_3\,[l_1 s_{23}\dot{q}_1\,(\dot{q}_2 + \dot{q}_3) + l_2 s_3 \dot{q}_3\,(\dot{q}_1 + \dot{q}_2)]$$

$$- \frac{m_3}{2}l_3\,(\dot{q}_1 + \dot{q}_2 + \dot{q}_3)\,[l_1 s_{23}\,(\dot{q}_2 + \dot{q}_3) + l_2 s_3 \dot{q}_3],$$

$$c_2 = \left(\frac{m_2}{2} + m_3\right) l_1 l_2 s_2 \dot{q}_1^2 + \frac{m_3}{2} l_3 \left[l_1 s_{23} \dot{q}_1^2 - l_2 s_3 \dot{q}_3 (\dot{q}_1 + \dot{q}_2)\right]$$

$$- \frac{m_3}{2} l_3 (\dot{q}_1 + \dot{q}_2 + \dot{q}_3) l_2 s_3 \dot{q}_3,$$

$$c_3 = \frac{m_3}{2} l_3 \left[l_1 s_{23} \dot{q}_1^2 + l_2 s_3 (\dot{q}_1 + \dot{q}_2)^2\right],$$

$$\Gamma_1 = \left(\frac{m_1}{2} + m_2 + m_3\right) l_1 s_1 + \left(\frac{m_2}{2} + m_3\right) l_2 s_{12} + \frac{m_3}{2} l_3 s_{123},$$

$$\Gamma_2 = \left(\frac{m_2}{2} + m_3\right) l_2 s_{12} + \frac{m_3}{2} l_3 s_{123}, \quad \Gamma_3 = \frac{m_3}{2} l_3 s_{123}.$$

(b) Making suitable assumptions, obtain expressions for joint control laws for controlling the manipulator by using the computed torque approach.

11

Robot Control

11.1 Introduction

Robot control is the engineering of how a controller is synthesized based on a representative model and implemeted, and how the robot is controlled while in operation. The robot-control problem is patently different from other applications of control engineering, in that no attempts are made to linearize the governing dynamical equations by the application of the small-perturbation method. Rather, the inherent nonlinear nature of robot dynamics is accepted as a *fait accompli* and dealt with quite directly. There is also the issue of how the robot-control software is written and its architecture, but this is not considered here as it has to do with computer software. A broad understanding of the robot-control problem is essential as it has an impact on the kinematic design of the mechanism used in the robot for the purpose of motion transmission.

A simplistic model of a robot is to look at it as a collection of *links* connected by *joints*. As the joints rotate or move and the links contract, expand, and reorient themselves, the tip of the robot's end-effector or *tool center point* (TCP) will change its position. To design a practical manipulator, it is not only of great importance to know the position of the TCP in world coordinates but it is also vital that it be predictable and controllable. Thus the actuators in the joints of the robot have to be controlled in a complex and deterministic manner to follow a prescribed sequence of set points or control commands.

Considering the classical nonlinear joint control problems associated with a robot, we observe that these are essentially adaptive because the robot link parameters are almost always not known accurately. We therefore consider first the various adaptive and model-based approaches to robot control.

11.1.1 Adaptive and Model-Based Control

A central idea in the control of a robot is the feedback-control loop, and a major feature of this loop is the controller that prescribes a control law relating the error between the desired output and the actual output on hand and the input to the

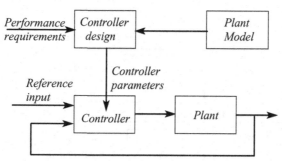

Figure 11.1. Principle of computer design of a controller and its relationship to a feedback controller.

actuator responsible for providing the control signal to a particular robot joint. The control law is mathematically synthesized by applying a series of performance criteria to its inputs and outputs. This process of synthesis or controller design, as it is usually referred to, is performed either manually or by use of appropriate computational tools, by employing a suitable mathematical model of the plant that is being controlled. The process is illustrated in Figure 11.1. In the context of computer-controlled systems, the concept of adaptive control arises from the desire to integrate both the computer computation of the controller as well as the computer control of the plant. A major breakthrough in the synthesis of adaptive controllers occurred when it was possible to design globally stable adaptive systems. The Lyapunov functional approach introduced in Chapter 7 is the prime technique that facilitates this. Thus, in its most basic form, an adaptively controlled system consists of a stabilizing controller, which itself could be adjusted or set by an adaptation mechanism. The latter uses its inputs, the performance requirements, and plant's inputs and outputs to adjust the controller parameters so as to meet the performance requirements in some optimal or goal-satisfying way. This process is illustrated in Figure 11.2.

In many situations involving robots, the plant has to perform tasks in an environment and the environmental parameters may have a direct or indirect bearing on the controller parameters. Thus it becomes essential that either the controller parameters or certain gains in the adjustment mechanism have to be varied in sympathy with the characteristics of the environment. This is usually done by

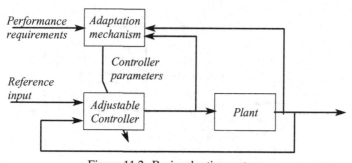

Figure 11.2. Basic adaptive system.

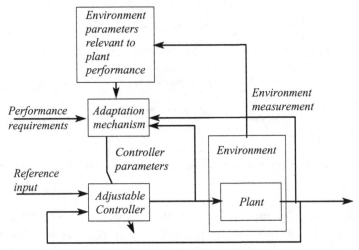

Figure 11.3. Principle of gain scheduling.

establishing a table directly relating a parameter or gain in the adjustment mechanism to the environmental measurements and is the principle of gain scheduling. It is illustrated in Figure 11.3, and in this example the parameters within the adaptation mechanism are themselves subject to a schedule as prescribed by a scheduler. The schedule itself is determined from a *control surface*, which is a relation among the control inputs, the model and environment parameters, the corresponding set points and plant outputs, and the performance parameters, if any. Given a particular set of model and environment parameters, the corresponding set points, and the plant outputs, the actual control inputs may then be determined.

Adaptive systems generally tend to be lot more complex than simple gain schedulers. In the adaptive system illustrated in Figure 11.4, the plant's performance is first estimated by use of its inputs and outputs and evaluation of a suitable metric. The estimate of the plant's performance metric is then compared with the corresponding desired standard value. Depending on the result of this comparison the

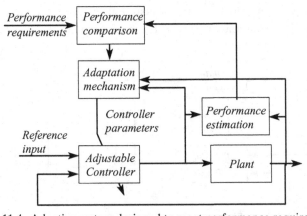

Figure 11.4. Adaptive system designed to meet performance requirements.

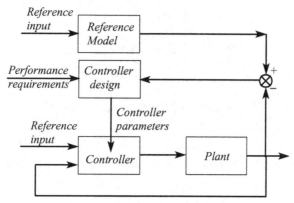

Figure 11.5. Principle of controller design using a reference model.

adaptation mechanism is altered so it is able to estimate the required controller parameters.

Adaptive systems are not really unique either in structure or in their implementation. One class of adaptive systems uses a reference model, so the adaptation mechanism can synthesize the controller parameters from the performance requirements and the error between the reference model's and the plant's output. The reference signal performs a number of useful functions in the context of adaptive systems. This type of adaptive system is illustrated in Figure 11.5. Of course, there is no guarantee in this case that the plant's output would follow the reference output.

In some situations there is often the additional explicit requirement that the plant's output follow the output of the reference model. In these cases, it is often essential to introduce the feedforward controller that uses the error between the reference model's and the plant's output as its input and provides an additional control input. The complete control input is then employed to control the plant as illustrated in Figure 11.6. The primary feedback-control loop serves as stabilizing controller whereas the feedforward controller ensures that the plant output tracks the output of the reference model. Model following by itself does not require that the feedback controller be adaptive. The feedforward controller by itself has the ability to ensure the model-following property. In particular, when a mathematical

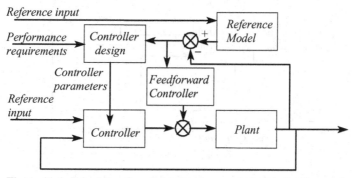

Figure 11.6. Principle of model-following control with feedforward.

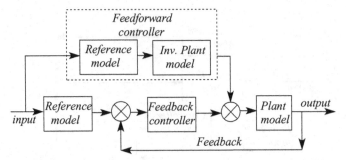

Figure 11.7. Model-following control using an inverse model.

representation of the plant is known and possesses a stable inverse in that the plant inputs can be synthesized from the plant outputs, the feedforward controller has a relatively simple structure, as illustrated in Figure 11.7. Following the controller partitioning approach, adaptive robot-control algorithms have been developed that can be broken down into two parts: a PD feedback part and a feedforward compensation part. The algorithm is computationally simple, but can excite unmodeled high-frequency dynamics if the gains are not carefully selected.

Figure 11.8 illustrates a typical adaptive system in which the adaptation mechanism uses not only the error between the reference model's and the plant's output, but also the input to and the output from the plant. This type of adaptive control is known as a model reference adaptive controller (MRAC). Typically a nonlinear plant is modeled, and a linear model is used as the reference model. The input signal is sent to the reference model, and the trajectory from the reference model is used as the desired trajectory for the adaptive controller. The adaptive controller adjusts the gains to force the plant to follow the desired linear model without the need to use a feedforward-control input. The MRAC has also been applied to robotic manipulators. It is known that the MRAC algorithm can perform well compared with nonadaptive controllers over a wide range of motions and payloads. The advantage

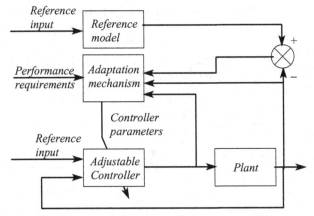

Figure 11.8. Model reference adaptive controller.

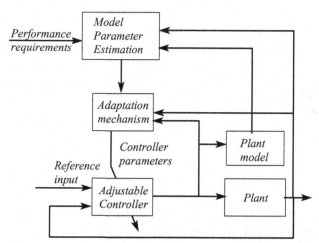

Figure 11.9. Adaptive control with parameter estimation or self-tuning (indirect adaptive control).

of this adaptive controller is that neither a detailed model of the system nor detailed information about the system parameters is needed.

The traditional approach to adaptive control known as indirect adaptive control involves the estimation of all the parameters of the assumed model. This process known as system identification requires only that a model structure be assumed. Once this is done all the parameters in the assumed model are estimated recursively. Once the model and its parameters are complete, the controller is synthesized, just as it would be in the case of an off-line design by use of an appropriate method of synthesis that would always result in a stabilizing controller. Following the synthesis, all the desirable parameters of the controller are obtained, and if these are significantly different from the actual parameters of the controller, the latter are altered so they are equal to the former. This type of adaptive control is said to be indirect as it is a two-stage process involving in the first instance the estimation of the assumed model parameters and then, in the second stage, the synthesis of the controller parameters. This type of adaptive controller is illustrated in Figure 11.9.

In the case of direct adaptive control, illustrated in Figure 11.10, the adaptation mechanism is programmed to directly synthesize the controller parameters without

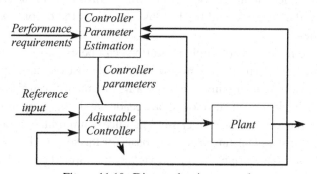

Figure 11.10. Direct adaptive control.

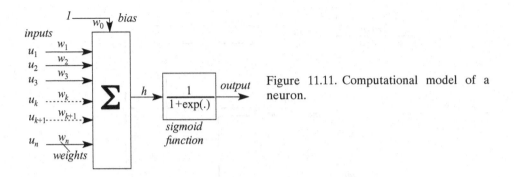

Figure 11.11. Computational model of a neuron.

the need to estimate the parameters of the assumed model and is similar to the basic adaptive system illustrated earlier in Figure 11.2.

A number of studies have been performed to implement adaptive control for robotic manipulators. An adaptive version of the computed torque method has been developed. The algorithm estimates parameters such as load mass, link mass, and friction, which appear in the nonlinear dynamic model of the system, and then uses the most recent parameter estimates to calculate the required servotorque. This control method is computationally intensive because it requires a complex and accurate model of the system to compute the necessary control torques.

One of the spin-offs following the development of adaptive control has been the implementation of artificial-neural-network-based neurocontrollers. Originally, artificial neural networks were used as adaptive control agents to model the dynamics of a plant. However, artificial neural networks were considered to be broader than specific models of traditional adaptive controllers, and their use quickly spread to other aspects of control theory. Artificial neural networks were based on a mathematical representation of the chemical action resulting in the firing of a signal in a biological neuron, as a nonlinear function, such as the *sigmoid*. A number of weights are used to linearly combine multiple inputs to a neuron before the combined scalar input is processed by an archetypal nonlinear function to generate the neuron's output. The complete model of each of the neurons (Figure 11.11) is thus a static nonlinear functional model with a vector input. The neurons could be networked to each other in layers, with the layers themselves being strung together in a sequence, resulting in an artificial neural network. The first of the layers in the sequence takes in the inputs to the network, and the last layer provides the outputs while all other layers are hidden.

Much of the neurocontrol implementations can be categorized into a few distinct approaches. It is particularly noteworthy that most of these approaches are based on supervised learning involving the adjustment of the weights. Reinforcement learning, having been developed a decade later than supervised learning, remains mostly the focus of the artificial-intelligence community's approach to robot control. Reinforcement learning is also the basis for the so-called intelligent control and involves either the learning of the previously mentioned control surface or the optimal performance metrics for synthesizing the control inputs.

Only the most common application of artificial neural networks to robot control is mentioned here: modeling an existing subsystem or a controller. This arose from early artificial-neural-network studies involving the newly "rediscovered" back-propagation algorithm, which bears a close relationship to the dynamic programming family of algorithms briefly discussed in Chapter 9, and the artificial neural network's keen ability to model a nonlinear function. The network can be trained to imitate any nonlinear mapping given sufficient resources in the form of hidden layers of cross-connected artificial neurons. The artificial neural network receives the same inputs as the subsystem it is trying to emulate and attempts to produce the same outputs. The error between the desired outputs, presented to the network by the supervisor, and the actual outputs are back-propagated through the net to adjust the weights.

An obvious question arises as to the utility of training such a network if we already have a mathematical representation of the existing subsystem. There are several reasons why we would require the artificial-neural-network-based model. In particular the artificial-neural-network models of plants do not have some of the undesirable features of mathematical models such as unstable zero dynamics. Furthermore they may be inverted with relative ease to construct an inverse model, which is an important component in the implementation of adaptive control. Finally, the custom implementation of artificial-neural-network models is possible, thus making it quite easy to encapsulate the entire adaptive controller in a particular robot joint.

Before our discussion of adaptive control closes, two important generalizations of it are mentioned; the first is the adaptive fuzzy-logic controller and the second is the model-predictive controller.

Traditional logic-based control systems are based on a simple control agent: the binary switch. This is a hardware realization of two-valued Boolean logic, and the resulting control outcome can always be reduced to an on–off-type two-state output; i.e., a firm on-state or off-state emerges as the output. Sophisticated programmed logic controllers are built around a family of these control agents by sequencing the switching operations according a predetermined time schedule. Fuzzy logic was invented by Lotfi Zadeh in the early 1960s to incorporate shades of meaning so the binary yes–no-type decisions could be replaced with "definitely yes," "probably yes," "maybe," "probably no," or "probably yes," Such decisions are much more representative of human decision making than simple Boolean logic is. Thus fuzzy logic emulates human decision making by using several levels of possibility in a number of uncertain characteristics. To quantify the meaning of fuzzy representations like "high," "medium," "low," and "zero," fuzzy logic uses the concept of grades or membership functions. This form of quantification is very application oriented.

For example, if one makes the statement "Tom's weight is high," the context is referring to the weight of a male human being. Thus associated with the qualifier "high" is a membership function. In the current context a weight of 75 kg may have a membership value of 0.0, 80 kg would correspond to membership value of 0.2, 85 kg to 0.4, 90 kg to 0.6, 95 kg to 0.8, and 100 kg to a membership of 1. For any weight

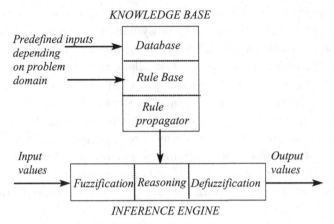

Figure 11.12. Principle of a fuzzy-logic controller.

below 75 kg and above 100 kg the membership is defined to be zero. Thus if Tom's actual weight is 92 kg the corresponding membership would be 0.68. On the other hand, if Peter's weight is 70 kg, the associated membership for the qualifier "high" would be zero, indicating that Peter's weight is not excessive.

In fuzzy logic the controller is specified as a set of rules that typically have the following form:

If the ERROR (E) is POSITIVE BIG AND the CHANGE IN ERROR (CE) is NEGATIVE SMALL the CONTROL (C) is NEGATIVE BIG.

The qualifiers POSITIVE BIG, NEGATIVE SMALL, and NEGATIVE BIG are all defined for each of the variables, ERROR, CHANGE IN ERROR, and CONTROL in terms of specific membership functions. The AND connectives are interpreted as the minimum operation on the corresponding membership values. In conventional fuzzy-logic control, the predicted system consequences as well as the model are implicit in the rules. Rules are obtained explicitly from an expert and presumably have been compiled as a sequence of inferences. In practice, there are several rules to process, and the OR connective is assumed to be implicit between any two or more rules. When two rules result in the same fuzzy-control consequence, the maximum value of the corresponding membership values is assigned to the fuzzy-control qualifier.

Figure 11.12 illustrates a typical fuzzy-logic controller, and the application of this typical controller to a DC servomotor is illustrated in Figure 11.13(a).

In the control of a typical DC motor the input variables are the position ERROR and the CHANGE IN ERROR. Before the control can be estimated, the numerical inputs are "fuzzified" in the sense that membership values are assigned corresponding to each qualifier such as POSITIVE BIG, POSITVE MEDIUM, POSITIVE SMALL, ZERO, NEGATIVE SMALL, NEGATIVE MEDIUM, and NEGATIVE BIG. Once this is done to all the inputs, each and every rule is applied. When a particular rule is applied, the result could be something like "CONTROL is NEGATIVE BIG" with a membership of 0.72. This would correspond to a specific numerical value that could be associated with the control generated by that rule with a grade of membership equal to 0.72. When all the

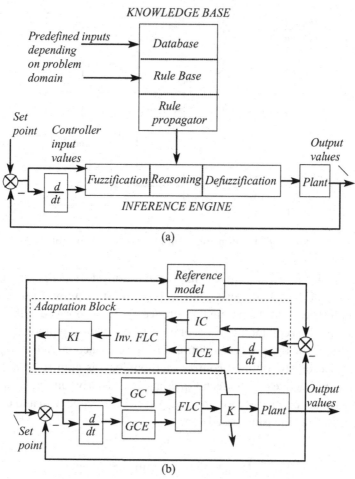

Figure 11.13. (a) Typical process plant with a fuzzy-logic controller in the loop, (b) reference-and inverse-model-based adaptive fuzzy-logic control using a fuzzy-logic inverse model.

rules are applied, the reasoning process is deemed to be complete and the result would be a spread of numerical values for the control, each associated with specific membership. The control input is now in a "fuzzy" form and must be "defuzzified." Typically, for example, one could choose the numerical control value with the highest associated membership. Alternatively the control value could be the weighted sum of the mid-value of each of the qualifiers contributing to the output weighted by its grade of membership. This results in a numerical control output. The cycle of computations is then recursively repeated, and the next set of numerical inputs is processed to compute the control input corresponding to the next time step.

Figure 11.13(b) illustrates a typical fuzzy-logic-based adaptive controller in which both the adaptation mechanism and the primary controller are based on fuzzy logic. The adaptation mechanism has as its input the error between the output of the reference model and the plant, and its output is used to modify the controller gain in the main control loop.

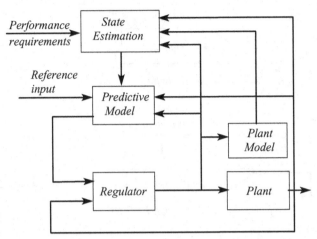

Figure 11.14. Principle of model-predictive control.

Figure 11.14 illustrates a typical implementation of *model-predictive control*, which is a type of digital adaptive control. In its most common form it is applied in process industries, and the nominal reference model is replaced with a predicted reference model. It is common to use a one- or even a two-step predictor and shape the control to follow the predicted output of the reference model. The prediction method is based on the optimal control methodology as applied to a state-space model. Model-predictive control techniques are particularly suitable in applications involving the visual control of motion as they are able to cope with and compensate for the process delay in vision-based control.

11.1.2 Taxonomies of Control Strategies

To understand the robot-control problem in greater depth, it is best to classify the different strategies of controlling a robot. The classification process itself may be achieved on different levels. First, considering a robot emulates a human, it is important to consider how humans approach the control problem. There are indeed several different techniques and systems that humans adopt in dealing with the problem of control. These are the bases for the human-centered approaches to classification of robot control.

Next we may classify robot control based on the tasks it needs to perform: coordination, redundancy management, force control, path following, and hybrid force and position control. Third, we may classify robot control on the basis of the implementation of the controller. Finally there is the actuator–control-sensor loop level classification of robot control.

11.1.3 Human-Centered Control Methods

Considering the role of a human as an operator, we may establish a model to conceptually capture the human's behavior in relation to the machine or vehicle the

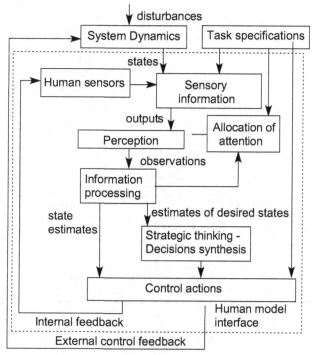

Figure 11.15. Role of the human operator.

human is operating. This is illustrated in Figure 11.15. Central to this model is the information-processing block and its relationship to the other blocks, which is shown in further detail in Figure 11.16.

Human behavioral models permit one to identify the principal human-centered approaches to robot control. However, before considering the human-centered approach to classification of robot control, we observe that there is a 1:1 duality between the basic human-centered approach and the traditional control engineering closed loop. This duality is illustrated in Figure 11.17. In traditional control engineering, a wide variety of sensors and actuators are used to control a process plant. But the synthesis of the appropriate control action is the primary function of the controller or control computer. This is in fact also the case of a human approach to the control problem.

One may identify four basic human-centered methods of control. These are given in Table 11.1. Also given in the table are the corresponding traditional control methods. *Reactive control* is a biologically inspired technique based on instant stimulus response that is characteristic of the autonomic nervous system and is responsible for fast and effective control of muscles by the motor nerves. Reactive control requires tight coupling of sensory inputs and effector outputs to allow the robot to respond rapidly to changing and unstructured environments. Reactive control requires very little memory and attention, relies very little on modeling, and there is no learning. In *deliberative control*, the robot assimilates all the information presented to it by its sensors, uses internal representations of the outside world and a set

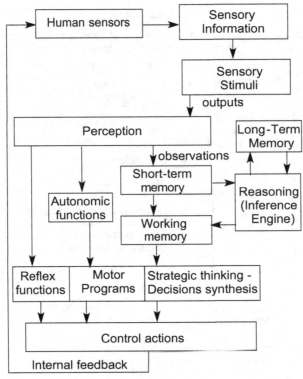

Figure 11.16. Structure of information-processing block.

of reasoning tools, and synthesizes algorithms to assemble an optimal strategic plan of action. For instance, in grasping a passive object, a deliberate and *intrinsically passive control* strategy is desirable. *Hybrid control* refers to a situation in which both reactive control and deliberate control operate in parallel. However, because reactive control is fast, it constitutes an inner loop whereas deliberate control acts

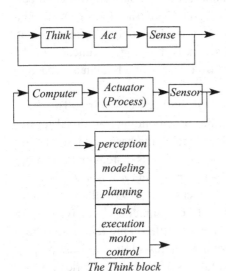

Figure 11.17. Duality between the basic human-centered approach and the traditional control engineering closed loop.

Table 11.1. *Human-centered approaches to robot control and their relations to traditional control methods*

Human-centered method	Traditional control method
Reactive control (autonomic)	Classical control loop with fixed gain control (fastest)
Deliberate control	Single-loop adaptive, optimizing control (slow)
Hybrid control	Dual-loop control – adaptive supervisory loop
Behavior-based control	Adaptive, predictive, and optimizing control
Learning and evolutionary approaches	Strategy learning and game theory

as a supervisory or *adaptive* outer loop. In fact, adaptive control may be used to implement such a supervisory attentional outer loop.

Although the control metaphor has been largely ignored in psychology, the preceding classification is completely in accordance with the *cognitive psychology* view. Thus reactive control involves either fully automatic processing, controlled by organized plans known as *schemas*, or partially automatic processing, without deliberate direction or conscious control but involving contention scheduling to resolve conflicts among schemas. In fact, the theory of *action slips* is based on the premise that action slips, the unintentional performance or nonperformance of actions, occur in part because of the existence of multiple modes of control. Action slips are the consequences of a failure to shift from automatic to attention-based deliberate control at the most appropriate instants of time.

For an organism to exert control over its environment, there must exist predictable relationships between an action and the resulting stimulation ("motor–sensory" relationships). *Behavior-based systems* have different "parts" or loops, with modular blocks representing behaviors, processes that take inputs and send outputs to each other, quite quickly. Models of behavior are then assembled to form the main "thinking" block in the control loop, as illustrated in Figure 11.18. Behavioral modeling involves using process inputs so as to optimize future *predicted* model behavior over a time interval known as the *prediction horizon*. Behavioral control exploits reliable properties of statistics, geometry, rules of optics, etc. Thus it has many similarities with model-predictive control in addition to being adaptive and optimizing.

Behavioral modeling is also used in the partitioning of robot-control software. Unlike a centralized architecture, modularity of robot software permits the implementation of distributed processing and thus allows the robot to respond sufficiently fast. This is the rationale behind behavior-model-based software partitioning known as the *subsumption* architecture. This also, quite logically, leads to the concept of *intelligent autonomous agents*, which are software modules that can exhibit behaviors. *Learning-control* systems are based on human learning of new or improved strategies to deal with a problem. Humans generally acquire certain skills and are then able to improve their ability to perform tasks, relying on these skills by repetitive execution of the tasks. This is the basis for learning.

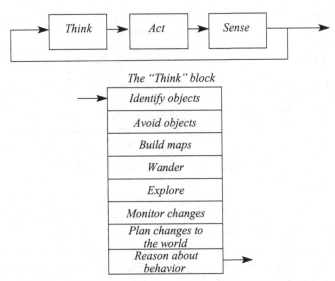

Figure 11.18. Behavioral modeling of the "think" block.

No discussion of biologically inspired control techniques can be complete without a mention of the cerebellum and its role in the control of the skeletomuscular system. The cerebellum ("little brain") is a distinct brain structure connected to the stem of the brain behind the cerebral hemispheres, at the base of the brain. It is known to be the place in the brain where movements learned by experience are stored. With this experiential knowledge available to it, the cerebellum exerts its control over the coordination of movements. It receives its sensory information from various other parts of the brain and the spinal cord. One such brain part is called the *inferior olive*, which itself receives the sensory information in the first instance and then relays it to the cerebellum.

It is, in fact, in the cerebellum, that data are analyzed and a plan of a course of action is established. To put this plan into action, information is relayed from the cerebellum to specialized nerve cells called the *Purkinje cells*. As each and every piece of information leaves the cerebellum via the Purkinje cells, they are able to exercise primary control over the execution of motor activities. It is thus indirectly responsible for human posture and the human's ability to maintain equilibrium. The cerebellum operates in sets of threes. There are three highways leading in and out of it, three main inputs, and three main outputs from three deep nuclei. One of the three nuclei, the *fastigial* nucleus, is primarily concerned with balance, and sends information to the vestibular nuclei in the inner ear and reticular nuclei in the eye, while the other two, the *dentate* and *interposed* nuclei, are concerned with voluntary movements. Functionally the cerebellum is split into three parts: the *archicerebellum*, which is responsible for the postures and vestibular reflexes; the *paleocerebellum*, which is responsible for the geometrical control and the flexing of the muscles; and the *neocerebellum*, which is responsible for the coordination and timing of movements.

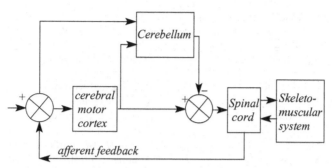

Figure 11.19. The cerebellum and motor control.

Whereas the cerebral motor cortex, a part of the layer of gray matter covering the brain, provides the motor control commands, the cerebellum functions also as a "feedforward" controller for these commands originating in the cerebrum, which is the seat of consciousness within the brain. From the experience gained based on models of the skeletomuscular behavior, it provides timing control of opposing muscles, as well as force and stiffness control. Although the spinal cord provides for independent control of muscle length and joint stiffness, "afferent" or sensed feedback relayed to it by the spinal cord allows the cerebellum to exercise control over it and indirectly control the skeletomuscular system as illustrated in Figure 11.19. An understanding of the role and functioning of the cerebellum has led to development of the *cerebellum model articulation controller* to act as a functional controller of robotic systems.

11.1.4 Robot-Control Tasks

The principal control-oriented tasks in a robot may be briefly summarized. When several robot manipulators, each with its own end-effector, operate on the same object, the actions of the end-effectors need to be coordinated. *Coordination* implies that the activities must be properly planned, programmed, and scheduled. Most serial robot manipulators are designed with excess free DOFs so as to give the end-effector adequate flexibility. This additional mobility or *redundancy* needs to be effectively managed. *Path planning* is a fundamental computational problem in robotics. In its simplest form, path planning is concerned with finding a collision-free motion path between a given starting point S to a final goal position F in a known obstacle-congested environment or space. Once a suitable path is synthesized, it is up to the robot's guidance system to ensure that the robot follows this prescribed path as closely as possible in some optimal way. This is the *path- or trajectory-following* task.

Path-planning algorithms are generally concerned with providing sequences of motion commands when little or no knowledge of the environment is available. Environmental modeling may involve the description of the trough of points traveled, areas with particular characteristics such as convex polygons, generalized

Table 11.2. *Task-based approaches to robot control and their relations to traditional control methods*

Task-based method	Traditional control method
Coordination	Planning and scheduling
Redundancy management	Multiloop control
Path following	Guidance and control
Force control	Servomechanism
Hybrid control	Two-input, two-output servo
Stiffness control	Adaptive control, self-tuning

cones, the generation of the locus of forbidden positions, and other related techniques. *Configuration space* planning that involves combined map- and sensor-based planning is based on the decomposition of a higher-dimensional configuration space and on planning optimally shaped trajectories in joint space by use of a variety of splines and other related curves. The planned trajectory is usually synthesized in segments and assembled together to establish the complete desired motion path in the workspace.

In most robot-control situations involving the end-effector, it is often necessary that the applied force or some attribute of it be controlled precisely. For example, in grasping a loaf of bread, a gripper must apply just the right level of force. This is the idea behind *force control*. *Hybrid control* refers to the combined control of both displacement and force, and *stiffness control* is an adaptive approach that seeks a displacement controller equivalent to the required force controller. The classification of robot-control systems based on the categories of tasks they are required to perform is illustrated in Table 11.2.

An important force-control paradigm is the need for near-static balance of all moving parts. A man carrying a heavy weight in one hand quite instinctively leans his own weight to the other side in order to balance the two moments that the two weights will exert about his point of support on the ground. This is an important principle, and it is therefore essential that mechanical links and joints be designed so the center of rotation is always close to the center of gravity, and consequently the links are effectively statically balanced. The net result of such a design is that the power required to drive the linkages is a near minimum.

11.1.5 Robot-Control Implementations

Robot-control system implementations vary from their rather simple forms in a typical DC servomotor that takes the form of one or two analog control loops to rather complex custom computer-based control loops involving sophisticated synthesis algorithms that are biologically inspired. The use of digital electronics and programmable Boolean logic is quite common and widespread in the manufacturing industry. A number of implementations that use custom fuzzy-logic-based hardware

Table 11.3. *Robot-control classification based on implementation*

Implementation approach	Minimal hardware/software
Analog control loop	Analog electronics
Digital control loop	Digital electronics, programmable logic
Fuzzy-logic (rule-based reasoning) control	Custom ware/microcontrollers
Artificial neural networks	Custom ware/microcomputers
Other biologically inspired algorithms (genetic algorithms)	Custom data structures/PC

are used for robot vision processing. There are also biologically inspired hardware implementations that mimic groups of synaptically connected sensory neurons, relay neurons (networking), pyramidal neurons (in the brain), and muscle-controlling motor neurons, forming an artificial neural network, to perform sensing, computing, and actuation-type control tasks. These methods are classified, and the categories are given in Table 11.3.

Unique among the biologically inspired implementation techniques are the genetic algorithms that are a class of learning or optimization algorithms based on the biologically motivated concept of organic evolution. Organic evolution is the change in a population of a species over a large number of generations that result in the formation of new species. The change implies "advance" or learning and adaptation to changing circumstances and environment. Unlike the classical concepts of control synthesis that assume that species of feasible control laws do not change as they are immutable, evolutionary systems are characterized by learning, adaptation, and mutation, inherently random processes. Implicit in classical control is the notion that the control laws are synthesized once and for all. However, the neo-Darwinian view is that the species of feasible control laws are modified by the inheritance of small changes in the *genotype*, representing changing circumstances, which have passed the test of *natural selection* or *fitness*. Alternatively the Lamarckian view is that the species of feasible control laws are modified by inheritance of changes in the *phenotype*, representing the changing environment and resulting in the use and misuse of certain aspects of the controller. The implementations of genetic algorithms have led to the development of custom data structures based on the genetic structure of *chromosomes* and a completely unusual approach to both controller synthesis and its subsequent optimization.

11.1.6 Controller Partitioning and Feedforward

A common manipulator controller paradigm known as *controller partitioning* splits the controller into two components: a servo-related component and a model-based controller component. Typically a DC servomotor is used to control a joint, and it is in turn controlled by a PD control loop. (The basic control algorithms for controlling a typical joint actuator are discussed in the Appendix.) This is also the structure of the servo-related component of the manipulator's controller. A PD control scheme emulates the effect of a linear spring and damper acting between the current state and the desired state. The model-based controller component is related to the *computed torque* controller paradigm and is usually synthesized by the application of the method of *feedback linearization* of the dynamics, a technique discussed in Chapter 10. The controller itself is so-called as it is computed from the dynamic model in terms of the joint coordinates, joint velocities, and joint torques.

Not all robot-control systems are built around a servomotor. Manipulators that are driven by stepper motors use a simple open-loop control system; i.e., one in which the controlling computer cannot sense the physical position or any other output it is controlling. A stepper motor is a motor that rotates through a small fixed angle in response to a single-pulse input. To drive such a motor, the computer simply transmits a fixed number of pulses and assumes that the motor gets there. Thus the motor could be *programmed* to perform a sequence of rotations. Although the motor cannot sense when something unexpected happens because of the absence of feedback, it is possible in principle for the motor to sense the occurrence of certain disturbing events and simply compensate for these by altering the number of pulses sent. This is the principle of programmed control with *feedforward*. The principle of feedback may be applied in one big loop by simply measuring the position (and orientation) of the end-effector and programming all other actuators to drive the position (and orientation) error at the end-effector to zero. On the other hand, one could introduce a number of inner feedback loops for groups of several links and ensure that these are displaced through a sequence of desired positions.

11.1.7 Independent Joint Control

Finally, each joint may be individually controlled by a servomotor to ensure that each of the joints is displaced through a sequence of prescribed displacements. Typically the joints themselves are controlled by a set of independent joint controller boards that communicate with each other by a dedicated bus, making a real-time operation rather difficult and expensive to achieve. Each joint controller is implemented as a simple PI (position and integral) controller around the joint velocity. Each joint must therefore be tuned, i.e., optimal values of the PI gains must be found for each of the joints. Considerably greater control may be exercised on the robot manipulator by this process. However, there is a price to pay in that the control system could be quite expensive. Generally one prescribes different modes of joint control in order realize the actual requirement, which may be in the form of a grasping specification, force control, impedance control, or a hybrid control mode. Each mode would possibly require a distinct and characteristic level of coordinated control in order to meet the specified requirement. The net result is a complex collection of coordination control tasks that must solved in real time. Controller classification based on the control-loop structure is summarized in Table 11.4.

Table 11.4. *Robot-control classification based on control-loop structure*

Level at which loop is implemented	Relevant classical controller structure
Joint level	Servo, multiple inner loops
Multilink or loop level	Closed loop (nonservo), multiple loops
End-effector level	Single, closed, outer loop control
Programmed feedforward	Feedforward, disturbance sensing
Programmed	Programmable logic (open loop)

11.2 HAL, Do You Understand JAVA?

Well, its time to talk to HAL. HAL, of course, was the legendary computer on board the spaceship in the movie version of Arthur C. Clarke's futuristic novel *2001: A Space Odyssey*, first published in the early 1950s. HAL could not only speak and understand conversations, but could also read lips purely from their visual images. Yet one could be sure that he would not understand JAVA. JAVA is a programming language that was created by Sun Microsystems, a Silicon Valley company specializing in the manufacturer of computer systems hardware and software based in California, long after the world premiere of *2001: A Space Odyssey* in 1968, which incidentally was attended both by Arthur C. Clarke and Alexei Leonov, the first man to walk in space. But JAVA, named after the beautiful island of Java in the Indonesian archipelago, is not understood by most Javanese as well and is not a robot programming language either. And, of course, HAL is no more.

Robot programming languages play an important role in being able to communicate and program a robot to perform a sequence of tasks. Programming languages designed to operate with robots are designed not only for communicating with them but also for instructing them to perform a structured sequence of tasks as defined in a "program." Robot programming languages may be either created from scratch or by modifying and adding to an existing computer programming language. Thus robot programming has evolved along with computer programming. Because programming is a specialized skill requiring considerable investment of resources, the current trend is to eliminate programming altogether. Thus it is believed that, by introducing robot simulators that are to directly create the necessary code to drive the robot, the need for programming could be eliminated.

Computers like HAL, which function as the *brain* of a robot, use an operating system or a so-called real-time kernel that acts as the interface between the programming language and the computer hardware. The kernel has direct access to all the hardware functions and makes these available in various forms to the operating system (OS) and the programming languages. An OS exploits the hardware resources offered by the processor or a processor board to provide a set of services to the user. Among other functions, it manages the secondary memory and input–output (I/O) devices for the user.

Most robot computer systems offer real-time computing facilities. Real-time computing may be defined as that type of computing in which the correctness depends not only on the logic and logical sequencing of computational operations but also on the time at which the operations are performed. A real-time system is one in which some or even one task is a real-time task. A real-time task is one that must meet a mandatory timing deadline (hard real-time task) or one in which it is highly desirable to meet a timing deadline and it is mandatory only that the task be completed (soft real time). A real-time system does not imply that it is fast; just *fast enough*. Another feature of a real-time system is that it responds to an interrupt on the basis of priority.

Robot OSs have a number of unique requirements over and above the requirements of an OS. Primarily they are required to operate in real time. Real-time operating systems (RTOSs) are deterministic in that they are able to perform tasks not only at the specified times but also in fixed predicable intervals of time. Therefore, provided it has the capacity, a RTOS can deterministically satisfy all user requests for services. Responsiveness is another feature of a RTOS and may be described as its ability to respond fast enough. User control is generally much broader in a RTOS than in a non-real-time OS. In a non-real-time OS the user has no control over the scheduling of multiple tasks whereas in a RTOS the user may exercise complete and direct control on the scheduling of all tasks. Of course, in the case of a RTOS the reliable and fail-soft[1] operation of the software modules is much more important than in a non-real-time OS, particular if the RTOS is used in a situation requiring high integrity, such as a hospital operating theater or an aircraft flight-control system. A real-time kernel is the heart of a RTOS and is responsible for providing all the basic facilities such as working with processes and threads, interprocess communication, and other features that are essential for real-time operations. Furthermore it provides a deterministic high performance and complete access for processor allocation, memory management, communication, and task and thread scheduling. Scheduling specifies how access to one or more reusable computational resource is granted while ensuring stability under transient overload. Thus the real-time kernel is a key component of the RTOS and is largely responsible in determining its real-time performance.

Robot software has evolved well beyond programming and simulation languages with the evolution of specific architectures for the partitioning of robot-control and vision software. The objective of using a proper architecture for the software is to provide an easier and flexible way of programming and controlling robot manipulators.

In a robot, a programming language, apart from offering facilities to control physical motions, to conduct operations in parallel, and to synchronize with external events and interrupts, also serves to communicate with and via the RTOS with the other hardware components such as the joint servomotors in a manipulator. Robot programming languages have become almost as varied as the robots they

---

[1] The term fail-soft is used to describe systems that are designed to terminate all nonessential processing whenever there are any failures. Thus systems that are in a fail-soft mode would be able to provide partial operational capability even after a failure in delivering a service.

are designed to communicate with. Languages available today may be classified in a multilevel structure, based on level of intelligence, task orientation, and structured organization, motion orientation, and point-to-point and microcomputer levels. Yet the International Standards Organization's open system interconnection (ISO/OSI) reference model for data communication protocols, with its seven-layer structure beyond the physical media (physical layer, data link layer, network layer, transport layer, session layer, presentation layer, application layer), is emerging as a basis for classifying robot application software including programming languages. Although not all of the seven layers are generally relevant, three of the seven layers may be considered important for robot-control communications. Robot-control communications software usually supports a simple network management protocol, which is conceptually in the application layer in terms of the ISO/OSI model. Moreover, the control software itself has a multilayered structure, with the lowest layer handling the servocontrol modules, the next layer providing primiting for such tasks as path generation, trajectory planning, and kinematic inversion, and higher layers dealing with tasks involving decision analysis and interpretation of high-level commands. Thus the control architecture is designed according to the principles of layers, each with different behavioral competencies. This not only allows the system to operate robustly but also facilitates the extension of the robot's control program at a later date simply by the addition of new levels on top of the existing ones. Adding new levels not only does not change the communication structure between the previous levels; it considerably enhances it. Through the suppression and inhibition of specific software elements in the layers below it is possible to modify previously implemented behavioral patterns without changing the existing architecture. This approach also allows for the robust execution of the program. If the control functions of a particular level fail, then the behavior patterns of the lower levels would continue to function properly.

11.3 Robot Sensing and Perception

Older texts in robotics often include a chapter on computer vision. But perception, whether in a human or in a robot, is not based purely on vision. Computer vision is an important and yet a relatively low-level function in a modern robot. A host of sensors is often used in building a robotic system, and some of these sensors are identified in the Table 11.5. Although the sensors provide the requisite sensation, what is more important is the perception of the robotic system. Perception involves the proper interpretation of sensed data. This aspect is substantially more important than mere sensing of information. A deeper understanding of robot perception may be gained by understanding human perception and cognition in some depth.

The human ear, illustrated in Figure 11.20, is an incredible natural "sensor" that is bombarded by stimuli of different kinds and from different directions. The human ear has three main regions; the outer, the middle, and the inner. On the one hand, although it is sensitive to sound waves in the 15-Hz–15-kHz frequency range, it is also the primary mechanism for providing humans with a sense of balance. As

Table 11.5. *Typical sensors used in building a robotic system*

Feature	Sensing principle or sensor type
Range	Time-of-flight (ultrasonic, laser, radio)
	Triangulation (stereo disparity or other stereoscopic)
	Coherent electromagnetic radiation
	From target size (if known)
	Optic flow
Proximity	Capacitive, ultrasonic, magnetic, or electromagnetic wave
Tactile	Binary, analog, tactile feelers or bumpers, distributed surface arrays
Forces and moments	Accelerometers
Speed	Doppler, odometers, anemometers (wind speeds)
Altitude	Barometers, altimeters, radar, sonar
Attitudes and attitude rates	Attitude gyroscopes, rate gyroscopes, rate-integrating gyroscopes
Positioning	Satellite-based positioning (GPS), IMUs

far as sound waves are concerned, they enter the outer ear, pass through the ear canal, and strike the eardrum. The drum's vibrations then travel down the middle ear and are amplified three times by a three-bone lever (the ossicles). The amplified pressure acts on a membrane, which in turn sends waves through the fluid in a snail-shaped organ, the cochlea. It is essentially a transducer, which passes on the signal and generates appropriate sensory neural firings to the auditory nerve for further transmittal to the brain. Sound perceived by the auditory nerve provides the primary

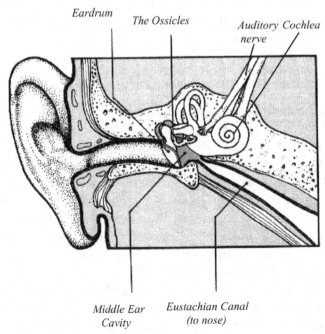

Figure 11.20. Illustration of the inner ear.

cues for the perception of spatial extent. These cues, the delay times associated with echoes, sound-frequency coloration, and the reflection density or reverberation are interpreted cognitively to give us a perception of the space in which a sound occurs.

The inner ear contains three semicircular canals, at right angles to each other and filled with fluid, that act as a three-axis sensor. The semicircular canals respond to angular acceleration at low frequencies and to angular velocity at high frequencies. This is because, while the semicircular canals respond to accelerations in their individual planes of rotation, they integrate these accelerations to report velocities at higher frequencies. The frequency-coded signals from the "ampullary" receptors of the semicircular canals are much more closely related to the angular velocity of the head than to its angular acceleration. They assist in maintaining body equilibrium.

The semicircular canals are also linked to the cochlea. Liquid inside the semicircular canals moves inside swellings or ampullae, thereby stimulating tiny hairlike cells capped by a cone of jelly (capulla). Head movement in the same plane as one of the canals causes the canal to move but not the liquid (endolymph) within it. The endolymph lags behind because of its inertia. The inertia of the endolymph swings the capulla, resulting in a distortion of the hair and hair cells. This results in a proportional signal to the auditory nerve. Thus the semicirculars work together to detect the head's rotational angular acceleration or angular velocity.

The semicircular canals have another important function in that they provide the vestibulo-ocular reflex (VOR) to the eye via the cerebellum. The function of the VOR is to stabilize the image on the retina of the eye while the head is subject to rotational or translational movements. The VORs are set-point command signals that stabilize the eye position in space during three-axis head rotations. The eye is the only human visual sensor and works like a stereo camera, imaging the real world in a form understandable to the human brain. The eye then generates a feedback signal, which is exactly the negative of the VOR. Thus the two together add up to zero, thereby indicating that the eye motion is completely synchronous with the head motion. When the two signals are not synchronous, there is conflict that results in so-called motion sickness.

The semicircular canals, together with two other sense organs, the otolith organs, the utricle and the saccule, illustrated in Figure 11.21, constitute the vestibular system. The otolith organs are embedded in the temporal bones on each side of the head near the inner ear. These organs are sensitive to gravity and linear acceleration of the head. They measure the linear acceleration that the head is subjected to as well as the orientation (tilt) of the linear acceleration vector relative to the local gravity vector, and on this basis the brain is able to compute how to maintain balance. The otoliths are mainly responsible for telling us which way is up as they act as physiological gravity perceptors. They are two membranous sacs, with hairlike structures embedded within, and act as accelerometers. It is at these hairlike structures where motion is transduced into a sensory neural firing.

Certain forced motions of the human body thus stimulate the sensory organs in the vicinity of the human ear, yielding cues, which are transduced into physiological

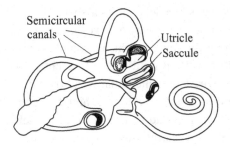
Semicircular canals
Utricle
Saccule

Figure 11.21. Illustration of the vestibular system in the ear.

neural signals. The physiological neural signals encode various motion- and orientation-related features. The CNS residing within the brain acts as a neural estimator, capable of accurately extracting motion and orientation parameters in a process almost similar to an electrical filter estimating a set of parameters from noisy measurements given a suitable model of the dynamics of the signal. The CNS also incorporates a multisensor fusion process whereby these various estimates of the motion and orientation parameters are combined by a weighting process followed by additional motor and cognitive processing to elicit reflexive and perceptual responses. Cognitive processing relies on a control strategy associated with internal models of the dynamics of the sensory organs, body dynamics, and cognitive responses. The models also include a smooth pursuit module on the basis of a model-following control strategy that is able to match eye responses and perceptual effects measured during various motion regimes in daylight and in darkness with visual motion cues held in memory. The processed outputs are the internal estimates of physical motion variables and servocommand reflex signals that govern the quasi-steady motions of the eye.

No robot perception would be complete without a provision for appropriate sensing and matching of audio–visual cues. Although processing of audio–visual information is performed within the robot's computer, the architecture of the sensing system is designed to resemble similar human systems. A computer vision system is a minimal requirement but may not be adequate for certain functions unless it is supplemented by appropriate motion and acoustic sensors. As a robot operator controls a robot, he/she continually views the outside world from his/her own moving reference. He or she does this to create a mental map of the neighborhood that provides visual cues to facilitate his or her primary task. The purpose of the synthetic displays of the outside world and other instruments is to provide the robot operator with the same visual cues he/she would obtain while performing the same task himself/herself.

There are two approaches to synthetic generation of outside-world displays or animation: key framing and motion capture. The key-framing technique requires that the animator specify limiting or key positions of the objects in the form of frames. A computer is used to fill in the missing frames, essential to creating a continuous animation, by smoothly interpolating between the positions captured by the key frames. To be able to effectively implement the technique, the animator must posses a detailed understanding of how moving objects behave over a time frame as well as the ability and talent to encode that information through key-framed config-urations. Motion capture, on the other hand, is based on recording and playing back the outside-world scenes in three dimensions. The key problems in such a computer generation of synthetic displays remain; i.e., the identification and optimal presen-tation of the most significant visual cues.

A realistic computer simulation of outside-world displays as seen by a robot operator is probably the best way of identifying the most significant and appropri-ate visual cues that may be provided to a trainee operator. Insight into the nature of visual cues essential for human perception may be gained by simulating the motions and perceptions of the human operator. Certain visual cues are essential to ensure that the simulation is realistic in that it incorporates the human's behavioral style and stance. Without these cues, the computer would still synthesize technically cor-rect movement that would appear unnatural.

To complete our understanding of visual cues, we examine how images are stored and retained in the human brain. Images have two components: a surface representation that is primarily superficial and a deep representation. The surface representation merely corresponds to one's experience of the image and embodies quasi-pictorial features. Surface representations are generated from the deep rep-resentations, which are elaborate knowledge structures stored in the brain's long-term memory. The deep representation of the image includes a list of propositions that encodes the relationships among the various properties on the image. Thus the surface representations may be actually reconstructed from component image elements with the aid of the deep representation. The elements of the image as well as the propositions that constitute the deep representations make up the visual cues associated with image. Visual cues may be broadly classified into two groups: object- or image-centered and observer-centered cues. Object- or image-centered cues include pictorial cues (perspective, interposition, height in the projected image plane, light, occlusion, shadow and diffuse interreflection, relative size, textural gra-dients, brightness/aerial perspective) and motion-parallax-related cues. Observer-related cues include motion feedback for ocular convergence, binocular disparity, stereo, and accommodation for the strains in the muscles controlling the optical lenses in the eyes.

The difficulty human operators experience in making precise judgments of dis-tance is due to the inadequacy of computer-generated imagery. The problem arises because a three-dimensional world is projected onto a two-dimensional image. Visual contact cues are important for simulations of larger-scale environments because they aid in achieving visual realism and in the perception of the position of objects in space. As one reaches to touch or grab an object, his or her hand positions itself appropriately in anticipation of contact with the object. Because this occurs prior to the physical contact, it necessarily must be cued by visual rather than haptic stimuli. Further, because light travels faster than sound, it can be said that spatial awareness is cued first by visual rather than by auditory stimuli, which only act to reinforce the visual stimuli and provide the VOR. The perception of spatial relations in computer-generated images is aided by six primary cues: perspective projection, relative motion, shadow, object texture, ground texture, and elevation. Visual motion cuing involves the rate of change of perspective and the streaming of points of contrast in an image.

In addition to vestibular proprioception, the human also receives kinesthetic proprioception stimulation. Kinesthesia is the awareness of the orientation and the rates of movement of different parts of the body arising from stimulation of receptors in the joints, muscles, and tendons. The focus here is on perception mediated exclusively or nearly so by variations in kinesthetic stimulation. Tactile perception refers to perception mediated solely by variations in cutaneous stimulation. Tactual perception may include sensing information tactilely (through the skin), kinesthetically (through the joints, muscles and tendons), or both. Haptic perception is a narrower term that refers to sensing information both tactilely and kinesthetically in which both the cutaneous sense and kinesthesis convey significant information about distal objects and events (distal refers to sites located away from the center or midline of the body). Auditory cues, in addition to being perceived by the ears, may be picked up by proprioception.

Finally, introducing realism into robot vision involves more than the realistic generation of motion and matching visual cues. The physiological sense of touch has two distinct aspects: the cutaneous sense, which refers to the ability to perceive textural patterns encountered by the skin surface; and the kinesthetic sense, which refers to the ability to perceive forces, moments, and their magnitudes. There is also a need for touch and feel sensing as well as the contact cues and feedback processes associated with these sensations. Visual contact cues have already been discussed; haptic contact cues associated with the sensations of touch and feel provide two typical feedbacks: kinesthetic feedback associated with kinesthesia, the sensation by which body weight, muscle tension, and movement are perceived, and propriocentric feedback, which is sensitive to changes in body position and movement. These feedbacks greatly influence the nature of the control that a human exerts over the interfaces to a robot as well as other manual controls. This establishes the need for at least *proximity, touch*, and *tactile* sensors. A proximity sensor is a device that detects the presence of an object without making physical contact with it. Touch and tactile sensors are devices that measure the forces of contact between the sensor and an object. This interaction obtained is confined to a small but predefined region. Touch sensing involves the detection and measurement of a contact force at a specific point. Output from basic proximity and touch sensors is in the form of binary information, namely an "on" signal if the object is present or to signify touch, and an "off" signal to signify the absence of the object or no touch. Tactile sensing involves the detection and measurement of the spatial distribution of forces perpendicular to a predetermined sensory area and the subsequent interpretation of the spatial information. Thus a tactile sensing array is assembled from a coordinated group of touch sensors. The measurement and the detection of the movement of an object relative to the sensor are processes of slip detection. These can be achieved either by a slip sensor designed specifically to measure slip or by the interpretation of data from a touch sensor or a tactile array.

EXERCISES

11.1. Reconsider the autonomously driven bicycle model developed in the preceding chapter.

    (a) Discuss the shortcomings of this dynamic model developed in the preceding chapter and explain how the rectification of these shortcomings may be effected.

    (b) Based on your experience in riding a bicycle, design a biomimetic controller capable of balancing the bicycle when in forward motion.

11.2. Demonstrate the feasibility of the bicycle's balancing controller designed in the preceding exercise by simulating the closed dynamics of a representative model of the bicycle.

11.3. Consider a system described by the following dynamical equations:

$$3\ddot{q}_1 + \ddot{q}_2 L \cos q_2 - \dot{q}_2^2 L \sin q_2 = u_1,$$
$$\ddot{q}_2 L^2 + \ddot{q}_1 L \cos q_2 + g L \sin q_2 = u_2.$$

    (a) Determine the inertia matrix, the Coriolis or centrifugal force vector, the gravitational force vector, and the vector of generalized external forces.

    (b) Describe briefly the computed torque control technique for the robot governed by the preceding dynamics and obtain the relevant control laws.

11.4. Reconsider the two-link manipulator illustrated in Figure 10.2.

    (a) Show that the equations of motion can be expressed in the form

$$\mathbf{I}_0\left(\mathbf{q}_0\right)\ddot{\mathbf{q}}_0 + \mathbf{C}_0\left(\mathbf{q}_0, \dot{\mathbf{q}}_0\right) - \boldsymbol{\Gamma}_{g0}\left(\mathbf{q}_0\right) = \boldsymbol{\Gamma},$$

where $\mathbf{q}_0$ is the two-vector of joint space variables in the frame of reference attached to the fixed link, $\mathbf{I}_0\left(\mathbf{q}_0\right)$ is the 2×2 inertia matrix, $\mathbf{C}_0\left(\mathbf{q}_0, \dot{\mathbf{q}}_0\right)$ represents the Coriolis and centrifugal force vectors, and $\boldsymbol{\Gamma}_{g0}$ is the vector of gravity torques acting at the joints.

    (b) Find an inverse model control law to transform the system into a double integrator system given by

$$\ddot{\mathbf{q}}_0 = \mathbf{v}.$$

    (c) Using the control law,

$$\mathbf{v} = \mathbf{r} - \left[\mathbf{K}_P\mathbf{q}_0 \quad \mathbf{K}_v\dot{\mathbf{q}}_0\right]^T,$$

where $\mathbf{r}$ is a reference input, select appropriate values for the control gains so the closed-loop system is decoupled, critically damped, and has the natural frequencies given by $\omega = \omega_0$.

11.5. A servocontrol loop for a robot is designed by incorporating a fuzzy-logic controller in a typical control loop. The rule matrix for this fuzzy controller module is shown in Table 11.6. The interpretation of each rule in this rule matrix takes the following form:

Table 11.6. *Rule matrix for the fuzzy controller*

	CHANGE IN ERROR							
ERROR	PB	PM	PS	PZ	NZ	NS	NM	NB
PB	PM	PM	PB	PB	PB	PB	PB	PB
PM	PZ	PS	PM	PM	PM	PM	PM	PM
PS	NZ	PZ	PZ	PS	PS	PS	PS	PS
PZ	NZ	NZ	PZ	PS	PZ	PZ	PZ	PZ
NZ	NZ	NZ	NZ	NZ	NS	NZ	PZ	PZ
NS	NS	NS	NS	NS	NS	NZ	NZ	PZ
NM	NM	NM	NM	NM	NM	NM	NS	NZ
NB	NB	NB	NB	NB	NB	NB	NM	NM

Notes: B, big; M, medium; S, small; Z, zero; P and N, positive and negative values of the position error and the change in position error.

If "position error is NM" and "change in position error is PS" then "control is NM."

The abbreviated linguistic terms are defined as follows: B, big; M, medium; S, small; Z, zero, while the hedges "P" and "N" refer to positive and negative values of the position error and the change in position error. The fuzzy sets associated with the rule matrix in Table 11.6 for the scaled, nondimensional position error, change in position error, and control input are shown in Figure 11.22.

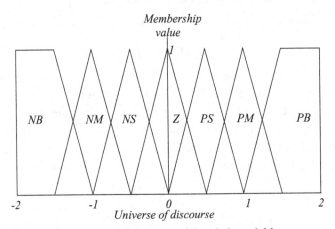

Figure 11.22. Definitions of linguistic variables.

(a) Given that the scaled nondimensional position error is -0.4 and that the scaled nondimensional change in position error is 1.2, identify the rule submatrix that would be applicable.

(b) Hence, find the control action, using an appropriate defuzzification method. Explain the method of defuzzification you have used and compare it with any other method of defuzzification you know.

11.6. (a) Using examples, discuss the principle of inverse-model- and internal-model-based neurocontrollers.

Give examples of at least two other neural-net-based controller structures.

(b) Explain the difference between supervised and unsupervised learning in artificial neural networks. Give an example of a neural network model that uses each procedure.

(c) Name and describe with the aid of a diagram at least one example of a neural network model in the category of feedforward neural networks and briefly discuss its applications to robot control.

12

Biomimetic Motive Propulsion

12.1 Introduction

In this chapter we consider some important benchmark problems involving biomimetic robots, such as the dynamics and balance of walking biped robots, dynamics and control of four-legged robotic vehicles, dynamics and control of robotic manipulators in free flight, dynamics and control of flapping propulsion in aerial vehicles, and the underwater propulsion of aquatic vehicles.

12.2 Dynamics and Balance of Walking Biped Robots

The dynamics and control of walking bipeds provides much insight into biomimetic robots. We consider a relatively simple model of a walking model that is capable of capturing the principal features of the kinematics and dynamics of coordinating walking that resemble a human gait. Generally a human can be modeled to operate in two-dimensional space, with the model having a head, a pair of arms attached by a shoulder joint to a torso, and two identical legs with knees as well as two ankles and feet. Such a model is capable of demonstrating uniform walking based on the stance and swing mechanism, dynamic balance of the torso and the head, the role of the ankle and feet in providing rolling contact with ground, and the role of the human arms that act as "stabilizers" while the robot is walking forward with uniform forward velocity. Such a planar model is illustrated in Figure 12.1.

12.2.1 Dynamic Model for Walking

To understand the basic kinematics of coordinated walking, a simplified seven-DOF model involving the head, torso, and legs is adequate. The legs by themselves form a classical five-bar linkage, and when the constraint of a single-leg support is included, the number of DOFs is reduced to just five. In addition, when the four moments at the leg joints are also specified to maintain steady walking, the model reduces to one with only a single DOF.

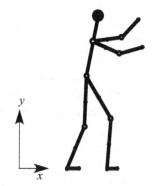

Figure 12.1. Thirteen-DOF planar model of a human walking, including the arms and feet to allow for balance and rolling contact with ground. The unfilled circles represent revolute joints, and the filled circles represent mass points.

The model is illustrated in Figure 12.2. In the first instance we establish a set of equations of motion that govern the dynamic evolution of the state of the model. We assume that the motion of the center of mass of the torso and head can be described by Cartesian coordinates (x_0, y_0) and that the total mass and moment of inertia of the torso and leg are m_0 and I_0, respectively. The orientation of the torso to the vertical is assumed to be described by the angle α.

The masses of the two upper legs from the pelvis to the knee joints are assumed to be m_1, and the masses of the two legs from the knee joints to the ankles are assumed to be m_2. The height of the center of mass of the torso from the pelvis is assumed to be r_0, and the distances of the masses m_1 and m_2 from the pelvis and the knee joints respectively are r_1 and r_2. The orientation angles of the leg links, the moments acting on these links, and the forces acting on the ankle when a leg is in contact with the ground are defined in Figure 12.2.

To establish the governing equations of motion and the appropriate constraint equations we need the Cartesian velocities of the centers of mass of the various component bodies as well as the angular velocity of the torso. The Cartesian coordinates defining the locations of the pelvic joint and the left and right knee joints are

$$x_{\text{pelvis}} = x_0 - r_0 \sin \alpha, \quad y_{\text{pelvis}} = y_0 - r_0 \cos \alpha, \tag{12.1}$$

$$x_{L\_\text{knee}} = x_0 - r_0 \sin \alpha - l_1 \sin (\alpha - \beta_L), \tag{12.2}$$

$$y_{L\_\text{knee}} = y_0 - r_0 \cos \alpha - l_1 \cos (\alpha - \beta_L), \tag{12.3}$$

Figure 12.2. Simplified seven-DOF model (excluding the single-leg or double-leg support constraints that could remove two or four DOFs) of a walking human.

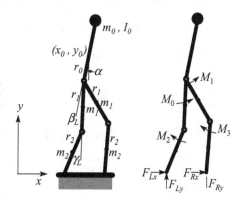

$$x_{R\_knee} = x_0 - r_0 \sin\alpha - l_1 \sin(\alpha - \beta_R), \qquad (12.4)$$

$$y_{R\_knee} = y_0 - r_0 \cos\alpha - l_1 \cos(\alpha - \beta_R), \qquad (12.5)$$

where l_1 is the length of the torso and head from the center of mass of the head to the pelvic joint. The Cartesian coordinates defining the locations of the centers of mass of the component bodies in the left and right legs are respectively given by

$$x_{L1} = x_0 - r_0 \sin\alpha - r_1 \sin(\alpha - \beta_L), \qquad (12.6)$$

$$y_{L1} = y_0 - r_0 \cos\alpha - r_1 \cos(\alpha - \beta_L), \qquad (12.7)$$

$$x_{L2} = x_0 - r_0 \sin\alpha - l_1 \sin(\alpha - \beta_L) - r_2 \sin(\alpha - \beta_L + \gamma_L), \qquad (12.8)$$

$$y_{L2} = y_0 - r_0 \cos\alpha - l_1 \cos(\alpha - \beta_L) - r_2 \cos(\alpha - \beta_L + \gamma_L), \qquad (12.9)$$

$$x_{R1} = x_0 - r_0 \sin\alpha - r_1 \sin(\alpha - \beta_R), \qquad (12.10)$$

$$y_{R1} = y_0 - r_0 \cos\alpha - r_1 \cos(\alpha - \beta_R), \qquad (12.11)$$

$$x_{R2} = x_0 - r_0 \sin\alpha - l_1 \sin(\alpha - \beta_R) - r_2 \sin(\alpha - \beta_R + \gamma_R), \qquad (12.12)$$

$$y_{R2} = y_0 - r_0 \cos\alpha - l_1 \cos(\alpha - \beta_R) - r_2 \cos(\alpha - \beta_R + \gamma_R). \qquad (12.13)$$

The Cartesian coordinates defining of the locations of the center of mass of the head and the points of contact of the left and right ankles with the ground are respectively given by

$$x_{\text{top}} = x_0 - (r_0 - l_0)\sin\alpha, \quad y_{\text{top}} = y_0 - (r_0 - l_0)\cos\alpha, \qquad (12.14)$$

$$x_{LG} = x_0 - r_0 \sin\alpha - l_1 \sin(\alpha - \beta_L) - l_2 \sin(\alpha - \beta_L + \gamma_L), \qquad (12.15)$$

$$y_{LG} = y_0 - r_0 \cos\alpha - l_1 \cos(\alpha - \beta_L) - l_2 \cos(\alpha - \beta_L + \gamma_L), \qquad (12.16)$$

$$x_{RG} = x_0 - r_0 \sin\alpha - l_1 \sin(\alpha - \beta_R) - l_2 \sin(\alpha - \beta_R + \gamma_R), \qquad (12.17)$$

$$y_{RG} = y_0 - r_0 \cos\alpha - l_1 \cos(\alpha - \beta_R) - l_2 \cos(\alpha - \beta_R + \gamma_R). \qquad (12.18)$$

The generalized coordinates are defined as

$$[q_1 \quad q_2 \quad q_3 \quad q_4 \quad q_5 \quad q_6 \quad q_7] = [x_0 \quad y_0 \quad \alpha \quad \beta_L \quad \beta_R \quad \gamma_L \quad \gamma_R]. \qquad (12.19)$$

The total kinetic energy of the bodies is given by

$$T = \frac{1}{2}\left[m_0(\dot{x}_0^2 + \dot{y}_0^2) + I_0\dot{\alpha}^2 + \sum_{i=0}^{2} m_i(\dot{x}_{Li}^2 + \dot{y}_{Li}^2 + \dot{x}_{Ri}^2 + \dot{y}_{Ri}^2) \right]. \qquad (12.20)$$

The total gravitational potential energy with reference to the ground is

$$V = m_0 g y_0 + \sum_{i=0}^{2} m_i g(y_{Li} + y_{Ri}). \qquad (12.21)$$

The virtual work done by the external forces acting at the ground contact points and moments acting at the pelvic and knee joints are respectively given by

$$
\begin{aligned}
\delta W_F &= \sum_{i=1}^{7} F_{Lx}\frac{\partial x_{LG}}{\partial q_i}\delta q_i + F_{Ly}\frac{\partial y_{LG}}{\partial q_i}\delta q_i + F_{Rx}\frac{\partial x_{RG}}{\partial q_i}\delta q_i + F_{Lx}\frac{\partial y_{RG}}{\partial q_i}\delta q_i \\
&= \sum_{i=1}^{7} Q_{Fi}\delta q_i,
\end{aligned}
\tag{12.22}
$$

and

$$
\delta W_M = M_0 \delta\beta_L + M_1 \delta\beta_R + M_2 \delta\gamma_L + M_3 \delta\gamma_R = \sum_{i=1}^{7} Q_{Mi}\delta q_i. \tag{12.23}
$$

The Euler–Lagrange equations are given by

$$
\frac{\partial}{\partial t}\frac{\partial T}{\partial \dot{q}_i} - \frac{\partial T}{\partial q_i} + \frac{\partial V}{\partial q_i} = Q_{Fi} + Q_{Mi}, i = 1, 2, 3 \ldots, 7. \tag{12.24}
$$

The basic mechanism of walking involves the stance and swing method discussed in Chapter 2. During the stance phase the robot is fully supported on either both legs or on one leg. When the robot is supported on both legs, the position coordinates of the ground contact points may be assumed stationary, and consequently this leads to four position constraint equations. These may be expressed as

$$
\begin{bmatrix} \cos(\alpha - \beta_L) \\ \sin(\alpha - \beta_L) \end{bmatrix} = \frac{1}{d(\gamma_L)}\Delta(\gamma_L)\begin{bmatrix} x_0 - r_0\sin\alpha \\ y_0 - r_0\sin\alpha \end{bmatrix}, \tag{12.25}
$$

$$
\begin{bmatrix} \cos(\alpha - \beta_R) \\ \sin(\alpha - \beta_R) \end{bmatrix} = \frac{1}{d(\gamma_R)}\Delta(\gamma_R)\begin{bmatrix} x_0 - x_{LR} - r_0\sin\alpha \\ y_0 - r_0\sin\alpha \end{bmatrix}, \tag{12.26}
$$

where

$$
d(\gamma) = l_1^2 + l_2^2 + 2l_1 l_2 \cos(\gamma), \tag{12.27}
$$

$$
\Delta(\gamma) = \begin{bmatrix} l_2\sin\gamma & (l_1 + l_2\cos\gamma) \\ (l_1 + l_2\cos\gamma) & -l_2\sin\gamma \end{bmatrix}, \tag{12.28}
$$

and x_{LR} is the distance of the right ground support point from the left.

Differentiating the position constraint equations results in the corresponding velocity and acceleration constraint equations. Thus when the robot is in this configuration the number of DOFs reduces from seven to three and explicit solutions may be obtained for the support forces, F_{Lx}, F_{Ly}, F_{Rx}, and F_{Ry}.

In the case of a single-leg support the position constraint equations are

$$
x_0 = r_0\sin\alpha + l_1\sin(\alpha - \beta_L) + l_2\sin(\alpha - \beta_L + \gamma_L), \tag{12.29}
$$

$$
y_0 = r_0\cos\alpha + l_1\cos(\alpha - \beta_L) + l_2\cos(\alpha - \beta_L + \gamma_L), \tag{12.30}
$$

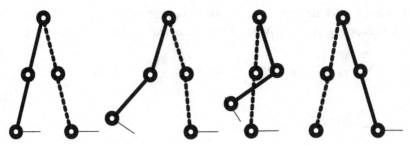

Figure 12.3. A typical sequence of the lower limb positions during the stance and swing mode of walking.

when the stance leg is the left leg and are

$$x_0 = x_{LR} + r_0 \sin \alpha + l_1 \sin (\alpha - \beta_R) + l_2 \sin (\alpha - \beta_R + \gamma_R), \qquad (12.31)$$

$$y_0 = r_0 \cos \alpha + l_1 \cos (\alpha - \beta_R) + l_2 \cos (\alpha - \beta_R + \gamma_R), \qquad (12.32)$$

when the stance leg is the right leg. Thus the position coordinates of the center of mass of the torso are no longer independent DOFs, which are now reduced to five.

A typical human gait half-cycle may be divided into four phases. The first phase is the double-support phase, when both feet are firmly on the ground. The next phase is the transition from the double-support to the single-support phase (or the takeoff phase), when the foot rolls smoothly as the support moves from the heel to the ball of the rear foot, which eventually leaves the ground. The third phase is the single-support phase or the swing phase, in which one foot is on the ground and the other foot is swinging across. The last phase is the transition from the single-support to the double-support phase or the landing phase, which is associated with the heel landing softly on the ground and the foot gradually rolling to the front as the support moves to the ball of the foot. Thus the robot returns to the double-support phase and continues to execute the four-phase cycle although the roles of two legs are interchanged. A second sequence of these four phases results in the robot returning to its original configuration.

The dynamic equations governing the motion of the robot during all the phases is made up of the ordinary differential equations previously discussed as well as the appropriate constraint equations in the support phases when the robot's kinematic topology changes from the double-support to the single-support phase and vice versa. In an ideal situation the transitions from one phase to the other can be modeled so they are governed by the conservation of the linear and the moment and momentum in appropriate directions.

In the walking mode, appropriate external moments at the leg joints M_0, M_1, M_2, and M_3 are commanded so the leg joint angles β_L, β_R, γ_L, and γ_R respond in a desired way. Consequently the legs are in periodic motion and an illustration of the legs in some typical positions is given in Figure 12.3. A typical set of approximate values of the leg joint angles during one walking cycle is shown in Table 12.1. In Table 12.1, the state "SLL" refers to support by the left leg, "SLR" refers to support by the right leg, and "DL" refers to double-leg support.

Table 12.1. *Commanded leg joint angles (in degrees) during one walking cycle*

Sl. no.	State	β_L	β_R	γ_L	γ_R
0	SLL	20	−20	0	20
1	SLL	0	30	0	90
2	SLL	−20	40	0	20
3	DL	−30	30	0	0
4	SLR	−20	20	20	0
5	SLR	30	0	90	0
6	SLR	40	−20	20	0
7	DL	30	−30	0	0
8	SLL	20	−20	0	20
9	SLL	0	30	0	90
10	SLL	−20	40	0	20
11	DL	−30	30	0	0
12	SLR	−20	20	20	0
13	SLR	30	0	90	0
14	SLR	40	−20	20	0
15	DL	30	−30	0	0

Notes: SLL, support by the left leg; SLR, support by the right leg; DL, double-leg support.

Consequently, walking in this cyclic manner implies that the four leg joint angles $\beta_L, \beta_R, \gamma_L$, and γ_R are constrained to behave in a certain way. This results in a further decrease in the number of independent DOFs, which are now reduced to one, i.e., the orientation of the torso α.

Although a number of different interpretations have been assigned to the nature of motion during the stance and swing mode of walking, the interpretation discussed in this section is the simplest one of its kind. Walking is generally "learned" by humans at a young age, and this suggests that walking patterns are stored in the brain and commanded as required. The net result is that the complex seven-DOF dynamic model is reduced to one with just one DOF.

12.2.2 Dynamic Balance during Walking: The Zero-Moment Point

We now turn our attention to the issue of dynamic balance during walking. First, we need to establish the condition for dynamic balance of a typical multibody kinematic chain of the type illustrated in Figure 12.1. The condition of dynamic balance may be stated in terms of two or more generic points. The first of these is the zero-moment point, which is defined as that point on the ground plane at which the net moment of the inertial and gravity forces acting on the kinematic chain has no components along the axes parallel to the ground plane. For a typical kinematic chain composed of a number of rigid bodies with both mass and mment of the inertia, the zero-moment point must satisfy the condition

$$\sum_i (\mathbf{r}_i - \mathbf{r}_{\text{zmp}}) \times m_i(\mathbf{a}_i - \mathbf{g}) + \mathbf{I}_i\alpha_i + \omega_i \times \mathbf{I}_i\omega_i = [0 \quad 0 \quad T_z]^T, \quad (12.33)$$

where $\mathbf{r}_i$ is the position vector of the center of mass of the ith body, $\mathbf{r}_{zmp}$ is the position vector of the zero-moment point on the ground plane, m_i and $\mathbf{I}_i$ are the mass and moment of inertia matrix of the ith body, ω_i and α_i are the angular velocity and angular acceleration vectors of the ith body, $\mathbf{g}$ is the gravity vector, and T_z may be an arbitrary number.

Considering our model of a walking robot in Figure 12.1, we find that the expression for the x coordinate of the zero-moment point is

$$
\begin{aligned}
x_{zmp} = {} & \frac{m_0 \left(\ddot{y}_0 x_0 - \ddot{x}_0 y_0 \right) - I_0 \ddot{\alpha}}{m_0 \ddot{y}_0 + m_0 g + \sum_{i=1}^{6} m_i \left(\ddot{y}_{Li} + \ddot{y}_{Ri} \right) + 2 m_i g} \\[2ex]
& + \frac{\sum_{i=1}^{6} m_i \left(\ddot{y}_{Li} x_{Li} - \ddot{x}_{Li} y_{Li} + \ddot{y}_{Ri} x_{Ri} - \ddot{x}_{Ri} y_{Ri} \right)}{m_0 \ddot{y}_0 + m_0 g + \sum_{i=1}^{6} m_i \left(\ddot{y}_{Li} + \ddot{y}_{Ri} \right) + 2 m_i g} \\[2ex]
& + \frac{g \left[m_0 x_0 + \sum_{i=1}^{6} m_i \left(x_{Li} + x_{Ri} \right) \right]}{m_0 \ddot{y}_0 + m_0 g + \sum_{i=1}^{6} m_i \left(\ddot{y}_{Li} + \ddot{y}_{Ri} \right] + 2 m_i g}.
\end{aligned}
\tag{12.34}
$$

In the case of the simplified model in Figure 12.2, the same expression holds but the summations run from $i = 1$ to $i = 2$.

To derive the condition the zero-moment point must satisfy for dynamic balance and observing that at the ground contact points the resultant reaction forces and moments act on the foot and the direction of the resultant moment is orthogonal to the ground plane. For simplicity we assume that the reaction moments are equal to zero. Thus the rate of change of moment of momentum in the case in which the robot is supported by two legs in contact with the ground plane must satisfy the moment equilibrium condition:

$$
\begin{aligned}
(\mathbf{r}_L - \mathbf{r}_{zmp}) &\times \mathbf{F}_L + (\mathbf{r}_R - \mathbf{r}_{zmp}) \times \mathbf{F}_R \\
&= \left[\sum_i (\mathbf{r}_i - \mathbf{r}_{zmp}) \times m_i (\mathbf{a}_i - \mathbf{g}) + \mathbf{I}_i \alpha_i + \omega_i \times \mathbf{I}_i \omega_i \right] \\
&= [0 \quad 0 \quad T_z]^T,
\end{aligned}
\tag{12.35}
$$

where $\mathbf{F}_L$ and $\mathbf{F}_R$ are the reaction force vectors in which the left and right feet are in contact with the ground and $\mathbf{r}_L$ and $\mathbf{r}_R$ are the corresponding position vectors of the ground contact points. Thus it follows that the zero-moment point must necessarily lie within the two footprints.

In the case in which the robot is supported by a single leg in rolling contact with the ground, the zero-moment point is not stationary but passes through the ground contact point of the supporting leg, so balance is regained when the robot returns to the double-leg-support configuration.

Last but not least, the arms are in out-of-phase motion relative to the legs, and this facilitates the synchronous motion of the zero-moment point relative to the points in contact with the ground so balance is continually maintained.

Assuming that the biped robot is walking with its body steady and with a uniform velocity, we require that

$$\dot{x}_{zmp} = V, \qquad (12.36)$$

where V is constant. The preceding requirement may be viewed as a constraint on $\dot{\alpha}$ and may be replaced with the simpler constraint equation given by

$$\dot{\alpha} = 0. \qquad (12.37)$$

Now all that remains is to determine the joint control torques, M_0, M_1, M_2, and M_3 so balance is always maintained. One approach to determining these control torques is to define the desired dynamic motion of x_{zmp}. With this in mind we define x_{SP} as the geometric center of the footprint polygon. In the case of a single supporting leg,

$$x_{SP} = x_{LG} \qquad (12.38)$$

when the supporting leg is the left leg;

$$x_{SP} = x_{RG} \qquad (12.39)$$

when the supporting leg is the right leg; and

$$x_{SP} = (x_{LG} + x_{RG})/2 \qquad (12.40)$$

when both legs are supporting the body. To obtain the joint control torques, we may choose the control law to minimize the performance index given by

$$J = \int_0^t (x_{zmp} - x_{SP})^2 dt + \int_0^t u^T(t)\,\mathbf{R}u(t)\,dt, \qquad (12.41)$$

where

$$u^T(t) = [\,M_1 \quad M_2 \quad M_3 \quad M_4\,]. \qquad (12.42)$$

Such a control law would ensure that x_{zmp} is always to x_{SP} and thus ensure balance of the walking robot.

12.2.3 Half-Model for a Quadruped Robot: Dynamics and Control

Quadruped mammals are endowed with articulated legs that are capable of executing a greater variety of walking and running gaits and, in particular, are able to retract the swing leg whenever there is the possibility of accidentally kicking the ground, leading to toe stubbing.

A relatively simple, planar, half-model known as the SCOUT has been developed by researchers at McGill University in Canada to demonstrate some of the basic features of the walking gaits of quadrupeds.

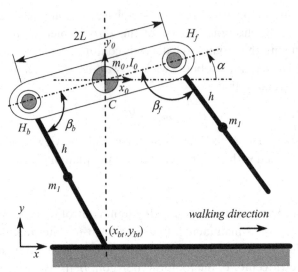

Figure 12.4. Planar kinematic chain half-model of a quadruped with five DOFs.

Understanding the kinematics and dynamics of such a half-model is the first step in synthesizing a more complex four-legged model involving multiple DOFs. This simple model is illustrated in Figure 12.4.

To establish the governing equations of motion and the appropriate constraint equations, we need the Cartesian velocities of the centers of mass of the two legs as well as the angular velocity of the main body. The Cartesian coordinates defining the locations of the hip joints are

$$x_{Hf} = x_0 + L\cos\alpha, \quad y_{Hf} = y_0 + L\sin\alpha, \tag{12.43}$$

$$x_{Hb} = x_0 - L\cos\alpha, \quad y_{Hb} = y_0 - L\sin\alpha. \tag{12.44}$$

The Cartesian coordinates defining of the locations of the centers of mass of the front and back legs are respectively given by

$$x_{fl} = x_0 + L\cos\alpha + r_1\cos(\beta_f + \alpha), \tag{12.45}$$

$$y_{fl} = y_0 + L\sin\alpha - r_1\sin(\beta_f + \alpha), \tag{12.46}$$

$$x_{bl} = x_0 - L\cos\alpha + r_1\cos(\beta_b - \alpha), \tag{12.47}$$

$$y_{bl} = y_0 - L\sin\alpha - r_1\sin(\beta_f - \alpha), \tag{12.48}$$

where r_1 is the distance of the leg center of mass from the hip joint. The Cartesian coordinates defining of the locations of toes of the front and back legs are respectively given by

$$x_{ft} = x_0 + L\cos\alpha + h\cos(\beta_f + \alpha), \tag{12.49}$$

$$y_{ft} = y_0 + L\sin\alpha - h\sin(\beta_f + \alpha), \tag{12.50}$$

$$x_{bt} = x_0 - L\cos\alpha + h\cos(\beta_b - \alpha), \tag{12.51}$$

$$y_{bt} = y_0 - L\sin\alpha - h\sin(\beta_f - \alpha), \tag{12.52}$$

where h is the height of each leg.

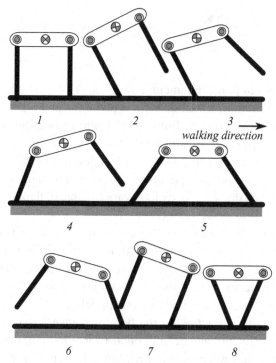

Figure 12.5. Sequence of leg postures to illustrate a quadruped half-model walking forward.

The dynamical equations and the constraints may now be established by the methods outlined in the preceding section. Moreover, because the robot is always supported on at least one leg, the DOFs reduce to α, β_f, and β_b. The latter two angles are specified when the robot is in locomotion, and consequently the only remaining DOF is α. Figure 12.5 depicts a sequence of leg postures as the robot is walking from an initial standing position.

The entire walking sequence consists of seven phases beyond the standing posture. In the figure, the robot starts from the initial standing position in the first phase. It then raises the front leg, swings it forward, lands the front leg on the ground, raises the back leg, swings it forward, lands the back leg, and finally repeats phases two through to eight in a rhythmic periodic manner to achieve forward locomotion at a uniform pace. The desired leg positions, determined by angles β_f and β_b, are acquired and stored as patterns by a process of learning and practice rather than by deliberate optimization every time the robot decides to walk.

12.3 Modeling Bird Flight: Robot Manipulators in Free Flight

The unique characteristics of bird flight are found in its motion dynamics, which to a large extent are influenced by aerodynamic lift and propulsive forces. Moreover, the dynamics of bird flight is unique on its own. The wings are made up of several hundred feathers that are controlled and coordinated so as to produce the desired aerodynamic lift and propulsive forces. Typically the resulting wing motion of a bird

Figure 12.6. Model of a bird in flight with articulated wing joints.

will generally consist of four fundamental motions: (1) flapping motion in a vertical
plane, (2) lead–lag motion, which denotes a posterior and anterior motion in the
horizontal plane, (3) feathering motion, which denotes a twisting motion of the wing
pitch, and (4) spanning motion, which denotes an alternately extending and contracting motion of the wing span. Flapping–translation involves a combination of
the first three modes. A simple model incorporating these capabilities is illustrated
in Figure 12.6, in which the wing–body joints are capable of flapping, feathering, and
lead–lag motion.

The motion of the body of the bird is a consequence of the reaction forces and
torques acting on the body. Thus the reactions of the wings that are attached to
the body by a number of revolute joints result in a coupling between the motion
of the bird's body and its wings. Thus it is often necessary that the net reaction
forces and torques acting on the body be completely balanced. The balance of the
reaction forces and torques acting on the body will ensure that the flight is steady
and uniform. This principle has been applied to manipulators attached to a free-
flying space vehicle, and we consider this particular application in some detail in
this section.

12.3.1 Dynamics of a Free-Flying Space Robot

We consider a typical free-flying space vehicle to which are attached at least two
or more manipulators or inertial controllers such as a momentum wheel, a reaction wheel, or a control moment gyroscope distributed symmetrically relative to the
space vehicle's body, as illustrated in Figure 12.7. The free-flying space vehicle and
the attached manipulators and controllers may be considered a multibody system.

The governing equations of motion may in general be expressed in the form

$$\begin{bmatrix} I_{bb}(\theta_i) & I_{bm}(\theta_i) & I_{bn}(\theta_i) \\ I_{bm}^T(\theta_i) & I_{mm}(\theta_i) & 0 \\ I_{bn}^T(\theta_i) & 0 & I_{nn}(\theta_i) \end{bmatrix} \begin{bmatrix} \ddot{x}_b \\ \ddot{\theta}_m \\ \ddot{\theta}_n \end{bmatrix} + \begin{bmatrix} c_b \\ c_m \\ c_n \end{bmatrix} = \begin{bmatrix} \mathcal{F}_b \\ \tau_m \\ \tau_n \end{bmatrix} + \begin{bmatrix} J_b^T \\ J_m^T \\ J_n^T \end{bmatrix} \mathcal{F}_e, \quad (12.53)$$

Figure 12.7. A free-flying space vehicle with attached manipulators.

where we choose the linear and angular velocity of the space vehicle's base body,

$$\dot{x}_b = \begin{bmatrix} v_b^T & \omega_b^T \end{bmatrix}^T, \qquad (12.54)$$

and the manipulator joint angles θ_i as the generalized coordinates. Furthermore, $I_{bb}(\theta_i)$ is the inertia matrix of the base, $I_{mm}(\theta_i)$ is the inertia matrix of the manipulators and controllers on one side of the space vehicle, which we refer to as the m group of manipulators, $I_{nn}(\theta_i)$ is the inertia matrix of the remaining manipulators and controllers, which we refer to as the n group of manipulators, $I_{bm}(\theta_i)$ and $I_{bn}(\theta_i)$ are the inertia coupling matrices responsible for linearly coupling the angular and linear acceleration of the m group and the n group of manipulators and controllers to that of the main body, $c_b(\theta_i, \dot{\theta}_j)$, $c_m(\theta_i, \dot{\theta}_j)$, and $c_n(\theta_i, \dot{\theta}_j)$ are the Coriolis and gyroscopic forces and torques in the body and manipulator equations of motion, and θ_m and θ_n are the joint angles corresponding to the m group and the n group of manipulators and controllers, respectively.

The space vehicle is generally in a freely floating mode for considerable lengths of time as, in this case, the external forces and torques acting on the space vehicle's body and those acting on the manipulator's end-effectors are assumed to be zero, i.e., $\mathcal{F}_b = 0$ and $\mathcal{F}_e = 0$. Thus the motion of the entire system is governed only by the internal forces and torques acting on the manipulator joints τ_m and τ_n. Consequently it follows that the linear and angular momenta of the system are constant. Hence,

$$\begin{bmatrix} \mathcal{P}^T & \mathcal{L}^T \end{bmatrix}^T = I_{bb}\dot{x}_b + I_{bm}\dot{\theta}_m + I_{bn}\dot{\theta}_n, \qquad (12.55)$$

where $\mathcal{P}$ is the linear momentum and $\mathcal{L}$ is the moment of momentum.

The upper set of these conservation equations defined in Equations (12.55) corresponds to the conservation of linear momentum. Generally, they may be integrated to obtain the equations determining the position of the origin of the body coordinate frame. The lower set of equations in Equations (12.55) corresponds to the conservation of moment of momentum. These equations cannot, in general, be integrated any further and provide a nonholonomic constraint. The linear velocity of the space vehicle's base body x_b may be eliminated from these equations and they may be expressed in terms of the angular velocity of the space vehicle's base body ω_b as

$$\mathcal{L}_0 = \bar{I}_{bb}\omega_b + \bar{I}_{bm}\dot{\theta}_m + \bar{I}_{bn}\dot{\theta}_n, \qquad (12.56)$$

where $\mathcal{L}_0$ is the initial moment of momentum. The first term is the moment of momentum of the space vehicle's base body, and the latter two terms are the contributions to $\mathcal{L}_0$ from the two groups of manipulators. Assuming that the angular velocity of the space vehicle's base body ω_b and the manipulator joint velocities $\dot{\theta}_m$ and $\dot{\theta}_n$ are evaluated initially,

$$\mathcal{L}_0 = \bar{I}_{bb}\omega_{b0} + \bar{I}_{bm}\dot{\theta}_{m0} + \bar{I}_{bn}\dot{\theta}_{n0}, \qquad (12.57)$$

any perturbations to the initial velocities ω_{b0}, $\dot{\theta}_{m0}$, and $\dot{\theta}_{n0}$ must satisfy the constraint

$$\bar{I}_{bb}\Delta\omega_b + \bar{I}_{bm}\Delta\dot{\theta}_m + \bar{I}_{bn}\Delta\dot{\theta}_n = 0. \qquad (12.58)$$

Similarly the linear velocity perturbations satisfy the constraints

$$I_{bb}\Delta\dot{x}_b + I_{bm}\Delta\dot{\theta}_m + I_{bn}\Delta\dot{\theta}_n = 0. \tag{12.59}$$

Constraint equations (12.58) and (12.59) are used to eliminate the variables $\Delta\dot{x}_b$ and $\Delta\omega_b$ from the kinematic and dynamic relations.

Consider, for example, the incremental change in the translational velocity of the end-effector that is due to changes in the space vehicle's base body velocity $\Delta\dot{x}_b$ and in the manipulator joint velocities $\Delta\dot{\theta}_m$ and $\Delta\dot{\theta}_n$, which can be expressed in terms of the Jacobian matrices as

$$\Delta\dot{x}_e = J_b\Delta\dot{x}_b + J_m\Delta\dot{\theta}_m + J_n\Delta\dot{\theta}_n. \tag{12.60}$$

Eliminating $\Delta\dot{x}_b$ from Equations (12.59) and (12.60) yields

$$\Delta\dot{x}_e = \left[\left(J_m - J_bI_{bb}^{-1}I_{bm}\right)\left(J_n - J_bI_{bb}^{-1}I_{bn}\right)\right]\begin{bmatrix}\Delta\dot{\theta}_m\\\Delta\dot{\theta}_n\end{bmatrix} = \mathbf{J}_G\begin{bmatrix}\Delta\dot{\theta}_m\\\Delta\dot{\theta}_n\end{bmatrix}. \tag{12.61}$$

The matrix $\mathbf{J}_G$ is known as the *free-flight Jacobian* or *generalized Jacobian* of the manipulator and allows one to directly express the end-effector's translational velocities in terms of the joint angles.

12.3.2 Controlling a Free-Flying Space Robot

One of the practical control problems associated with these types of space-vehicle-based manipulators is the ability to move the manipulator's arms in space without affecting the attitude of the space vehicle's base body. In this case we require that

$$\Delta\omega_b = 0. \tag{12.62}$$

Consequently constraint (12.58) reduces to

$$\bar{I}_{bm}\Delta\dot{\theta}_m + \bar{I}_{bn}\Delta\dot{\theta}_n = 0. \tag{12.63}$$

When constraint equation (12.63) is satisfied exactly, we require that

$$\Delta\dot{\theta}_m = -\bar{I}_{bm}^{-1}\bar{I}_{bn}\Delta\dot{\theta}_n, \tag{12.64}$$

provided $\bar{I}_{bm}$ is invertible. When

$$\bar{I}_{bn} = \bar{I}_{bm}\mathbf{G}, \tag{12.65}$$

constraint (12.63) may be expressed as

$$\Delta\dot{\theta}_m = -\mathbf{G}\Delta\dot{\theta}_n, \tag{12.66}$$

which has the structure of a feedback-control law.

Alternatively one could minimize a performance index defined as

$$J = \int_0^t (\bar{I}_{bm}\Delta\dot{\theta}_m + \bar{I}_{bn}\Delta\dot{\theta}_n)^2 dt + \int_0^t u^T(t)\mathbf{R}u(t)\,dt, \tag{12.67}$$

where $u(t)$ is the vector of control torques acting at the manipulator joints. When this is done the space vehicle's base body angular velocity remains almost constant but is not exactly maintained at a steady value.

12.4 Flapping Propulsion of Aerial Vehicles

Probably the most well-known application of flapping propulsion is the Ornithopter, an aircraft that is heavier than air and flies by flapping its wings. The discovery that flapping wings produce both thrust and lift was made in the early part of the 20th century. It was first predicted by considerations of the physical phenomenon of the vortices shed by a wing in an aerodynamic flow and subsequently verified experimentally. The first theoretical prediction of the thrust or drag production that is due to the appearance of a vortex street behind a flapping wing appeared in the 1930s. It was postulated that a clockwise rotation structure of the row of vortices above the middle plane and an anticlockwise rotation structure of the row of vortices below the middle plane, when the flow is going from left to right, was characteristically associated with drag production. On the other hand, an anticlockwise rotation structure of the row of vortices above the middle plane and clockwise rotation structure of the row of vortices below the middle plane was characteristically associated with thrust production. The vorticity is necessary to generate thrust by transferring momentum from the aerofoil to the fluid. In the late 1930s, shortly after compact formulas for estimating the lift force and aerodynamic moment acting on a thin aerofoil in a low-speed flow were developed in terms of the celebrated Theodorsen function $C(k)$, where k is the nondimensional reduced frequency of oscillation of the aerofoil, a theory for the calculation of the associated thrust was proposed by Garrick.[1] Garrick was able to show that, in an inviscid incompressible flow, a positive thrust is generated for all plunging motions with the efficiency of energy conversion from that needed to maintain the oscillations to a propulsive thrust being 100% at low frequencies approaching zero. However, Garrick also found the magnitude of the thrust to be proportional to the square of the frequency, although the propulsive efficiency was found to reduce asymptotically to 50% as the frequencies approached infinity.

The wings of an Ornithopter behave differently in the downstroke and in the upstroke. The periodic variation of the lift normal to the equivalent total velocity results in a small mean component of thrust in the direction of flight. The net mean thrust increases when more thrust is produced during the downstroke relative to the drag in the upstroke.

The laminar flow aerodynamics of aerial vehicles with flapping wings resembles that of birds and insects more than the traditional human-piloted aircraft we are familiar with. The small size of insects, which do not glide like large birds, causes them instead to flap with considerable change of wing shape during a single

---

[1] Garrick, I. E. (1937). Propulsion of a Flapping and Oscillating Aerofoil, NACA Report No. 567. National Advisory Committee for Aeronautics.

flapping cycle. Insects and small birds overcome a deteriorating aerodynamic performance under steady flow conditions at low characteristic Reynolds numbers by
using unsteady mechanisms of flapping and flexible wings. In bird and insect flight,
flapping motion does not take place in a plane. In the case of an insect, when the
flapping is in fact most efficient, the wings are attached to the thorax. The thorax is
packed with a number of flight muscles together with various other mechanisms for
coupling and operating the wings. An insect is able to control the flapping mechanism from here to carry out almost any aerial maneuver. The thorax can be considered simply as an elastic box with a lid (called the tergal plate) on top.

Two types of flapping mechanisms, *direct* and *indirect*, may be identified in most
flying insects. The direct-flapping mechanism is found in dragonflies and grasshop-
pers, and in this case the muscles driving the flapping wings are attached directly
to a pair of wing levers on each side of a pivot. Alternating contraction of these
levers is responsible for the generation of the flapping oscillations. In such a mechanism, the amplitude of flapping oscillation can be varied independently for each
wing. The indirect-flapping mechanism is found in bees and flies, and in this case
there are two pairs of muscles or links connected to the walls of the thorax. They
are not attached near the wing's base as in a direct mechanism. Rather, the flapping
motion is brought about by the up-and-down motion of the tergal plate. In both
cases the mechanisms are spatial in nature, and consequently, during one cycle of
flapping, the wings undergo spanwise twisting and lead–lag motion, which modifies
the relative angle of attack so as to continuously generate optimum ratios of lift and
thrust.

The thrust-producing capabilities of aerofoils oscillating relative to a flow have
led to proposals for energy extraction such as the Wells turbine for extracting
energy from the ocean waves and the McKinney–DeLaurier "wingmill" for extracting energy from the wind. In the Wells turbine the basic idea, as a consequence
of the Katzmayr or Knoller–Betz effect, was to extract energy from an oscillating
stream. The Wells turbine, named after its inventor Alan Wells, is a special type of
turbine, capable of maintaining constant direction of rotation although the airflow
passing through it is oscillating. The latter proposal was, however, based on the phenomenon of flutter in aerofoils, which is well known to hydronautical and aeronautical engineers. When an aerofoil is free to oscillate in both plunging and pitching or
feathering modes in a flow, then, under certain conditions, power is transferred from
the flow to the aerofoil, as it is capable of performing sustained oscillations in the
flow. In the case of an aircraft this transfer of power to an oscillating wing can lead
to sustained and undesirable fluttering oscillations that in turn can lead to disastrous
structural failure. In a wingmill this phenomenon is exploited to draw power from
the flow with efficiencies that are comparable to those of conventional windmills.

Theodorsen[2] was able provide closed-form expressions for the frequency-
dependent aerodynamic lift and aerodynamic moment acting on a flat-plate aerofoil

---

[2] Theodorsen, T. (1935). General theory of aerodynamic instability and the mechanism of flutter, NACA Report No. 496. National Advisory Committee for Aeronautics.

Figure 12.8. Illustration of a typical aerofoil section, showing the plunging and pitching DOFs at
the pivot point.

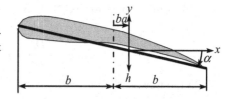

oscillating sinusoidally in both plunge and pitch modes and shedding vortices in
accordance with Kelvin's theorem in fluid mechanics. In Theodorsen's analysis it
was assumed that the flow was both inviscid and incompressible so the unsteady
velocity distribution can be expressed as the gradient of a potential, which satisfies a Laplace equation. The oscillations were assumed to be of small amplitude,
so the expressions for the pressure, obtained from the Bernoulli equation for total
energy, can be linearized. The flow was assumed to be fully attached during the
oscillations. To ensure there are no discontinuities at the trailing edge, the Kutta
condition requiring a smooth flow at the trailing edge is imposed. As this analysis
is the basis for the derivation of the thrust that is due to a flapping wing, its main
features are now highlighted.

12.4.1 Unsteady Aerodynamics of an Aerofoil

Consider a typical section of a two-dimensional wing extending in both directions
to infinity. The DOFs and geometry of the aerofoil, which is idealized as a flat plate,
are shown in Figure 12.8.

Theodorsen's most significant contribution was the fact that he modeled the
shed vortex street behind the trailing edge as a continuous, plane, two-dimensional,
vortex sheet with distributed vorticity and stretching from the trailing edge to an
infinite distance behind the wing. The shape of the vortex sheet is assumed to be
steady. To simplify the analysis, the vortex sheet is constrained to remain on the
horizontal plane along its entire length and the shape of the vortex-sheet model is
not allowed to evolve in response to the velocities self-induced by the vortex sheet.

The aerofoil is assumed to be suspended in a free stream with the flow velocity
being in the positive x direction far ahead of the aerofoil, uniform, steady, and equal
to U. It is assumed to be at an angle of attack α to the flow field, and the plunge
displacement is denoted as h, as illustrated in Figure 12.8. The semichord of the
aerofoil is assumed to be b, and the pivot point is assumed to be a semichords aft
of the midchord. The reference frame is assumed to be attached to the aerofoil at
midchord. The velocity vector of the flow field is defined by

$$\mathbf{V} = (U + u)i + vj + wk, \tag{12.68}$$

where u, v, and w are the local perturbation velocity components of the uniform
far-field flow velocity, U. A perturbation velocity potential is then defined by

$$(u \quad v \quad w) = \nabla\phi = \left(\frac{\partial}{\partial x} \quad \frac{\partial}{\partial y} \quad \frac{\partial}{\partial z} \right)\phi, \tag{12.69}$$

and the continuity relation for the conservation of mass in the case of the unsteady flow is

$$\frac{D\rho}{Dt} + \rho \nabla \cdot (\nabla \phi) = 0. \tag{12.70}$$

When the flow is incompressible, continuity equation (12.70) is

$$\nabla \cdot (u \quad v \quad w) = \nabla \cdot \nabla \phi = \frac{\partial^2}{\partial x^2}\phi + \frac{\partial^2}{\partial y^2}\phi + \frac{\partial^2}{\partial z^2}\phi = \nabla^2 \phi = 0. \tag{12.71}$$

Compressible flows, based on the unsteady Bernoulli relation, in the absence of any significant body forces, may be expressed as

$$\frac{\partial \phi}{\partial t} + \frac{1}{2}[\mathbf{V} \cdot \mathbf{V} - U^2] + \int_{p_\infty}^{p} \frac{dp}{\rho} = 0. \tag{12.72}$$

The last integral in Equation (12.72) is simplified by recognizing that the square of the speed of sound in the medium that is characterized by a unique pressure–density relationship in the entire flow field is,

$$a^2 = \left(\frac{\partial p}{\partial \rho}\right)_{s=c} = \frac{dp}{d\rho} = \frac{\gamma p}{\rho}. \tag{12.73}$$

Hence it follows that

$$d(a^2) = \gamma d\left(\frac{p}{\rho}\right) = (\gamma - 1)\frac{dp}{\rho}. \tag{12.74}$$

Thus the last integral in Bernoulli relation (12.72) may then be evaluated and expressed as

$$\int_{p_\infty}^{p} \frac{dp}{\rho} = \int_{a_\infty^2}^{a^2} \frac{d(a^2)}{(\gamma - 1)} = \frac{1}{(\gamma - 1)}\left(a^2 - a_\infty^2\right). \tag{12.75}$$

It may also be shown that

$$\frac{D}{Dt}\int_{p_\infty}^{p} \frac{dp}{\rho} = \frac{a^2}{\rho}\frac{D\rho}{Dt}. \tag{12.76}$$

Thus from Equation (12.70) the equation for the velocity potential may be expressed as

$$\nabla^2 \phi - \frac{1}{a^2}\left[\frac{\partial^2 \phi}{\partial t^2} + \frac{\partial}{\partial t}(\mathbf{V} \cdot \mathbf{V}) + \frac{1}{2}\mathbf{V} \cdot \nabla (\mathbf{V} \cdot \mathbf{V})\right] = 0. \tag{12.77}$$

In the case of incompressible flows, the perturbation pressure differential with reference to the free-stream (far-field) pressure p_∞ satisfies the Bernoulli relation

$$\int_{p_\infty}^{p} \frac{dp}{\rho} = -\frac{\partial \phi}{\partial t} - \frac{1}{2}[\mathbf{V} \cdot \mathbf{V} - U^2],$$

and may be expressed as

$$\frac{p - p_\infty}{0.5\rho U^2} = \frac{\Delta p}{0.5\rho U^2} = -\frac{1}{U^2}\left(\frac{\partial \phi}{\partial t} + U\frac{\partial \phi}{\partial x}\right). \tag{12.78}$$

The pressure is integrated around the aerofoil surface to obtain the forces and moments. In the case of a flat-plate-type aerofoil surface, the aerodynamic forces and moments are also obtained by evaluating the pressure distribution across the aerofoil surface.

The solution for the velocity potential is essential for obtaining the pressure distribution across the wing. The boundary conditions that must be imposed on the velocity distributions play a vital role in determining the pressure distribution. The principal boundary conditions are as follows:

1. The normal component of the flow velocity across the aerofoil surface must necessarily be equal to zero as any flow across the surface is physically impossible.
2. The Kutta–Joukowsky condition requires that the flow velocity be finite at the trailing edge. This ensures that the pressure differential across the wake is zero.

Assuming the flow to be incompressible implies that $a = \infty$, and this reduces the governing equation for the velocity potential to the Laplace equation,

$$\nabla^2 \phi = 0. \tag{12.79}$$

One approach to solving for the velocity potential and hence the velocity distribution is to assume that the flat-plate aerofoil is replaced with a sheet of distributed vorticity. The vorticity is assumed to extend behind the wing to infinity, where it represents the wake that is also modeled as a planar vortex sheet, as explained earlier. A sheet of vortices can support a jump in tangential velocity (i.e., a force), while the normal velocity is continuous. This is the main reason for using a vortex sheet to represent a lifting surface.

Vortex flow is also as a form of potential flow, as it satisfies the Laplace equation. We may express the potential that is due to any such point vortex (in a cylindrical reference frame centered at the vortex source) at point j as

$$\phi_j = (\gamma_j / 2\pi)\theta. \tag{12.80}$$

Then, we can find the velocity induced by the vortex at any point in the flow field by taking the gradient of the potential:

$$\mathbf{v}_j = \nabla \phi_j = (\hat{k} / 2\pi r_j)\gamma_j, \tag{12.81}$$

where r_j is the distance from the center of the vortex and γ_j is the vortex strength. This type of source has proved particularly useful for approximating the flow over aerofoils because it automatically satisfies the far-field boundary condition of Laplace's equation, which allows us to decompose the potential into two components: the potential that is due to the interaction of all the bound vortices and the potential at infinity. We also know that a two-dimensional vortex singularity satisfies Laplace's equation (i.e., a point vortex). Thus all that remains to be satisfied is the boundary condition on the surface of the aerofoil.

The aerofoil boundary condition can then be expressed as the sum of the induced velocity that is due to a sheet of vortices bound to the aerofoil surface, approximated as a flat plate, and the induced velocity on the aerofoil surface that is due to the free vortices in the wake must equal the aerofoil surface velocity. Thus,

$$w_a = w_b + \lambda_{\text{wake}}, \tag{12.82}$$

where

$$w_b = \frac{1}{2\pi} \int_{-b}^{b} \frac{\gamma_b}{x - \xi} d\xi, \quad \lambda_{\text{wake}} = \frac{1}{2\pi} \int_{b}^{\infty} \frac{\gamma_b}{x - \xi} d\xi, \tag{12.83}$$

w_a is the aerofoil surface velocity, γ_b is the strength of the distributed vortices that satisfies

$$\gamma_b(x, t) = -2\phi_x(x, 0+, t) = -2u(x, 0+, t), \tag{12.84}$$

and $\phi(x, z, t)$ is the corresponding velocity potential that satisfies the Laplace equation, (12.79).

Assuming that $u \approx U$, the velocity of the free stream far ahead of the aerofoil, and that $v \cong 0$, we obtain the nondimensional relation on the aerofoil surface, $z = z_a(x, y, t)$,

$$w_a = \frac{w}{U}\bigg|_{z=z_a} = \left(\frac{\partial}{\partial t} + U\frac{\partial}{\partial x}\right) \frac{z_a(x, y, t)}{U}. \tag{12.85}$$

Thus,

$$\frac{1}{2\pi} \int_{-b}^{b} \frac{\gamma_b}{x - \xi} d\xi = w_a - \lambda_{\text{wake}}. \tag{12.86}$$

Integrating the vortex flux from $x = -b$ to x, we obtain the partial circulation given by

$$\Gamma(x, t) = \int_{-b}^{x} \gamma_b(\xi, t) \, d\xi = -2 \int_{-b}^{x} \frac{\partial}{\partial \xi} \phi(\xi, z, t)\big|_{z=z_a=0+} d\xi$$

and

$$\Gamma(x, t) = -2\phi(x, 0+, t). \tag{12.87}$$

The velocity potential $\phi(x, z, t)$ is related to the nondimensional pressure coefficient by

$$C_p = \frac{p}{(1/2)\rho U^2} = -2\left(\frac{\partial}{\partial x} + \frac{1}{U}\frac{\partial}{\partial t}\right) \phi(x, z, t). \tag{12.88}$$

Thus, across the wake, as one would expect no pressure differential, for $x \geq b$, $z = +0$, and

$$C_p = \frac{p}{(1/2)\rho U^2} = -2\left(\frac{\partial}{\partial x} + \frac{1}{U}\frac{\partial}{\partial t}\right) \phi(x, z, t) = 0. \tag{12.89}$$

It follows that, on the wake surface,

$$\phi(x, 0+, t) = \phi[b, 0+, t - (x - b)/U], \tag{12.90}$$

and consequently on the wake surface,

$$\Gamma(x, t) = \Gamma[b, t - (x - b)/U]. \tag{12.91}$$

Equation (12.91) reduces to the full circulation $\Gamma(b, t)$ when $x = b$, and for $x \geq b$, $z = +0$,

$$\gamma_b(x, t) = -\frac{1}{U} \times \frac{\partial}{\partial t} \Gamma[b, t - (x - b)/U]. \qquad (12.92)$$

Hence, observe that the wake-induced velocity may be expressed as

$$\lambda_{\text{wake}} = \frac{1}{2\pi} \int_b^\infty \frac{\gamma_b}{x - \xi} d\xi = \frac{-1}{2\pi U} \int_b^\infty \frac{d\xi}{x - \xi} \times \frac{\partial}{\partial t} \Gamma[b, t - (\xi - b)/U]. \qquad (12.93)$$

If we assume that the full circulation is sinusoidal, i.e.,

$$\Gamma(b, t) = \hat{\Gamma}(b) \exp(i\omega t), \Gamma(x, t) = \Gamma(b, t) \exp[-i\omega(x - b)/U], \qquad (12.94)$$

then it follows that

$$\begin{aligned}
\lambda_{\text{wake}} &= \frac{-1}{2\pi U} \int_b^\infty \frac{1}{x - \xi} \times \frac{\partial}{\partial t} \Gamma[b, t - (x - b)/U] d\xi \\
&= \frac{-i\omega b}{U} \frac{\Gamma(b, t)}{2\pi b} \int_b^\infty \frac{\exp\{-i(\omega b/U)[(\xi - b)/b]\}}{x - \xi} d\xi.
\end{aligned} \qquad (12.95)$$

To evaluate $\hat{\Gamma}(b)$, consider the integral equation solution for γ_b obtained from Equations (12.83) and given by

$$\gamma_b = \frac{2}{\pi} \sqrt{\frac{b - x}{b + x}} \int_{-b}^{+b} \sqrt{\frac{b + \xi}{b - \xi}} \frac{w_a - \lambda_{\text{wake}}}{x - \xi} d\xi \qquad (12.96)$$

and

$$\Gamma(b, t) = \int_{-b}^b \gamma_b(\xi, t) d\xi = \frac{2}{\pi} \int_{-b}^b \left(\sqrt{\frac{b + \xi}{b - \xi}}\right) (w_a - \lambda_{\text{wake}}) \int_{-b}^{+b} \sqrt{\frac{b - x}{b + x}} \frac{1}{x - \xi} dx d\xi,$$

or

$$\Gamma(b, t) = \int_{-b}^b \gamma_b(\xi, t) d\xi = 2 \int_{-b}^b \sqrt{\frac{b + \xi}{b - \xi}} (w_a - \lambda_{\text{wake}}) d\xi. \qquad (12.97)$$

If we let

$$Q = \frac{1}{\pi} \int_{-b}^b \sqrt{\frac{b + \xi}{b - \xi}} w_a d\xi, \qquad (12.98)$$

then it follows from the solution for λ_{wake} that

$$2\pi Q = \Gamma(b, t) + 2 \int_{-b}^b \sqrt{\frac{b + \xi}{b - \xi}} \lambda_{\text{wake}} d\xi = \Gamma(b, t) \left(1 - \frac{2ikI}{b}\right), \qquad (12.99)$$

where the integral I is

$$I = \frac{1}{2\pi} \int_b^\infty \exp\{-i(\omega b/U)[(\xi - b)/b]\} \left(\int_{-b}^{+b} \sqrt{\frac{b - x}{b + x}} \frac{1}{x - \xi} dx\right) d\xi, \xi > b,$$

which may be evaluated as

$$I = \frac{b}{2ik} \left\{ 1 + \frac{\pi i k}{2} \exp{(ik)} \left[H_1^{(2)}(k) + i H_0^{(2)}(k) \right] \right\}.$$

Hence,

$$\Gamma(b,t) = -\frac{4 \exp{(-ik)} Q}{ik \left[H_1^{(2)}(k) + i H_0^{(2)}(k) \right]}. \tag{12.100}$$

The solution for the distributed vortex strength γ_b is now known, and the solution for the nondimensional pressure coefficient may be obtained from Equations (12.87) and (12.89). If we let the amplitude of the downwash velocity be sinusoidal,

$$w_a(x,t) = \hat{w}_a(x) \exp{(i\omega t)}, \tag{12.101}$$

we may show that[3]

$$\begin{aligned}
\Delta C_p(x) = {} & \frac{4}{\pi} \sqrt{\frac{b-x}{b+x}} \int_{-b}^{b} \sqrt{\frac{b+x_1}{b-x_1}} \frac{\hat{w}_a(x_1)}{x-x_1} dx_1 \\
& + \frac{4i\omega}{\pi U} \sqrt{(b^2-x^2)} \int_{-b}^{b} \sqrt{\frac{1}{b^2-x_1^2}} \frac{\hat{p}_a(x_1)}{x-x_1} dx_1 \\
& + \frac{4}{\pi \times b} [1 - C(k)] \sqrt{\frac{b-x}{b+x}} \int_{-b}^{b} \sqrt{\frac{b+x_1}{b-x_1}} \hat{w}_a(x_1) dx_1,
\end{aligned} \tag{12.102}$$

where the reduced frequency k is related to the circular frequency of sinusoidal oscillations by the relation $k = \omega b / U$, and

$$\hat{p}_a(x) = \int_{-b}^{x} \hat{w}_a(x_1) dx_1, \quad C(k) = \frac{\int_1^\infty \frac{\xi}{\sqrt{\xi^2-1}} e^{-ik\xi} d\xi}{\int_1^\infty \frac{\xi+1}{\sqrt{\xi^2-1}} e^{-ik\xi} d\xi},$$

which reduces to,

$$C(k) = \frac{K_1(ik)}{K_1(ik) + K_0(ik)} = \frac{H_1^{(2)}(k)}{H_1^{(2)}(k) + i H_0^{(2)}(k)}. \tag{12.103}$$

Hence the amplitudes of the nondimensional lift and pitching moment about the midchord line are

$$\frac{\hat{L}}{\frac{1}{2}\rho U^2 b} = -\frac{4C(k)}{b} \int_{-b}^{b} \sqrt{\frac{b+x_1}{b-x_1}} \hat{w}_a(x_1) dx_1 - \frac{4ik}{b^2} \int_{-b}^{b} \sqrt{b^2-x_1^2} \hat{w}_a(x_1) dx_1, \tag{12.104}$$

---

[3] See, for example, Ashley, H. and Landahl, M. T. (1965). *Aerodynamics of Wings and Bodies*, Addison-Wesley, Reading, MA, Chapter 10. [Also available as a Dover paperback from Dover Publications, New York.]

Table 12.2. *Values of the Theodorsen function, $C(k) = F(k) + i G(k)$*

K	$F(k)$	$-G(k)$
0	1.000	0.0
0.05	0.909	0.130
0.10	0.832	0.172
0.20	0.728	0.189
0.30	0.665	0.179
0.40	0.625	0.165
0.50	0.598	0.151
0.60	0.579	0.138
0.80	0.554	0.116
1.00	0.539	0.100
1.20	0.530	0.088
1.50	0.521	0.0736
2.00	0.513	0.0577
4.00	0.504	0.0305
6.00	0.502	0.0206
10.00	0.501	0.0124
∞	0.5	0.0

$$\frac{\hat{M}}{0.5\rho U^2 (2b^2)} = -\frac{1}{b} \int_{-b}^{b} \left(\sqrt{\frac{b+x_1}{b-x_1}} - \frac{2}{b}\sqrt{b^2 - x_1^2} \right) \hat{w}_a (x_1)\, dx_1$$

$$+ \frac{C(k)}{b} \int_{-b}^{b} \sqrt{\frac{b+x_1}{b-x_1}} \hat{w}_a (x_1)\, dx_1 \qquad (12.105)$$

$$- \frac{i\omega}{U} \int_{-b}^{b} \left(\sqrt{\frac{1}{b^2 - x_1^2}} - \frac{2}{b}\sqrt{b^2 - x_1^2} \right) \hat{p}_a (x_1)\, dx_1,$$

respectively.

Theodorsen was able to evaluate the effect of the wake on a wing in a very concise manner, which he expressed in terms of the function $C(k)$, where the parameter k is referred to as the reduced frequency. It arises from the circulation terms resulting from the wake and is a complex function of the scaled or reduced frequency of vibration.

Theodorsen, who introduced the function, expressed it in terms of complex Bessel functions of the first and second kinds as well as in terms of Hankel functions. Separating the real and imaginary parts, we may write $C(k)$ as

$$C(k) = F(k) + iG(k), \qquad (12.106)$$

and typical numerical values are tabulated in Table 12.2. The reduced frequency k is also related to the Strouhal number, which is defined as

$$St = fA/U = \omega A/2\pi U = k(A/2\pi b), \qquad (12.107)$$

where A is the amplitude of motion of the aerofoil either at the trailing edge or at any other significant point on the aerofoil surface.

In practice it is quite common to use a rational function approximation of the Theodorsen function, given by

$$C(k) = \frac{0.5\,(ik)^2 + 0.2813\,ik + 0.01365}{(ik)^2 + 0.3455\,ik + 0.01365}. \tag{12.108}$$

12.4.2 Generation of Thrust

An important parameter in the design of flapping mechanisms is the efficiency of the conversion of the average work done in unit time to maintain oscillations against the generalized force and moments acting on the aerofoil into thrust or propulsive energy. We therefore consider the calculation of this average work done in unit time and the propulsive thrust generated by a flapping wing. To generalize the original expressions derived by Garrick for the propulsive thrust, the previously mentioned average work done in unit time, and the rate of kinetic energy transfer to the wake, we consider an aerofoil oscillating in a generalized mode shape. The aerofoil motion is described in terms of a finite number of assumed mode shapes, $\phi_i(\xi), i = 2, 3, \ldots, N$, where ξ is a nondimensional coordinate directed toward the trailing edge of the flap and the corresponding modal displacement amplitudes q_i. The vertical downward displacement of any mass point in the section is then given as

$$v = \sum_{i=1}^{N} q_i(t)\, z_i(\xi) = b \sum_{i=1}^{N} q_i(t)\, \phi_i(\xi), \tag{12.109}$$

where the first two modes are assumed to be the plunging and pitching modes defined as

$$\phi_1(\xi) = 1, \quad \phi_2(\xi) = \xi - a. \tag{12.110}$$

Although nondimensional, the parameters that determine the pressure coefficient are not nondimensional. To nondimensionalize all the parameters that influence the preceding expression for $\Delta C_p(x)$, we define nondimensional quantities in terms of $\xi = x/b$, which is itself nondimensional. Thus the downwash amplitude in the ith mode is defined by

$$\hat{w}_i(\xi_1) = \hat{w}_i(x_1/b) = \hat{w}_a(x_1). \tag{12.111}$$

The integral quantity

$$\hat{p}_a(x) = \int_{-b}^{x} \hat{w}_a(x_1)\, dx_1 = b \int_{-1}^{x/b} \hat{w}_a(x_1)\, d\left(\frac{x_1}{b}\right) = b \int_{-1}^{x/b} \hat{w}_i(x_1/b)\, d\left(\frac{x_1}{b}\right),$$

i.e.,

$$\hat{p}_a(x) = b\hat{p}_i(\xi), \quad \text{with } \hat{p}_i(\xi) = \int_{-1}^{\xi} \hat{w}_i(\xi_1)\, d\xi_1. \tag{12.112}$$

The quantity $\hat{w}_i(\xi_1)$ representing the nondimensional downwash is then related to the physical modal displacement z_i and the nondimensional modal displacement

$\phi_i(\xi)$ by

$$\hat{w}_i(\xi) = \frac{1}{U}\left(i\omega + U\frac{\partial}{\partial x}\right)z_i = \left[\frac{i\omega b}{U} + \frac{\partial}{\partial(x/b)}\right]\frac{z_i}{b} = \left(ik + \frac{d}{d\xi}\right)\phi_i(\xi).$$

(12.113)

The generalized modal pressure coefficient in the jth mode is

$$\begin{aligned}
\Delta C_p^j(\xi) = {} & \frac{4}{\pi}\sqrt{\frac{1-\xi}{1+\xi}}\int_{-1}^{1}\sqrt{\frac{1+\xi_1}{1-\xi_1}}\frac{\hat{w}_j(\xi_1)}{\xi-\xi_1}d\xi_1 \\
& + \frac{4ik}{\pi}\sqrt{1-\xi^2}\int_{-1}^{1}\sqrt{\frac{1}{1-\xi_1^2}}\frac{\hat{p}_j(\xi_1)}{\xi-\xi_1}d\xi_1 \\
& + \frac{4}{\pi}[1-C(k)]\sqrt{\frac{1-\xi}{1+\xi}}\int_{-1}^{1}\sqrt{\frac{1+\xi_1}{1-\xi_1}}\hat{w}_i(\xi_1)\,d\xi_1.
\end{aligned}$$

(12.114)

To estimate the average work done in unit time to maintain oscillations against the generalized force acting on the aerofoil, we use the expression for the generalized modal pressure coefficient and compute the generalized forces acting on the aerofoil corresponding to each of the assumed modes of oscillation. The generalized modal force coefficients may be defined in terms of the nondimensional integral expression:

$$G_{ij} = G_{ij}(\phi_i,\phi_j) = \frac{1}{2\pi}\int_{-1}^{1}\Delta C_p^j(\xi_1)\,\phi_i(\xi_1)\,d\xi_1.$$

(12.115)

Performing the outer integrals first and removing the singular part of the principal-value integrals reduces (12.115) to the double-integral expression given by

$$\begin{aligned}
G_{ij} = G_{ij}(\phi_i,\phi_j) = {} & \frac{2}{\pi^2}\int_{-1}^{1}\sqrt{\frac{1}{1-\xi_1^2}}\hat{w}_j(\xi_1)\,I_1(\xi_1)\,d\xi_1 \\
& + \frac{2ik}{\pi^2}\int_{-1}^{1}\sqrt{\frac{1}{1-\xi_1^2}}\hat{p}_j(\xi_1)\,I_2(\xi_1)\,d\xi_1 + [1-C(k)] \\
& \times \left[\frac{1}{\pi}\int_{-1}^{1}\sqrt{\frac{1}{1-\xi_1^2}}\hat{w}_j(\xi_1)(1+\xi_1)\,d\xi_1\right]\left[\frac{2}{\pi}\int_{-1}^{1}\sqrt{\frac{1}{1-\xi^2}}\phi_i(\xi)(1-\xi)\,d\xi\right],
\end{aligned}$$

(12.116)

where

$$\begin{aligned}
I_1(\xi_1) = {} & \phi_i(\xi_1)\left(1-\xi_1^2\right)\int_{-1}^{1}\frac{1}{\sqrt{1-\xi^2}}\frac{d\xi}{\xi-\xi_1} \\
& + (1+\xi_1)\left\{\int_{-1}^{1}\sqrt{\frac{1}{1-\xi^2}}\frac{[\phi_i(\xi)(1-\xi)-\phi_i(\xi_1)(1-\xi_1)]}{\xi-\xi_1}d\xi\right\},
\end{aligned}$$

and

$$I_2\{\xi_1, c_i\} = \phi_i(\xi_1)\left(1 - \xi_1^2\right) \int_{c_i}^1 \frac{1}{\sqrt{1-\xi^2}} \frac{d\xi}{\xi - \xi_1}$$
$$+ \left\{ \int_{c_i}^1 \sqrt{\frac{1}{1-\xi^2}} \frac{\left[\phi_i(\xi)\left(1-\xi^2\right) - \phi_i(\xi_1)\left(1-\xi_1^2\right)\right]}{\xi - \xi_1} d\xi \right\}.$$

The total nondimensional lift and pitching moments about the midchord acting on the airfoil are evaluated independently by use of

$$\frac{L}{\pi \rho U^2 b} = -\frac{2}{\pi} \left[\int_{-1}^1 g(\xi_1)\,\hat{w}_j(\xi_1)\left(1+\xi_1\right) d\xi_1 + ik \int_{-1}^1 g(\xi_1)\,\hat{w}_j(\xi_1)\left(1-\xi_1^2\right) d\xi_1 \right]$$
$$+ \frac{2}{\pi} \left\{ [1 - C(k)] \int_{-1}^1 g(\xi_1)\,\hat{w}_j(\xi_1)\left(1+\xi_1\right) d\xi_1 \right\} \tag{12.117}$$

and

$$\frac{M_0}{\pi \rho U^2 b^2} = \frac{2}{\pi} \left[\int_{-1}^1 g(\xi_1)\,\hat{w}_j(\xi_1)\left(1-\xi_1^2\right) d\xi_1 + ik \int_{-1}^1 g(\xi_1)\left(\frac{1}{2}-\xi_1^2\right) \hat{p}_j(x_1)\, dx_1 \right]$$
$$- \frac{2}{\pi} \left\{ \frac{1}{2}\,[1 - C(k)] \int_{-1}^1 g(\xi_1)\,\hat{w}_j(\xi_1)\left(1+\xi_1\right) d\xi_1 \right\}, \tag{12.118}$$

where

$$g(\xi_1) = 1/\sqrt{1 - \xi_1^2}. \tag{12.119}$$

When the generalized displacements are sinusoidal, the generalized force in the ith mode may be expressed as

$$\mathbf{G}_i = \pi \rho U^2 b^2 \sum_{j=1}^N G_{ij} q_j(t) = \pi \rho U^2 b^2 \sum_{j=1}^N G_{ij} q_{j0} \exp(i\omega t). \tag{12.120}$$

We assume that the generalized displacements are of the form

$$q_j(t) = q_{j0} \sin(\omega t + \varphi_j). \tag{12.121}$$

The corresponding expression for the generalized force in the ith mode is then given by

$$\mathbf{G}_i = \pi \rho U^2 b^2 \sum_{j=1}^N [\mathrm{Re}\,(G_{ij}) \sin(\omega t + \varphi_j) + \mathrm{Im}\,(G_{ij}) \cos(\omega t + \varphi_j)]\, q_{j0}.$$

The average work done in unit time to maintain oscillations against the generalized force acting on the aerofoil is

$$\bar{W} = -\frac{\omega}{2\pi} \int_0^{2\pi/\omega} \left[\sum_{i=1}^N \mathbf{G}_i \dot{q}_i(t) \right] dt.$$

The integral may be expressed as

$$\bar{W} = \frac{\pi\rho\omega}{4\pi}(\omega Ub)^2 \int_0^{2\pi/\omega} \left(\sum_{i=1}^{N} \sum_{j=1}^{N} w_{ij} q_{j0} q_{i0} \right) dt, \qquad (12.122)$$

where

$$w_{ij} = -2\left[\mathrm{Re}\,(G_{ij}) \sin(\omega t + \varphi_j) + \mathrm{Im}\,(G_{ij}) \cos(\omega t + \varphi_j) \right] \cos(\omega t + \varphi_i).$$

Performing the integrations, we may write the average work done in unit time to maintain oscillations against the generalized force acting on the aerofoil as

$$\bar{W} = \left(\frac{\pi\rho\omega^3 b^4}{k^2} \right) \left(\sum_{i=1}^{N} \sum_{j=1}^{N} W_{ij} q_{j0} q_{i0} \right) = \left(\frac{\pi\rho\omega^3 b^4}{k^2} \right) C_W, \qquad (12.123)$$

where

$$W_{ij} = \frac{1}{2\pi} \int_0^{2\pi} w_{ij}\, d(\omega t) = \left[\mathrm{Re}\,(G_{ij}) \sin(\varphi_{ij}) - \mathrm{Im}\,(G_{ij}) \cos(\varphi_{ij}) \right],$$

with $\varphi_{ij} = \varphi_i - \varphi_j$.

We now consider the kinetic energy of the wake far away from the aerofoil, which we obtain by integrating the kinetic energy of an element of mass over a control volume that encloses the far-field wake. Thus the kinetic energy of the wake is give by the volume integral,

$$E = \frac{1}{2} \int_v \int \int \rho \left[\left(\frac{\partial\phi}{\partial x} \right)^2 + \left(\frac{\partial\phi}{\partial z} \right)^2 \right] dx\, dz = \frac{1}{2} \int_s \rho\phi \frac{\partial\phi}{\partial n}\, dx, \qquad (12.124)$$

which is reduced to the surface integral over the top and bottom surfaces of the wake within the control volume. Thus at a point along the x axis in the far-field wake,

$$E_1 = 0.5\rho w [\phi(x, 0^+, t) - \phi(x, 0^-, t)], \qquad (12.125)$$

where w is the induced vertical velocity there. The quantity w is given by

$$w = \frac{1}{2\pi} \int_{-b}^{b} \frac{\gamma_b}{x - \xi}\, d\xi - \frac{i\omega b}{U} \frac{\Gamma(b, t)}{2\pi b} \int_b^{\infty} \frac{\exp\{-i(\omega b/U)[(\xi - b)/b]\}}{x - \xi}\, d\xi,$$

where γ_b is given by Equation (12.96) and

$$\phi(x, 0^+, t) - \phi(x, 0^-, t) = -\int_{-b}^{x} \gamma_b(\xi, t)\, d\xi = -\Gamma(x, t).$$

Hence the average increase in energy in the wake per unit time and the average work done by the propulsive force P per unit time are respectively given by

$$\bar{E} = \frac{\omega U}{2\pi} \int_0^{2\pi/\omega} E_1\, dt, \qquad PU = \bar{W} - \bar{E}. \qquad (12.126)$$

Thus it can be shown that the propulsive force or thrust is

$$P = \pi\rho U^2 b S^2 + \frac{1}{2}\rho U^2 \int_{-b}^{b} C_p \frac{\partial z_a(x, y, t)}{\partial x}\, d\xi, \qquad (12.127)$$

where the first term represents the thrust that is due to the component of the suction at the leading edge, directed upstream, that does not contribute to the lift. It can be shown that the quantity S related to the vortex strength γ_b and defined by

$$S \equiv \underset{x \to -b}{Lt} \left(\frac{\sqrt{b+x}}{2\sqrt{b}} \gamma_b \right),$$

is given by

$$S = \frac{\sqrt{2}}{\pi\sqrt{b}} \underset{x \to -b}{Lt} \int_{-b}^{+b} \sqrt{\frac{b+\xi}{b-\xi}} \frac{(w_a - \lambda_{\text{wake}})}{x-\xi} d\xi. \tag{12.128}$$

As easy way of evaluating S is to note that the pressure coefficient ΔC_p and twice the negative strength of the vorticity distribution $-2\gamma_b$ approach one another at the leading edge $x = -b$ because $\hat{\Gamma}(-1) = 0$. Hence

$$S = \underset{x \to -b}{Lt} \left[0.5 \left(\sqrt{1+x/b} \right) \gamma_b \right] = \underset{x \to -b}{Lt} \left[0.25 \left(\sqrt{1+x/b} \right) \Delta C_p \right]. \tag{12.129}$$

Thus we obtain a nondimensional generalization of the quantity S introduced by Garrick for plunging, feathering, and flap modes, given as

$$S = \frac{\sqrt{2}}{\pi} \int_{-b}^{b} \sqrt{\frac{b+x_1}{b-x_1}} \frac{\hat{w}_a(x_1)}{x-x_1} dx_1 + \frac{\sqrt{2}}{\pi \times b} [1-C(k)] \int_{-b}^{b} \sqrt{\frac{b+x_1}{b-x_1}} \hat{w}_a(x_1) dx_1.$$

The expression for S in a linear combination of modes defined by

$$\tilde{w}(x,t) = \sum_{i=1}^{N} q_i(t) \hat{w}_i(x) = \sum_{i=1}^{N} q_{j0} \exp(i\omega t) \hat{w}_i(x) \tag{12.130}$$

is then given by the sum of the nondimensional integrals as

$$S = \frac{\sqrt{2}}{\pi} \int_{-1}^{1} \sqrt{\frac{1+\xi_1}{1-\xi_1}} \frac{\tilde{w}(\xi_1,t)}{\xi-\xi_1} d\xi_1 + \frac{\sqrt{2}}{\pi} [1-C(k)] \int_{-1}^{1} \sqrt{\frac{1+x_1}{1-x_1}} \tilde{w}(\xi_1,t) d\xi_1. \tag{12.131}$$

By adopting the same procedure as the one adopted for evaluating $\bar{W}$, we may express the average thrust generated $\bar{P}$ when the generalized displacements are of the form $q_j(t) = q_{j0} \sin(\omega t + \varphi_j)$ as

$$\bar{P} = \left(\frac{\pi \rho \omega^2 b^3}{k^2} \right) \left(\sum_{i=1}^{N} \sum_{j=1}^{N} P_{ij} q_{j0} q_{i0} \right) = \left(\frac{\pi \rho \omega^2 b^3}{k^2} \right) C_T. \tag{12.132}$$

We consider the special case of a flat-plate aerofoil with a trailing-edge flap oscillating in plunging, feathering about an axis, and the first rigid flap mode. In this case we obtain the same expressions for $\bar{W}$ and $\bar{P}$ as Garrick did.

In the case of the aerofoil oscillating in plunging (first mode) and feathering about an axis (second mode), they are given by

$$C_W = Fk \left(\frac{h_0}{b} \right)^2 + \left\{ \frac{k}{2} \left(\frac{1}{2} - a \right) - \left(\frac{1}{2} + a \right) \left[Fk \left(\frac{1}{2} - a \right) + G \right] \right\} \alpha_0^2$$
$$+ \left[\left(\frac{k}{2} - 2akF + G \right) \cos\varphi_{21} + (F - kG) \sin\varphi_{21} \right] \alpha_0 \frac{h_0}{b} \tag{12.133}$$

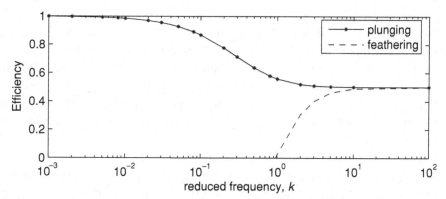

Figure 12.9. Comparison of plunging and feathering efficiencies.

and

$$
\begin{aligned}
C_T = {} & |C(k)|^2 \left(\frac{h_0 k}{b}\right)^2 + |C(k)|^2 \left[1 + \left(\frac{1}{2} - a\right)^2 k^2\right]\alpha_0^2 \\
& + \left\{\frac{k^2}{4}(1 - 2a) - \left[1 + \left(\frac{1}{2} - a\right)k^2\right]F - \left(\frac{1}{2} + a\right)Gk\right\}\alpha_0^2 \\
& + \left[2|C(k)|^2\left(\frac{1}{2} - a\right)k^2 + \frac{k^2}{2} - Fk^2 + kG\right]\alpha_0 \frac{h_0}{b}\cos\varphi_{21} \\
& + \left[2|C(k)|^2 k + \frac{k^2}{2} - Fk - Gk^2\right]\alpha_0 \frac{h_0}{b}\sin\varphi_{21},
\end{aligned}
\tag{12.134}
$$

where φ_{21} is the phase lead of the feathering oscillation relative to the plunging oscillation. It follows that the efficiency in the conversion of the average work done in unit time to maintain oscillations against the generalized force acting on the aerofoil to propulsive thrust is given by

$$
e = \bar{P}U/\bar{W} = C_T/kC_W.
\tag{12.135}
$$

Figure 12.9 compares the energy conversion efficiency e for the case of pure flapping and pure feathering about midchord. We observe that the efficiency of pure feathering about the trailing edge never exceeds the corresponding efficiency of pure plunging. It is worth noting that, for this reason, insects often use impulsive feathering more as a switching mechanism to control the generation of thrust. The influence of in-plane motion is not considered as in this case the second effects are relatively as important as the linear motion, and consequently the linear theory is not representative of the physical situation.

12.4.3 Controlled Flapping for Flight Vehicles

From the analysis presented in the previous subsection it is apparent that pure plunging always leads to a positive thrust coefficient that increases with frequency. Furthermore, although the efficiency of energy conversion is equal to 100% at low frequencies of flapping, it tends to fall to 50% at higher frequencies approaching infinity. Although this finding means that there is no prospect of improving the

propulsive efficiency at low frequencies, there is the possibility of improving the
propulsive efficiency well beyond 50% at high frequencies by combining pure flapping with other modes of oscillation, such as feathering about a fixed axis and lead–
lag motions in the plane of the aerofoil. In most practical situations some feathering
is unavoidable. It is natural to try to maximize the efficiency of energy conversion
to thrust at all frequencies of flapping while holding the average work done in unit
time to maintain oscillations against the generalized force acting on the aerofoil as
a constant.

Considering only the feathering and plunging modes, and assuming that the
feathering angle is a variable, we maximize the thrust with respect to the feathering
angle by setting the derivative of the thrust coefficient with respect to α_0 to zero, i.e.,

$$
\frac{\partial}{\partial \alpha_0} C_T = 2|C(k)|^2 \left[1 + \left(\frac{1}{2} - a \right)^2 k^2 \right] \alpha_{0c}
$$

$$
+ 2 \left\{ \frac{k^2}{4}(1 - 2a) - \left[1 + \left(\frac{1}{2} - a \right) k^2 \right] F - \left(\frac{1}{2} + a \right) Gk \right\} \alpha_{0c}
$$

$$
+ \left[2|C(k)|^2 \left(\frac{1}{2} - a \right) k^2 + \frac{k^2}{2} - Fk^2 + kG \right] \frac{h_{0c}}{b} \cos \varphi_{21}
$$

$$
+ \left[2|C(k)|^2 k + \frac{k^2}{2} - Fk - Gk^2 \right] \frac{h_{0c}}{b} \sin \varphi_{21} = 0.
$$

(12.136)

Hence it follows that

$$
\alpha_{0c} = -\left[(K_1 \cos \varphi_{21} + K_2 \sin \varphi_{21})/2C_{T,\alpha\alpha} \right] (h_{0c}/b),
$$

(12.137)

where α_{0c} is the amplitude of the feathering component in the optimum mode and
h_{0c} is the displacement of the corresponding plunging component in the optimum
mode:

$$
C_{T,\alpha\alpha} = |C(k)|^2 \left[1 + \left(\frac{1}{2} - a \right)^2 k^2 \right]
$$

$$
+ \left\{ \frac{k^2}{4}(1 - 2a) - \left[1 + \left(\frac{1}{2} - a \right) k^2 \right] F - \left(\frac{1}{2} + a \right) Gk \right\},
$$

(12.138)

$$
K_1 = 2|C(k)|^2 \left(\frac{1}{2} - a \right) k^2 + \frac{k^2}{2} - Fk^2 + kG,
$$

(12.139)

$$
K_2 = 2|C(k)|^2 k + \frac{k^2}{2} - Fk - Gk^2.
$$

(12.140)

The corresponding coefficient of the average work done in unit time to maintain
oscillations against the generalized force acting on the aerofoil is

$$
C_{Wc} = Fk \left(\frac{h_{0c}}{b} \right)^2 + \left\{ \frac{k}{2} \left(\frac{1}{2} - a \right) - \left(\frac{1}{2} + a \right) \left[Fk \left(\frac{1}{2} - a \right) + G \right] \right\} \alpha_{0c}^2
$$

$$
+ \left[\left(\frac{k}{2} - 2akF + G \right) \cos \varphi_{21} + (F - kG) \sin \varphi_{21} \right] \alpha_{0c} \frac{h_{0c}}{b}.
$$

(12.141)

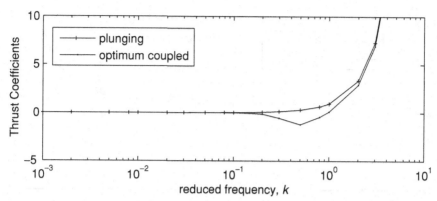

Figure 12.10. Comparison of thrust in plunging and in the optimally coupled case.

To compare with the case of flapping, we set this coefficient equal to the corresponding coefficient for the case of pure plunging with unit amplitude to give

$$C_{Wc} = 0.5 C_{Whh} \equiv Fk. \tag{12.142}$$

We obtain the optimum thrust coefficient by evaluating the thrust coefficient for $\alpha_0 = \alpha_{0c}$ and $h_0 = h_{0c}$, where we obtain h_{0c} by solving Equation (12.142). The thrust coefficient developed in the case of plunging with optimum feathering is compared with the case of pure plunging in Figure 12.10 and is seen to approach the case of pure flapping when the reduced frequency exceeds unity.

The practical implementation of this type of plunging coupled with optimum feathering can be realized in practice by appropriate aeroelastic tailoring. Even with appropriate aeroelastic tailoring, the maximum possible thrust can be realized at only one specific reduced frequency. The methodology presented here opens up the possibility of further increasing the thrust by use of a trailing-edge flap in which the flap is forced to satisfy a maximizing control law.

We may estimate the theoretical maximum thrust that can be developed by using a controlled trailing-edge flap by the preceding method. The maximum thrust is compared with that developed in pure plunging for a trailing-edge flap hinged at midchord in Figure 12.11, which shows an increase in the thrust over a small range of reduced frequencies. As a percentage increase, it is quite significant as it represents more than a 100% increase over a range of frequencies. If we now assume that the aerofoil also feathers in the opposite direction to the flap and by half the flap angle, so the mean camber line is always horizontal, the corresponding thrust coefficients are as in Figure 12.12. The figure shows that any change in the local angle of attack at the leading edge tends to have a detrimental effect on the thrust generated.

12.5 Underwater Propulsion and Its Control

Whereas flapping propulsion in birds involves the generation of thrust accompanied by a significant lift force, fish normally operate in a state of neutral buoyancy and not only generate thrust but also minimize both skin friction and form drag. In 1936 Sir

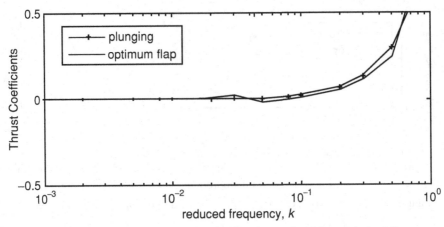

Figure 12.11. Comparison of thrust in plunging and with an optimized flap.

James Gray was able to arrive at the conclusion that, from a rudimentary estimate of the physiological power of a dolphin in a turbulent fluid flow, this power was mathematically insufficient for the dolphin to achieve the speeds mariners had observed them to reach. The boundary layer over a rigid, smooth surface of a similarly shaped body in the same Reynolds number range is known to be turbulent. The power required for maintaining the forward velocity estimated on the basis of the turbulent skin friction is generally known to be far in excess of that which the dolphin uses. Gray's Paradox, as it came to be known, was resolved some 25 years later when the biologist Kramer hypothesized that compliance of dolphin skin would delay water flowing over the skin from separating and becoming turbulent, thus dramatically reducing drag and solving Gray's Paradox. Like earlier scientific announcements that found horses' legs too weak to support their bodies and yet able to carry them and bumblebees anatomically unable to take to the air and yet able to fly effortlessly, the dolphin was found to be able to swim at relatively high speeds not just by generating thrust but by cleverly reducing both the skin friction and the form drag.

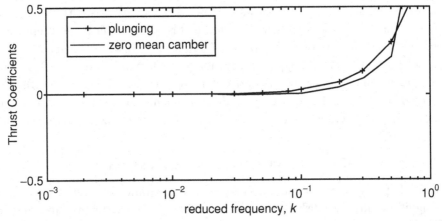

Figure 12.12. Comparison of thrust in plunging and in a mode with varying camber with mean camber line horizontal.

Fish as a class seem to display two distinct mechanisms of thrust generation: body and/or caudal fin (BCF) propulsion and median and/or paired fin (MPF) propulsion. The latter is quite similar to flapping propulsion, whereas the former is based on a different mechanism. Thus we focus primarily on the first of the preceding two mechanisms, which relies on a wavelike displacement traveling downstream in the body, faster than the forward velocity of motion, to generate vorticity in the wake. The vorticity in the wake is responsible for transferring momentum to the flow field, resulting in a net forward thrust. Moreover, one interesting observation is that for relatively shallow but rigid waves the integrated viscous wall shear or viscous drag is substantially less than that for an equivalent flat surface.

We consider a flexible flat-plate aerofoil with a displacement of the form

$$z = z_0 \exp\left[i\omega\left(t - \lambda x/U\right)\right], \tag{12.142}$$

where the ratio of the free-stream velocity to the velocity of the wave is λ. The aerofoil surface velocity or downwash is then given by

$$w_a = \frac{1}{U}\left(i\omega + U\frac{\partial}{\partial x}\right)z = \frac{i\omega b}{U}\left(1 - \lambda\right)\frac{z_0}{b}\exp\left[i\omega\left(t - \lambda x/U\right)\right], \tag{12.143}$$

which may be expressed in nondimensional variables as

$$w_a = w_0 \exp\left(-i\lambda k\xi\right)\exp\left(i\omega t\right), \quad w_0 = \frac{i\omega b}{U}\left(1 - \lambda\right)\frac{z_0}{b}. \tag{12.144}$$

The nondimensional pressure coefficient is

$$\Delta C_p^j(\xi) = 4w_0\left[I^0\left(\xi, \lambda k\right) + I^1\left(\xi, \lambda k\right)\right]\sqrt{\frac{1-\xi}{1+\xi}}$$

$$-\frac{4w_0}{\lambda}I^0\left(\xi, \lambda k\right)\sqrt{1-\xi^2} + 4w_0\left[1 - C\left(k\right)\right]\left[J_0\left(\lambda k\right) - iJ_1\left(\lambda k\right)\right]\sqrt{\frac{1-\xi}{1+\xi}}, \tag{12.145}$$

where the integrals,

$$I^n\left(\xi, \lambda k\right) = \frac{1}{\pi}\int_{-1}^{1}\frac{\xi_1^n}{\sqrt{1-\xi_1^2}}\frac{\exp\left(-i\lambda k\xi_1\right)}{\xi - \xi_1}d\xi_1,$$

are evaluated by use of the series expansion

$$\exp\left(-ik\xi\right) = \exp\left(-ik\cos\theta\right) = J_0\left(k\right) + 2\sum_{n=1}^{\infty}\left(-i\right)^n J_n\left(k\right)\cos\left(n\theta\right).$$

With the notation $U(0) = 0$, $U(1) = 1$, the integrals are

$$I^n\left(\xi, \lambda k\right) = \frac{J_0\left(\lambda k\right)}{\pi}U\left(n\right)\xi - 4\sum_{m=1}^{\infty}\left(-1\right)^m J_m\left(\lambda k\right)\frac{\sin m(\cos^{-1}\xi)}{\sin(\cos^{-1}\xi)}\xi^n, \quad n = 0, 1.$$

The average work done in unit time to maintain oscillations against the generalized force acting on the aerofoil $\bar{W}$ may be found by averaging over a cycle

$$W = -\frac{1}{2}\rho U^2\int_{-b}^{b}C_p\frac{\partial z_a\left(x, y, t\right)}{\partial t}d\xi. \tag{12.146}$$

The average propulsive force may be found by averaging the expression for P given by Equation (12.127). The evaluations must necessarily be performed by use of appropriate numerical methods. The details of the evaluations of the lift and the pitching moment, $\bar{P}$ and $\bar{W}$, are left as an exercise for the reader. The results show that $\bar{P}$ and $\bar{W}$ vanish at $k = 0$, and, for large values of k, the magnitudes of $\bar{P}$ and $\bar{W}$ behave like k^2 and k^3, respectively.

The basic mechanism of propulsion and its control is identical to the case of the flapping aerofoil and is a consequence of the momentum gain that is due the vortex wake that results in the fluid's being expelled behind the trailing edge of the aerofoil. The magnitude of the thrust generated is not only a function of the reduced frequency and the traveling-wave amplitude but also of the traveling-wave velocity. In most fish a further increase in the thrust is possible by use of natural lubricants to reduce the skin friction drag. Fish mucus is secreted by cells of the epidermis that reduce the water resistance up to 60%, thereby effectively increasing the thrust. However, during an impulsive start, the frictional drag represents only a small percentage of the total drag and it is the total mass of the system, including the virtual mass of the fluid that makes a dominating contribution to it.

Drag-reducing mechanisms are invoked whenever it is necessary to gain acceleration, and these involve a reduction of mass that is not essential. These reductions include control of nonmuscle tissue density, which reduces the total dead weight that must be accelerated relative to the essential mass of the muscle motor. This reduction results in a increase in the power-to-load ratio, which is a typical metric that is indicative of the reduction in resistance. Although all density-saving adaptations are assumed to be due to the effects of neutral buoyancy, which will contribute to the increase in acceleration, mass-reducing mechanisms in the skin are not due to the forces of buoyancy. Thus fish are able to exercise control over the propulsive thrust by a variety of mechanisms.

EXERCISES

12.1. Consider the dynamic walking model described in Subsection 12.2.1. Assuming the case in which the support is provided by the left leg, show that the position constraint equations are

$$x_0 = r_0 \sin \alpha + l_1 \sin (\alpha - \beta_L) + l_2 \sin (\alpha - \beta_L + \gamma_L),$$
$$y_0 = r_0 \cos \alpha + l_1 \cos (\alpha - \beta_L) + l_2 \cos (\alpha - \beta_L + \gamma_L).$$

Hence obtain the velocity and acceleration constraint equations.

12.2. Consider the dynamic walking model described in Subsection 12.2.1 and assume that the support is provided by both legs. Show that the position constraints are given by Equations (12.25) and (12.26) and hence obtain the velocity and acceleration constraint equations.

12.3. Consider the "compass gait" model used for gait simulation illustrated in Figure 12.13.

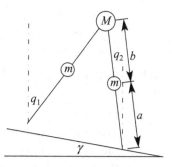

Figure 12.13. The compass gait model.

The dynamic equations governing the dynamics may be expressed as

$$\mathbf{I}(\mathbf{q})\,\ddot{\mathbf{q}} + \mathbf{C}(\mathbf{q}, \dot{\mathbf{q}})\,\dot{\mathbf{q}} - g\Gamma_g(\mathbf{q}) = 0,$$

where $\mathbf{q} = [q_1 \quad q_2]^T$ is the vector of the swing and support leg angles, respectively, measured relative to the inertial vertical for both links, which are assumed to be identical.

(a) Show that

$$I(\mathbf{q}) = \begin{bmatrix} mb^2 & -m(a+b)b\cos(2\alpha) \\ -m(a+b)b\cos(2\alpha) & (m+M)(a+b)^2 + ma^2 \end{bmatrix},$$

$$\mathbf{C}(\mathbf{q}, \dot{\mathbf{q}}) = \begin{bmatrix} 0 & m(a+b)b\sin(2\alpha)\,\dot{q}_2 \\ m(a+b)b\sin(2\alpha)\,\dot{q}_1 & 0 \end{bmatrix},$$

$$\Gamma_g(\mathbf{q}) = \begin{bmatrix} -mb\sin(q_1) \\ ((M+m)(a+b) + ma)\sin(q_2) \end{bmatrix}, \quad \text{where, } \alpha = \frac{q_2 - q_1}{2}.$$

(b) The foot collision results in an instantaneous change of velocity governed by the conservation of angular momentum around the point of impact. Assuming that the swing foot can swing through the ground, show that the conservation relation may be expressed as

$$\mathbf{Q}^+(\alpha)\,\dot{\mathbf{q}}^+ = \mathbf{Q}^-(\alpha)\,\dot{\mathbf{q}}^-,$$

where

$$\mathbf{Q}^+(\alpha) = \mathbf{Q}_1^+ + \mathbf{Q}_2^+, \mathbf{Q}^-(\alpha) = \mathbf{Q}_1^- + \mathbf{Q}_2^-,$$

$$\mathbf{Q}_1^+ = -mab \begin{bmatrix} 1 & 1 \\ 0 & 1 \end{bmatrix},$$

$$\mathbf{Q}_2^+ = (a+b)\cos(2\alpha) \begin{bmatrix} 0 & M(a+b) + 2ma \\ 0 & 0 \end{bmatrix},$$

$$\mathbf{Q}_1^- = -mb(a+b)\cos(2\alpha) \begin{bmatrix} 1 & 1 \\ 0 & 1 \end{bmatrix},$$

$$\mathbf{Q}_2^- = \begin{bmatrix} mb^2 & (M+m)(a+b)^2 + ma^2 \\ mb^2 & 0 \end{bmatrix}.$$

(c)   Numerically simulate the response of the compass gait model by using a typical numerical integration method and demonstrate the existence of stable limit cycles.

12.4. Consider the half-model of the quadruped robot described in Subsection 12.2.3.

(a)   Derive the governing dynamical equations of motion.

(b)   The foot collision results in an instantaneous change of velocity governed by the conservation of angular momentum around the point of impact. Obtain the relevant conservation relations.

(c)   Numerically simulate the response of the model by using a typical numerical integration method and demonstrate the gait patterns.

12.5. A certain remotely piloted vehicle is modeled (Figure 12.14) after an aquatic animal as a central two-dimensional body (point mass) that propels itself by an articulated three-link mechanism. The final link in the mechanism is acted on by a single force, F.

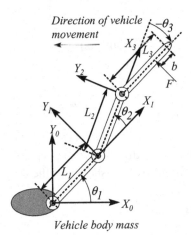

Figure 12.14. Remotely piloted vehicle modeled after an aquatic cetacean animal (the bottlenose dolphin, *Tursiops truncatus*).

(a)   Assume that the mass of the central body is M and that the masses of each of the links (assumed uniform) are m_i. Ignore all gravitational effects and obtain expressions for the kinetic energy and the Lagrangian dynamical equations.

(b)   For purposes of thrust generation, assume that the angle θ_2 is constant and equal to zero. Assume that the body and the trailing link represent two-dimensional aerofoils capable of flapping at a high frequency and obtain an expression for the thrust developed by the prototype model.

(c)   Suitably modify the equations of motion obtained in part (a) to include the mean thrust and simulate the motion of the prototype vehicle. Include the aerodynamic forces and moments of the mean motion of aerofoils in your dynamic model and simulation.

12.6. Consider the pressure acting on the surface of a flexible flat-plate aerofoil performing traveling-wave oscillations and show that the lift is given by

$$L(k) = 4\pi \left(\frac{1}{2}\rho U^2\right) bw_0 K(k, \lambda) \exp(i\omega t),$$

where

$$K(k, \lambda) = [J_0(\lambda k) - iJ_1(\lambda k)] C(k) + (i/\lambda) J_1(\lambda k).$$

Answers to Selected Exercises

Chapter 3

3.1. $[10.366 \quad 2.098 \quad 11 \quad 1]^T$; **3.3.** $[-1 \quad 2 \quad -2]^T$;

3.4. If $\mathbf{T} = \begin{bmatrix} \mathbf{R} & \mathbf{d} \\ \mathbf{0} & 1 \end{bmatrix}$, $\mathbf{R}_1 = \mathbf{R}_2 = \begin{bmatrix} 0 & 1 & 0 \\ 0 & 0 & -1 \\ -1 & 0 & 0 \end{bmatrix}$, $\mathbf{d}_1 = \begin{bmatrix} 6 \\ 8 \\ 7 \end{bmatrix}$, $\mathbf{d}_2 = \begin{bmatrix} 8 \\ -7 \\ -6 \end{bmatrix}$;

3.5. Second; **3.6.** $\begin{bmatrix} 0 & 0 & -1 & 3 \\ 1 & 0 & 0 & -2 \\ 0 & -1 & 0 & 1 \\ 0 & 0 & 0 & 1 \end{bmatrix}$; **3.7.** $\begin{bmatrix} 0 & 1 & 0 & 6 \\ 0 & 0 & -1 & 9 \\ -1 & 0 & 0 & 2 \\ 0 & 0 & 0 & 1 \end{bmatrix}$;

3.8. $\begin{bmatrix} \cos\theta_3 & -\sin\theta_3 & 0 & L\cos\theta_3 + L\cos\theta_1 \\ \sin\theta_3 & \cos\theta_3 & 0 & L\cos\theta_3 + L\sin\theta_1 \\ 0 & 0 & 1 & 0 \\ 0 & 0 & 0 & 1 \end{bmatrix}$, $\theta_3 = \theta_1 + \theta_2$;

3.9. $\begin{bmatrix} c_4c_5c_6 - s_4s_6 & -c_4c_5s_6 - s_4c_6 & c_4s_5 & d_6c_4s_5 \\ s_4c_5c_6 + c_4s_6 & -s_4c_5s_6 + c_4c_6 & s_4s_5 & d_6s_4s_5 \\ -s_5c_6 & s_5s_6 & -c_5 & -d_6c_5 \\ 0 & 0 & 0 & 1 \end{bmatrix}$;

3.10. $\begin{bmatrix} 0 & -1 & 0 & 1 \\ 1 & 0 & 0 & -1 \\ 0 & 0 & 1 & 1 \\ 0 & 0 & 0 & 1 \end{bmatrix}$, $\sqrt{3}, 0°, [0 \quad 0 \quad 0]^T$;

3.11. $\frac{1}{\sqrt{3}}[1 \quad -1 \quad 1]^T,$ $\begin{bmatrix} 0 & 1 & 0 & x \\ 0 & 0 & -1 & y \\ -1 & 0 & 0 & z \\ 0 & 0 & 0 & 1 \end{bmatrix};$

3.12. (a) $\frac{1}{\sqrt{2}}[1 \quad 1 \quad 0]^T,$ (b) 4.

Chapter 4

Answers to the exercises of this chapter will depend on your choice of coordinates.

4.1. $\begin{bmatrix} c_{123} & -s_{123} & 0 & L_3 c_{123} + L_2 c_{12} + L_1 c_1 \\ s_{123} & c_{123} & 0 & L_3 s_{123} + L_2 s_{12} + L_1 c_1 \\ 0 & 0 & 1 & 0 \\ 0 & 0 & 0 & 1 \end{bmatrix};$

4.2. $\begin{bmatrix} 0 & 0 & 1 & d_2 \\ -1 & 0 & 0 & 0 \\ 0 & -1 & 0 & d_1 \\ 0 & 0 & 0 & 1 \end{bmatrix};$ **4.3.** $\begin{bmatrix} c_1 & 0 & -s_1 & a_1 c_1 - d_2 s_1 \\ s_1 & 0 & c_1 & d_2 c_1 + a_1 s_1 \\ 0 & c_2 & 0 & 0 \\ 0 & 0 & 0 & 1 \end{bmatrix};$

4.4. $\begin{bmatrix} c_{13} & -s_{13} & 0 & a_3 c_{13} - d_2 s_1 + a_1 c_1 \\ s_{13} & c_{13} & 0 & a_3 s_{13} + d_2 c_1 + a_1 s_1 \\ 0 & 0 & 1 & 0 \\ 0 & 0 & 0 & 1 \end{bmatrix};$

4.7. $\begin{bmatrix} c_{123} & s_{123} & 0 & a_3 c_{123} + a_2 c_{12} + a_1 c_1 \\ s_{123} & -c_{123} & 0 & a_3 s_{123} + a_2 s_{12} + a_1 s_1 \\ 0 & 0 & 1 & d_2 + h - d_4 \\ 0 & 0 & 0 & 1 \end{bmatrix}.$

Chapter 5

Answers to the exercises of this chapter will depend on your choice of coordinates.

5.1. $q_1 = \cos\left(\frac{\theta_4}{2}\right) \sin\left(\frac{\theta_5}{2}\right) \sin\left(\frac{\theta_6}{2}\right) - \sin\left(\frac{\theta_4}{2}\right) \sin\left(\frac{\theta_5}{2}\right) \cos\left(\frac{\theta_6}{2}\right),$

$q_2 = \cos\left(\frac{\theta_4}{2}\right) \sin\left(\frac{\theta_5}{2}\right) \cos\left(\frac{\theta_6}{2}\right) + \sin\left(\frac{\theta_4}{2}\right) \sin\left(\frac{\theta_5}{2}\right) \sin\left(\frac{\theta_6}{2}\right),$

$q_3 = -\cos\left(\frac{\theta_4}{2}\right) \cos\left(\frac{\theta_5}{2}\right) \sin\left(\frac{\theta_6}{2}\right) - \sin\left(\frac{\theta_4}{2}\right) \cos\left(\frac{\theta_5}{2}\right) \cos\left(\frac{\theta_6}{2}\right),$

$$q_4 = \cos\left(\frac{\theta_4}{2}\right)\cos\left(\frac{\theta_5}{2}\right)\cos\left(\frac{\theta_6}{2}\right) - \sin\left(\frac{\theta_4}{2}\right)\cos\left(\frac{\theta_5}{2}\right)\sin\left(\frac{\theta_6}{2}\right),$$

$$\tan\theta_6 = -\frac{T_{32}}{T_{31}} = \frac{q_1 q_4 - q_2 q_3}{q_1 q_3 + q_2 q_4}, \tan(\theta_4) = \frac{T_{23}}{T_{13}} = \frac{q_1 q_4 + q_2 q_3}{q_1 q_3 - q_2 q_4},$$

$$\cos(\theta_5) = q_3^2 + q_4^2 - q_1^2 - q_2^2;$$

5.2. 4.1:
$$\begin{bmatrix} x_{03} \\ y_{03} \\ z_{03} \end{bmatrix} = \begin{bmatrix} a_3 c_3 c_{12} + a_2 c_2 c_1 + a_1 c_1 \\ a_3 c_3 s_{12} + a_2 c_2 s_1 \\ 0 \end{bmatrix};$$

4.2:
$$\begin{bmatrix} x_{03} \\ y_{03} \\ z_{03} \end{bmatrix} = \begin{bmatrix} a_1 c_1 - d_2 s_1 \\ a_1 s_1 + d_2 c_1 \\ 0 \end{bmatrix}; \quad 4.3: \begin{bmatrix} x_{02} \\ y_{02} \\ z_{02} \end{bmatrix} = \begin{bmatrix} a_1 c_1 - d_2 s_1 \\ a_1 s_1 + d_2 c_1 \\ 0 \end{bmatrix};$$

4.4:
$$\begin{bmatrix} x_{03} \\ y_{03} \\ z_{03} \end{bmatrix} = \begin{bmatrix} a_3 c_{13} + a_1 c_1 - d_2 s_1 \\ a_3 s_{13} + a_1 s_1 - d_2 c_1 \\ 0 \end{bmatrix};$$

4.6:
$$\begin{bmatrix} x \\ y \\ z \end{bmatrix} = \begin{bmatrix} a_4 c_{1234} + a_3 c_{123} + a_2 c_{12} + a_1 c_1 \\ a_4 s_{1234} + a_3 s_{123} + a_2 s_{12} + a_1 s_1 \\ 0 \end{bmatrix};$$

5.4. (a) $b_i^2 = (x_{Bi} - x_{Pi})^2 + (y_{Bi} - y_{Pi})$, (x_{Bi}, y_{Bi}), are the base triangle and platform triangle corner coordinates, respectively,

$$\begin{bmatrix} x_{P1} & x_{P2} & x_{P3} \\ y_{P1} & y_{P2} & y_{P3} \\ 1 & 1 & 1 \end{bmatrix} = \begin{bmatrix} c\phi & s\phi & x_d \\ -s\phi & c\phi & y_d \\ 0 & 0 & 1 \end{bmatrix}\begin{bmatrix} 0 & a/2 & -a/2 \\ a/\sqrt{3} & a/2\sqrt{3} & a/2\sqrt{3} \\ 1 & 1 & 1 \end{bmatrix},$$

and

$$\begin{bmatrix} x_{B1} & x_{B2} & x_{B3} \\ y_{B1} & y_{B2} & y_{B3} \\ 1 & 1 & 1 \end{bmatrix} = \begin{bmatrix} 0 & L/2 & -L/2 \\ L/\sqrt{3} & L/2\sqrt{3} & L/2\sqrt{3} \\ 1 & 1 & 1 \end{bmatrix}.$$

Chapter 7

7.4. $\mathbf{J} = \begin{bmatrix} \mathbf{z}_0 & \mathbf{z}_0 & \mathbf{z}_0 \\ \mathbf{z}_0 \times \mathbf{r}_3 & \mathbf{z}_0 \times (\mathbf{r}_3 - \mathbf{r}_1) & \mathbf{z}_0 \times (\mathbf{r}_3 - \mathbf{r}_2 - \mathbf{r}_1) \end{bmatrix},$

where

$$\mathbf{z}_0 = \begin{bmatrix} 0 \\ 0 \\ 1 \end{bmatrix}, \ \mathbf{r}_1 = \begin{bmatrix} a_1c_1 \\ a_1s_1 \\ 0 \end{bmatrix}, \ \mathbf{r}_2 = \begin{bmatrix} a_2c_{12} \\ a_2s_{12} \\ 0 \end{bmatrix}$$

and

$$\mathbf{r}_3 = \begin{bmatrix} a_1c_1 + a_2c_{12} + a_3c_{123} \\ a_1s_1 + a_2s_{12} + a_3s_{123} \\ 0 \end{bmatrix}.$$

7.6. $\begin{bmatrix} \mathbf{0} & \mathbf{0} \\ \mathbf{z}_2 & \mathbf{z}_1 \end{bmatrix}, \ \mathbf{z}_1 = \begin{bmatrix} 1 \\ 0 \\ 0 \end{bmatrix}, \ \mathbf{z}_2 = \begin{bmatrix} 0 \\ 1 \\ 0 \end{bmatrix},$

$$\begin{bmatrix} \mathbf{z}_0 & \mathbf{0} \\ (a_1 + a_2)\,\mathbf{z}_0 \times \mathbf{z}_d & \mathbf{z}_d \end{bmatrix}, \ \mathbf{z}_0 = \begin{bmatrix} 0 \\ 0 \\ 1 \end{bmatrix}, \ \mathbf{z}_d = \begin{bmatrix} c_1 \\ s_1 \\ 1 \end{bmatrix};$$

7.8. $\begin{bmatrix} \mathbf{z}_0 & \mathbf{0} & \mathbf{z}_0 \\ \mathbf{z}_0 \times \mathbf{r}_1 & \mathbf{z}_d & \mathbf{z}_0 \times \mathbf{r}_3 \end{bmatrix}, \ \mathbf{z}_0 = \begin{bmatrix} 0 \\ 0 \\ 1 \end{bmatrix}, \ \mathbf{z}_d = \begin{bmatrix} c_1 \\ s_1 \\ 1 \end{bmatrix},$

$$\mathbf{r}_1 = \begin{bmatrix} a_1c_1 + a_2c_1 + a_3c_{12} \\ a_1s_1 + a_2s_1 + a_3s_{12} \\ 0 \end{bmatrix}, \ \mathbf{r}_3 = \begin{bmatrix} a_3c_{12} \\ a_3s_{12} \\ 0 \end{bmatrix}.$$

Chapter 8

8.7. (b) $T = \dfrac{1}{2}mL^2\left[\dot{\theta}^2 + \dfrac{\dot{\phi}^2}{3} + \dot{\phi}\dot{\theta}\cos(\theta - \phi)\right],$

$$V = mg\left[L(1 - \cos q_1) + \dfrac{L}{2}(1 - \cos q_2)\right];$$

8.7. (c) The two Euler–Lagrange equations of motion are

$$mL^2\left[\ddot{q}_1 + \dfrac{\ddot{q}_2}{2}\cos(q_1 - q_2) + \dfrac{\dot{q}_2^2}{2}\sin(q_1 - q_2)\right] + mgL\sin q_1 = 0,$$

$$mL^2\left[\dfrac{\ddot{q}_1}{2}\cos(q_1 - q_2) + \dfrac{\ddot{q}_2}{3} - \dfrac{\dot{q}_1^2}{2}\sin(q_1 - q_2)\right] + \dfrac{mgL}{2}\sin q_2 = 0;$$

8.7. (d) To linearize, we set $\cos(q_1 - q_2) \approx 1$, $\sin(q_1 - q_2) \approx 0$, $\sin q_2 \approx q_2$, and $\sin q_1 \approx q_1$. The linearized equations are

$$mL^2\left(\ddot{q}_1 + \frac{\ddot{q}_2}{2}\right) + mgLq_1 = 0, \; mL^2\left(\frac{\ddot{q}_1}{2} + \frac{\ddot{q}_2}{3}\right) + \frac{mgL}{2}q_2 = 0,$$

$$\frac{mL^2}{12}\begin{bmatrix} 12 & 6 \\ 6 & 4 \end{bmatrix}\begin{bmatrix} \ddot{q}_1 \\ \ddot{q}_2 \end{bmatrix} + \frac{mgL}{2}\begin{bmatrix} 2 & 0 \\ 0 & 1 \end{bmatrix}\begin{bmatrix} q_1 \\ q_2 \end{bmatrix} = \begin{bmatrix} 0 \\ 0 \end{bmatrix};$$

8.8. (a) $T = \frac{2\pi L}{a}\sqrt{\frac{2h}{3g}}$; (b) $T = 2\pi\sqrt{\frac{h}{g}}$.

Chapter 10

10.1. $L = T - V, T = \frac{1}{2}mL^2\left[\dot{\theta}^2 + \frac{\dot{\phi}^2}{3} + \dot{\phi}\dot{\theta}\cos(\theta - \phi)\right]$,

$$V = mg\left[L(1 - \cos q_1) + \frac{L}{2}(1 - \cos q_2)\right];$$

10.2. $V = g\Gamma_{11}\sin\theta_1 + g\Gamma_{22}\sin(\theta_1 + \theta_2)$,

where

$$\Gamma_{11} = (m_1 L_{1cg} + m_2 L_1 + ML_1), \; \Gamma_{22} = (m_2 L_{2cg} + ML_2).$$

$$T = \frac{1}{2}(I_{11} - I_{21})\dot{\theta}_1^2 + \frac{1}{2}I_{22}(\dot{\theta}_1 + \dot{\theta}_2)^2 + I_{21}\dot{\theta}_1(\dot{\theta}_1 + \dot{\theta}_2),$$

where

$$I_{11} = m_1\left(L_{1cg}^2 + k_{1cg}^2\right) + (m_2 + M)L_1^2 + (m_2 L_{2cg} + ML_2)L_1\cos(\theta_2),$$

$$I_{21} = (m_2 L_{2cg} + ML_2)L_1\cos(\theta_2), \; I_{22} = m_2\left(L_{2cg}^2 + k_{2cg}^2\right) + ML_2^2,$$

$$I_{12} = m_2\left(L_{2cg}^2 + k_{2cg}^2\right) + ML_2^2 + (m_2 L_{2cg} + ML_2)L_1\cos(\theta_2) = I_{22} + I_{12}.$$

The Lagrangian is defined as $L = T - V$.

Chapter 12

12.1. Velocity constraints:

$$
\begin{bmatrix} \dot{x}_0 \\ \dot{y}_0 \end{bmatrix} + \begin{bmatrix} l_1 \cos(\alpha - \beta_L) + l_2 \cos(\alpha - \beta_L + \gamma_L) & 0 \\ -l_1 \sin(\alpha - \beta_L) - l_2 \sin(\alpha - \beta_L + \gamma_L) & 0 \end{bmatrix} \begin{bmatrix} \dot{\beta}_L \\ \dot{\beta}_R \end{bmatrix}
$$

$$
\times \begin{bmatrix} -r_0 \cos\alpha - l_1 \cos(\alpha - \beta_L) - l_2 \cos(\alpha - \beta_L + \gamma_L) \\ r_0 \sin\alpha + l_1 \sin(\alpha - \beta_L) + l_2 \sin(\alpha - \beta_L + \gamma_L) \end{bmatrix} \dot{\alpha}
$$

$$
+ \begin{bmatrix} -l_2 \cos(\alpha - \beta_L + \gamma_L) & 0 \\ l_2 \sin(\alpha - \beta_L + \gamma_L) & 0 \end{bmatrix} \begin{bmatrix} \dot{\gamma}_L \\ \dot{\gamma}_R \end{bmatrix} = 0.
$$

Acceleration constraints:

$$
\begin{bmatrix} \ddot{x}_0 \\ \ddot{y}_0 \end{bmatrix} + \begin{bmatrix} l_1 \cos(\alpha - \beta_L) + l_2 \cos(\alpha - \beta_L + \gamma_L) & 0 \\ -l_1 \sin(\alpha - \beta_L) - l_2 \sin(\alpha - \beta_L + \gamma_L) & 0 \end{bmatrix} \begin{bmatrix} \ddot{\beta}_L \\ \ddot{\beta}_R \end{bmatrix}
$$

$$
\times \begin{bmatrix} -r_0 \cos\alpha - l_1 \cos(\alpha - \beta_L) - l_2 \cos(\alpha - \beta_L + \gamma_L) \\ r_0 \sin\alpha + l_1 \sin(\alpha - \beta_L) + l_2 \sin(\alpha - \beta_L + \gamma_L) \end{bmatrix} \ddot{\alpha}
$$

$$
+ \begin{bmatrix} -l_2 \cos(\alpha - \beta_L + \gamma_L) & 0 \\ l_2 \sin(\alpha - \beta_L + \gamma_L) & 0 \end{bmatrix} \begin{bmatrix} \ddot{\gamma}_L \\ \ddot{\gamma}_R \end{bmatrix}
$$

$$
= \begin{bmatrix} -r_0 \dot{\alpha}^2 \sin\alpha - l_1(\dot{\alpha} - \dot{\beta}_L)^2 \sin(\alpha - \beta_L) - l_2(\dot{\alpha} - \dot{\beta}_L + \dot{\gamma}_L)^2 \sin(\alpha - \beta_L + \gamma_L) \\ -r_0 \dot{\alpha}^2 \cos\alpha - l_1(\dot{\alpha} - \dot{\beta}_L)^2 \cos(\alpha - \beta_L) - l_2(\dot{\alpha} - \dot{\beta}_L + \dot{\gamma}_L)^2 \cos(\alpha - \beta_L + \gamma_L) \end{bmatrix}.
$$

12.2. Velocity constraints:

$$
\mathbf{L} \begin{bmatrix} \dot{x}_0 \\ \dot{y}_0 \\ \dot{\beta}_L \\ \dot{\beta}_R \end{bmatrix} + \begin{bmatrix} -r_0 \cos\alpha - l_1 \cos(\alpha - \beta_L) - l_2 \cos(\alpha - \beta_L + \gamma_L) \\ r_0 \sin\alpha + l_1 \sin(\alpha - \beta_L) + l_2 \sin(\alpha - \beta_L + \gamma_L) \\ -r_0 \cos\alpha - l_1 \cos(\alpha - \beta_R) - l_2 \cos(\alpha - \beta_R + \gamma_R) \\ r_0 \sin\alpha + l_1 \sin(\alpha - \beta_R) + l_2 \sin(\alpha - \beta_R + \gamma_R) \end{bmatrix} \dot{\alpha}
$$

$$
+ \begin{bmatrix} -l_2 \cos(\alpha - \beta_L + \gamma_L) & 0 \\ l_2 \sin(\alpha - \beta_L + \gamma_L) & 0 \\ 0 & -l_2 \cos(\alpha - \beta_R + \gamma_R) \\ 0 & l_2 \sin(\alpha - \beta_R + \gamma_R) \end{bmatrix} \begin{bmatrix} \dot{\gamma}_L \\ \dot{\gamma}_R \end{bmatrix} = 0,
$$

where

$$
\mathbf{L} = \begin{bmatrix} 1 & 0 & l_1 \cos(\alpha - \beta_L) + l_2 \cos(\alpha - \beta_L + \gamma_L) & 0 \\ 0 & 1 & -l_1 \sin(\alpha - \beta_L) - l_2 \sin(\alpha - \beta_L + \gamma_L) & 0 \\ 1 & 0 & 0 & l_1 \cos(\alpha - \beta_R) + l_2 \cos(\alpha - \beta_R + \gamma_R) \\ 0 & 1 & 0 & -l_1 \sin(\alpha - \beta_R) - l_2 \sin(\alpha - \beta_R + \gamma_R) \end{bmatrix}
$$

Acceleration constraints:

$$
\mathbf{L}
\begin{bmatrix} \ddot{x}_0 \\ \ddot{y}_0 \\ \ddot{\beta}_L \\ \ddot{\beta}_R \end{bmatrix}
+
\begin{bmatrix}
-r_0 \cos \alpha - l_1 \cos(\alpha - \beta_L) - l_2 \cos(\alpha - \beta_L + \gamma_L) \\
r_0 \sin \alpha + l_1 \sin(\alpha - \beta_L) + l_2 \sin(\alpha - \beta_L + \gamma_L) \\
-r_0 \cos \alpha - l_1 \cos(\alpha - \beta_R) - l_2 \cos(\alpha - \beta_R + \gamma_R) \\
r_0 \sin \alpha + l_1 \sin(\alpha - \beta_R) + l_2 \sin(\alpha - \beta_R + \gamma_R)
\end{bmatrix} \ddot{\alpha}
$$

$$
+
\begin{bmatrix}
-l_2 \cos(\alpha - \beta_L + \gamma_L) & 0 \\
l_2 \sin(\alpha - \beta_L + \gamma_L) & 0 \\
0 & -l_2 \cos(\alpha - \beta_R + \gamma_R) \\
0 & l_2 \sin(\alpha - \beta_R + \gamma_R)
\end{bmatrix}
\begin{bmatrix} \ddot{\gamma}_L \\ \ddot{\gamma}_R \end{bmatrix}
$$

$$
=
\begin{bmatrix}
-r_0 \dot{\alpha}^2 \sin \alpha - l_1(\dot{\alpha} - \dot{\beta}_L)^2 \sin(\alpha - \beta_L) - l_2(\dot{\alpha} - \dot{\beta}_L + \dot{\gamma}_L)^2 \sin(\alpha - \beta_L + \gamma_L) \\
-r_0 \dot{\alpha}^2 \cos \alpha - l_1(\dot{\alpha} - \dot{\beta}_L)^2 \cos(\alpha - \beta_L) - l_2(\dot{\alpha} - \dot{\beta}_L + \dot{\gamma}_L)^2 \cos(\alpha - \beta_L + \gamma_L) \\
-r_0 \dot{\alpha}^2 \sin \alpha - l_1(\dot{\alpha} - \dot{\beta}_R)^2 \sin(\alpha - \beta_R) - l_2(\dot{\alpha} - \dot{\beta}_R + \dot{\gamma}_R)^2 \sin(\alpha - \beta_R + \gamma_R) \\
-r_0 \dot{\alpha}^2 \cos \alpha - l_1(\dot{\alpha} - \dot{\beta}_R)^2 \cos(\alpha - \beta_R) - l_2(\dot{\alpha} - \dot{\beta}_R + \dot{\gamma}_R)^2 \cos(\alpha - \beta_R + \gamma_R)
\end{bmatrix}.
$$

Appendix: Attitude and Quaternions

A.1 Defining Attitude: Frames of Reference

The motion of a rigid body can be specified by its position and attitude. The first quantity describes the translational motion of the center of mass of the body. The latter quantity describes the rotational motion of the body about an axis passing through the center of mass. In general, the position and attitude are independent. The attitudinal motion can be considered to consist of two aspects: attitude kinematics and attitude dynamics. Attitude kinematics is the fundamental description of changes in attitude with time without any consideration of the torques applied to the body. On the other hand, attitude dynamics refers to the motion of a body in response to applied torques. The main aspect in the kinematics of attitude is the description of the attitude or orientation of a body in space, which is the subject of this appendix. To define the orientation of the body in space, we begin by defining three mutually perpendicular axes fixed in the body at its center of gravity. The *body axes* are a right-handed triple of orthogonal axes. The orientation or attitude of a rigid body is then defined as the orientation of the body axes relative to a set of reference axes. A frame fixed in space defines a reference axes system and is said to be an inertial frame. The *reference axes* are a right-handed triple of orthogonal axes. Typically a set of three mutually orthogonal axes constitutes a right-handed reference frame when there is a right-handed clockwise screw rotation of the x axis toward the y axis points to the z axis. All the reference frames considered are assumed to be right-handed reference frames. A number of reference frames are used to define the position and orientation of a rigid body.

A.1.1 Inertial and Noninertial Frames

The inertial reference frame is usually an orthogonal frame consisting of three mutually perpendicular axes fixed in space and with its origin coinciding with the center of the Earth. By convention, the x axis of this frame points to a fixed point in space known as the first point in Aries, which lies in the equatorial plane of the Earth as well as in the plane containing the Greenwich meridian. The Earth-fixed reference frame is fixed in the Earth and also has its origin at the center of the Earth. The Earth-fixed geographic reference frame has its axes pointing in the actual local north, east, and downward directions, respectively. The origin of the frame is fixed locally at a point on the Earth's surface defined by the local latitude and longitude. In most robotic applications, however, the Earth is assumed to be nonrotating, and it follows that any Earth-fixed geographic frame is also an inertial frame fixed in space.

A.2 Rotating Frames of Reference

In the description of the kinematics of robot manipulators, one always needs to consider the rotation of one rigid body relative to another. When a reference frame fixed in a body is rotating relative to another or relative to a fixed reference frame such as a local inertial frame, the relationship between the two reference frames or *coordinate* frames may be described by a *transformation*. In the study of robot manipulators it is most important to be able to synthesize this transformation and to parameterize it with a minimum number of independent quantities. To this end we consider the resolution of a vector in two and three dimensions.

A.2.1 Resolution of a Position Vector in Two and Three Directions

Considering the position vector, we find that the components along the x_0 and y_0 axes are

$$R_x = |\mathbf{R}| \cos\theta, \quad R_y = |\mathbf{R}| \sin\theta = |\mathbf{R}| \cos\left(\frac{\pi}{2} - \theta\right),$$

where $|\mathbf{R}|$ is the magnitude of the vector $\mathbf{R}$. They be expressed in general as

$$R_x = |\mathbf{R}| l, \quad R_y = |\mathbf{R}| m.$$

The *direction cosines* l and m are then defined as

$$l = \frac{R_x}{|\mathbf{R}|} = \cos\theta, \quad m = \frac{R_y}{|\mathbf{R}|} = \cos\left(\frac{\pi}{2} - \theta\right)$$

and satisfy the condition

$$l^2 + m^2 = 1.$$

Thus, given unit vectors $\mathbf{i}_0$ and $\mathbf{j}_0$ in the direction of the Cartesian axes x_0 and y_0, the position vector itself may be expressed as a vector sum of its components:

$$\mathbf{R} = R_x\mathbf{i}_0 + R_y\mathbf{j}_0 = |\mathbf{R}| l\mathbf{i}_0 + |\mathbf{R}| m\mathbf{j}_0.$$

The definition of direction cosines may be extended to the case of a three-dimensional Cartesian reference frame. Thus, in this case, the position vector may be expressed in terms of unit vectors in the three mutually perpendicular directions as

$$\mathbf{R} = R_x\mathbf{i}_0 + R_y\mathbf{j}_0 + R_z\mathbf{k}_0 = |\mathbf{R}| l\mathbf{i}_0 + |\mathbf{R}| m\mathbf{j}_0 + |\mathbf{R}| n\mathbf{k}_0,$$

and it follows that the *direction cosines l, m,* and *n* are then defined as

$$l = \frac{R_x}{|\mathbf{R}|}, \quad m = \frac{R_y}{|\mathbf{R}|}, \quad n = \frac{R_z}{|\mathbf{R}|}$$

and satisfy the condition

$$l^2 + m^2 + n^2 = 1.$$

A.2.2 Rotations in Two Dimensions

The rotation of a two-dimensional reference frame in the plane of a paper is considered first. The angle of rotation is denoted by ϕ. Let the position vector be denoted by $\mathbf{R}$, the Cartesian component vectors in the reference frame (x_0, y_0) be denoted as R_{x0} and R_{y0}, and the Cartesian component vectors in the rotated reference frame (x_1, y_1) be denoted as R_{x1} and R_{y1}. The components of the position vector in the rotated reference frame change because of the rotation of the reference frame. The two components after rotation are

$$\begin{bmatrix} R_{1x} \\ R_{1y} \end{bmatrix} = |\mathbf{R}| \begin{bmatrix} \cos(\theta - \varphi) \\ \sin(\theta - \varphi) \end{bmatrix} = \begin{bmatrix} |\mathbf{R}| (\cos\theta \cos\varphi + \sin\theta \sin\varphi) \\ |\mathbf{R}| (\sin\theta \cos\varphi - \cos\theta \sin\varphi) \end{bmatrix}.$$

When $\varphi = 0$,

$$\begin{bmatrix} R_{0x} \\ R_{0y} \end{bmatrix} = |\mathbf{R}| \begin{bmatrix} \cos\theta \\ \sin\theta \end{bmatrix}.$$

Hence the components of the position vector in the rotated reference frame are given by

$$R_{1x} = R_{0x}\cos\theta + R_{0y}\sin\theta, \quad R_{1y} = -R_{0x}\sin\theta + R_{0y}\cos\theta,$$

which could be expressed in matrix form as

$$\begin{bmatrix} R_{1x} \\ R_{1y} \end{bmatrix} = \begin{bmatrix} \cos\varphi & \sin\varphi \\ -\sin\varphi & \cos\varphi \end{bmatrix} \begin{bmatrix} R_{0x} \\ R_{0y} \end{bmatrix}.$$

The angular velocity of the reference frame is the time derivative of the rotation angle $d\varphi/dt$ and is a vector in a direction perpendicular to both the unit vectors $\boldsymbol{i}_0$ and $\boldsymbol{j}_0$.

If a unit vector in this direction is denoted by $\boldsymbol{k}_0$, then the two-dimensional rotational transformation of a three-dimensional vector, with components R_{x0}, R_{y0}, and R_{z0} in the (x_0, y_0, z_0) reference frame, for a rotation ϕ about the z_0 axis may be expressed as

$$R_{1x} = R_{0x}\cos\theta + R_{0y}\sin\theta, \quad R_{1y} = -R_{0x}\sin\theta + R_{0y}\cos\theta, \quad R_{1z} = R_{0z},$$

where R_{x1}, R_{y1}, and R_{z1} are the components in the rotated reference frame. Adopting a simpler notation for the components of the position vector in each of the two frames, i.e., as

$$\mathbf{R} = R_{x0}\mathbf{i}_0 + R_{y0}\mathbf{j}_0 + R_{z0}\mathbf{k}_0 \equiv x_0\mathbf{i}_0 + y_0\mathbf{j}_0 + z_0\mathbf{k}_0$$

and

$$\mathbf{R} = R_{x1}\mathbf{i}_1 + R_{y1}\mathbf{j}_1 + R_{z1}\mathbf{k}_1 \equiv x_1\mathbf{i}_1 + y_1\mathbf{j}_1 + z_1\mathbf{k}_1,$$

the relationship between the components of the position vector in the rotated reference frame and the components of the position vector in the original reference frame is given in matrix notation as

$$\begin{bmatrix} x_1 \\ y_1 \\ z_1 \end{bmatrix} = \begin{bmatrix} \cos\varphi & \sin\varphi & 0 \\ -\sin\varphi & \cos\varphi & 0 \\ 0 & 0 & 1 \end{bmatrix} \begin{bmatrix} x_0 \\ y_0 \\ z_0 \end{bmatrix} = \mathbf{R}_{10} \begin{bmatrix} x_0 \\ y_0 \\ z_0 \end{bmatrix},$$

where $\mathbf{R}_{10}$ is the rotational transformation matrix defined as

$$\mathbf{R}_{10} = \begin{bmatrix} \cos\varphi & \sin\varphi & 0 \\ -\sin\varphi & \cos\varphi & 0 \\ 0 & 0 & 1 \end{bmatrix}.$$

A.2.3 Axis Transformations by Direction Cosines

It has already been stated that when a reference frame fixed in a body is rotating relative to another or relative to a fixed reference frame such as a local inertial frame, the relationship between the two reference frames or coordinate frames may be described by a transformation. Such a transformation may also be expressed in terms of the direction cosines defined earlier.

From the direction cosines defined earlier, one may express each of the components of a vector in the rotated reference axes in terms of the sum of the contributions of each of the components of the same vector in the original axes. Thus, considering just the x_0 component, the transformed vector in the rotated reference frame is

$$\mathbf{R}|_x = x_0\,l_{11}\,\mathbf{i}_0 + x_0\,l_{21}\,\mathbf{j}_0 + x_0\,l_{31}\,\mathbf{k}_0.$$

Applying the same concept to the y_0 and z_0 components,

$$\mathbf{R}|_y = y_0\,l_{12}\,\mathbf{i}_0 + y_0\,l_{22}\,\mathbf{j}_0 + y_0\,l_{32}\,\mathbf{k}_0,$$

$$\mathbf{R}|_z = z_0\,l_{13}\,\mathbf{i}_0 + z_0\,l_{23}\,\mathbf{j}_0 + z_0\,l_{33}\,\mathbf{k}_0,$$

and collecting together the contributions to the vector components along each of three rotated reference axes, we obtain

$$x_1 = l_{11}\,x_0 + l_{12}\,y_0 + l_{13}\,z_0, \quad y_1 = l_{21}\,x_0 + l_{22}\,y_0 + l_{23}\,z_0,$$

$$z_1 = l_{31}\,x_0 + l_{32}\,y_0 + l_{33}\,z_0,$$

where the direction cosines satisfy

$$l_{1n}^2 + l_{2n}^2 + l_{3n}^2 = 1, \quad n = 1, 2, 3.$$

The relationship between the components of the position vector in the rotated reference frame and the components of the position vector in the original reference frame is given in matrix notation as

$$
\begin{bmatrix} x_1 \\ y_1 \\ z_1 \end{bmatrix} = \begin{bmatrix} l_{11} & l_{12} & l_{13} \\ l_{21} & l_{22} & l_{23} \\ l_{31} & l_{32} & l_{33} \end{bmatrix} \begin{bmatrix} x_0 \\ y_0 \\ z_0 \end{bmatrix} = \mathbf{R}_{10} \begin{bmatrix} x_0 \\ y_0 \\ z_0 \end{bmatrix},
$$

where $\mathbf{R}_{10}$ is the rotational transformation matrix defined as

$$
\mathbf{R}_{10} = \begin{bmatrix} l_{11} & l_{12} & l_{13} \\ l_{21} & l_{22} & l_{23} \\ l_{31} & l_{32} & l_{33} \end{bmatrix}.
$$

A.3 Synthesis of Rotational Transformations

Three-dimensional transformations of reference frames can be defined as a sequence of successive rotations about different axes. Typically the reference frame fixed rigidly to the body (x_B, y_B, z_B) is rotated about the roll axis x_B by an angle $-\phi$. Thus we have a new reference frame (x_1, y_1, z_1) in the rotated position. This frame is then rotated about the new pitch axis y_1 by an angle $-\theta$. As a result of this next rotation we obtain the (x_2, y_2, z_2) frame. This frame is then rotated once more about the resulting yaw axis by an angle $-\psi$. The reference frame produced by the three successive rotations can be aligned parallel to an inertial reference frame (x_I, y_I, z_I). The relationships among the four reference frames can be written as

$$
\begin{bmatrix} x_1 \\ y_1 \\ z_1 \end{bmatrix} = \begin{bmatrix} 1 & 0 & 0 \\ 0 & \cos\phi & -\sin\phi \\ 0 & \sin\phi & \cos\phi \end{bmatrix} \begin{bmatrix} x_B \\ y_B \\ z_B \end{bmatrix}, \quad \begin{bmatrix} x_2 \\ y_2 \\ z_2 \end{bmatrix} = \begin{bmatrix} \cos\theta & 0 & \sin\theta \\ 0 & 1 & 0 \\ -\sin\theta & 0 & \cos\theta \end{bmatrix} \begin{bmatrix} x_1 \\ y_1 \\ z_1 \end{bmatrix},
$$

$$
\begin{bmatrix} x_I \\ y_I \\ z_I \end{bmatrix} = \begin{bmatrix} \cos\psi & -\sin\psi & 0 \\ \sin\psi & \cos\psi & 0 \\ 0 & 0 & 1 \end{bmatrix} \begin{bmatrix} x_2 \\ y_2 \\ z_2 \end{bmatrix}.
$$

The transformation relating the body-fixed reference frame to the inertial reference frame is given by the triple product of the preceding transformation matrices in the proper sequence. It must be emphasized that the transformations are not commutative. Each of the preceding transformation matrices as well as the triple product is an orthogonal matrix, i.e., the inverse of the matrix is also equal to its transpose.

We may obtain the same transformation by starting with the inertial reference frame, performing the yaw, pitch, and roll sequence of positive rotations (or the 3–2–1 sequence), and aligning the resulting frame with a body-fixed reference frame, preferably one that is aligned with the principal axes of the body. This procedure has emerged as a standard for defining the body-fixed axes.

A.3.1 Euler Angle Sets

The rotational angles defined in the preceding axes transformations are usually referred to as Euler angles. These transformations are defined in terms of the Euler angles. The $yaw(\psi)$–$pitch(\theta)$–$roll(\phi)$ or 3–2–1 *sequence* is an example of an Euler angle sequence in robot-manipulator dynamics. The i–j–k Euler angle rotation means that the first rotation by angle ψ is about the i axis, the second rotation by angle θ is about the j axis, and the third rotation by angle ϕ is about the k axis. They give rise to a basic set of coordinates for defining the attitude of a rigid body.

The transformation relating the body-fixed reference axes to the space-fixed inertial reference axes is

$$[x_I \quad y_I \quad z_I]^T = \mathbf{T}_{IB}[x_B \quad y_B \quad z_B]^T,$$

where

$$\mathbf{T}_{IB} = \begin{bmatrix} \cos\psi & -\sin\psi & 0 \\ \sin\psi & \cos\psi & 0 \\ 0 & 0 & 1 \end{bmatrix} \begin{bmatrix} \cos\theta & 0 & \sin\theta \\ 0 & 1 & 0 \\ -\sin\theta & 0 & \cos\theta \end{bmatrix} \begin{bmatrix} 1 & 0 & 0 \\ 0 & \cos\phi & -\sin\phi \\ 0 & \sin\phi & \cos\phi \end{bmatrix}.$$

The inverse transformation is

$$[x_B \quad y_B \quad z_B]^T = \mathbf{T}_{IB}^{-1}[x_I \quad y_I \quad z_I]^T = \mathbf{T}_{BI}[x_I \quad y_I \quad z_I]^T,$$

where

$$\mathbf{T}_{BI} = \begin{bmatrix} 1 & 0 & 0 \\ 0 & \cos\phi & \sin\phi \\ 0 & -\sin\phi & \cos\phi \end{bmatrix} \begin{bmatrix} \cos\theta & 0 & -\sin\theta \\ 0 & 1 & 0 \\ \sin\theta & 0 & \cos\theta \end{bmatrix} \begin{bmatrix} \cos\psi & \sin\psi & 0 \\ -\sin\psi & \cos\psi & 0 \\ 0 & 0 & 1 \end{bmatrix}.$$

The convention that is followed is that $\mathbf{T}_{BI}$ is a matrix that transforms representations in the reference inertial coordinate system to representations in the body-fixed coordinate system.

Another popular Euler angle sequence, which we shall refer to as the "alternative Euler angle sequence" or as the 3–2–3 sequence, is a rotation about the body-fixed z_B axis by an angle α followed by a rotation by an angle β about the rotated y_1 axis, followed by a rotation by an angle γ about the new z_2 axis of the frame obtained after the two previous rotations.

The transformation relating the body-fixed reference frame to the inertial reference frame for this alternative sequence is given by

$$[x_I \quad y_I \quad z_I]^T = \mathbf{T}_{IB}[x_B \quad y_B \quad z_B]^T,$$

where

$$\mathbf{T}_{IB} = \begin{bmatrix} \cos\gamma & -\sin\gamma & 0 \\ \sin\gamma & \cos\gamma & 0 \\ 0 & 0 & 1 \end{bmatrix} \begin{bmatrix} \cos\beta & 0 & \sin\beta \\ 0 & 1 & 0 \\ -\sin\beta & 0 & \cos\beta \end{bmatrix} \begin{bmatrix} \cos\alpha & -\sin\alpha & 0 \\ \sin\alpha & \cos\alpha & 0 \\ 0 & 0 & 1 \end{bmatrix}.$$

The inverse transformation is

$$[x_B \quad y_B \quad z_B]^T = \mathbf{T}_{IB}^{-1}[x_I \quad y_I \quad z_I]^T = \mathbf{T}_{BI}[x_I \quad y_I \quad z_I]^T,$$

where

$$\mathbf{T}_{BI} = \begin{bmatrix} \cos\alpha & \sin\alpha & 0 \\ -\sin\alpha & \cos\alpha & 0 \\ 0 & 0 & 1 \end{bmatrix} \begin{bmatrix} \cos\beta & 0 & -\sin\beta \\ 0 & 1 & 0 \\ \sin\beta & 0 & \cos\beta \end{bmatrix} \begin{bmatrix} \cos\gamma & \sin\gamma & 0 \\ -\sin\gamma & \cos\gamma & 0 \\ 0 & 0 & 1 \end{bmatrix}.$$

A.3.2 Geometric Interpretations

A fundamental property of these orthogonal matrix transformations is the invariance of the magnitude of a vector after transformation. There also exists a set of vectors that are unchanged in direction as well after the application of the transformation. Each transformation may be reduced to a single rotation about each of these vectors. There is at least one such for any orthogonal transformation vector.

According to *Euler's theorem*, the general displacement of a rigid body with one point fixed is equivalent to a single rotation about some axis through that point. In other words, for any rotation there exists a given axis of rotation in one coordinate system that remains invariant in another reference coordinate system. Any rotation may be expressed as a rotation through some angle about some axis. This parameterization of the attitude is called the Euler axis/angle parameterization.

The geometric interpretations of these transformation matrices are of fundamental importance in robot kinematics. Irrespective of the orthogonality properties, the singularities in the transformations and the lack of uniqueness of the representations are important shortcomings in the kinematics of the transformations. The geometric interpretation as a single rotation about some axis is the basis for an alternative representation of the transformation by use of four unique parameters. The need for such an alternative representation arises because of certain pitfalls of Euler angle representations, which are discussed in the next subsection.

A.3.3 Pitfalls of Rotational Sequences

The sequence of transformations just discussed is not a unique sequence, i.e., the transformation relating the body-fixed and space-fixed inertial reference frames can be established by use of a sequence of rotations about different sets of axes. One may use a pitch–roll–yaw sequence, a pitch–yaw–roll sequence, or a yaw–roll–yaw sequence.

The transformation relating the body-fixed reference frame to the inertial reference frame for a yaw, pitch, and roll sequence applied to the inertial frame was already discussed. When the pitch angle is set equal to ±90° in this transformation we find that the roll angle and the yaw angle are not unique. This feature is the cause of the phenomenon that is known as "gimbal lock" in inertial measuring systems designed to estimate the position and orientation of a body from measurements of acceleration and attitude. It is a direct consequence of the Euler angle representation and a major pitfall of the Euler angle approach.

A.3.4 Kinematics of Rotational Transformations

We can obtain the angular velocity vector of the body-fixed reference frames by simply adding up the individual angular velocity contribution from each successive rotation. In the preceding example it is the vector sum of the yaw rate about the z axis, the pitch rate about the y_1 axis, and the roll rate about the x_2 axis. The angular velocity may then be resolved to the body-fixed reference frame by use of the transformations previously developed relating the unit vector in the various reference frames. The body components of the angular velocity are related to the attitude rates for the first Euler angle sequence by

$$
\begin{bmatrix} p_B \\ q_B \\ r_B \end{bmatrix} = \begin{bmatrix} \omega_1 \\ \omega_2 \\ \omega_3 \end{bmatrix} = \begin{bmatrix} 1 & 0 & -\sin\theta \\ 0 & \cos\phi & \sin\phi\cos\theta \\ 0 & -\sin\phi & \cos\phi\cos\theta \end{bmatrix} \begin{bmatrix} \dot\phi \\ \dot\theta \\ \dot\psi \end{bmatrix},
$$

and the inverse relations are

$$
\begin{bmatrix} \dot\phi \\ \dot\theta \\ \dot\psi \end{bmatrix} = \begin{bmatrix} 1 & \sin\phi\tan\theta & \cos\phi\tan\theta \\ 0 & \cos\phi & -\sin\phi \\ 0 & \sin\phi/\cos\theta & \cos\theta/\cos\theta \end{bmatrix} \begin{bmatrix} p_B \\ q_B \\ r_B \end{bmatrix},
$$

where the overdots $(\cdot)$ represent the time differentiation operator d/dt. For the alternative Euler angle sequence or the 3–2–3 sequence, the corresponding relations are

$$
\begin{bmatrix} p_B \\ q_B \\ r_B \end{bmatrix} = \begin{bmatrix} \omega_1 \\ \omega_2 \\ \omega_3 \end{bmatrix} = \begin{bmatrix} 0 & \sin\alpha & -\sin\beta\cos\alpha \\ 0 & \cos\alpha & \sin\beta\sin\alpha \\ 1 & 0 & \cos\beta \end{bmatrix} \begin{bmatrix} \dot\alpha \\ \dot\beta \\ \dot\gamma \end{bmatrix},
$$

and the inverse relations are

$$
\begin{bmatrix} \dot\alpha \\ \dot\beta \\ \dot\gamma \end{bmatrix} = \begin{bmatrix} \cos\alpha/\tan\beta & -\sin\alpha/\tan\beta & 1 \\ \sin\alpha & \cos\alpha & 0 \\ -\cos\alpha/\sin\beta & \sin\alpha/\sin\beta & 0 \end{bmatrix} \begin{bmatrix} p_B \\ q_B \\ r_B \end{bmatrix}.
$$

It may be noted that the transformation matrix itself satisfies a differential equation, the coefficients of which are dependent on the angular velocities. Thus, if the angular velocities are measured, the differential equation could be integrated in a computer to obtain the transformation in real time. This is an essential feature of strapped-down mechanization of an inertial measurement system using accelerometers and rate gyros for the measurement of accelerations and angular velocities in three mutually perpendicular directions. However, in view of the pitfalls of the Euler angle representation, the transformation matrix may be represented by the four-parameter representation or quaternion.

A.3.5 Kinematics of the Direction Cosine Matrix

Consider a rotation by an angle φ and the associated rotational transformation about the z_0 axis:

$$\mathbf{R}_{10} = \begin{bmatrix} \cos\varphi & \sin\varphi & 0 \\ -\sin\varphi & \cos\varphi & 0 \\ 0 & 0 & 1 \end{bmatrix}.$$

Taking the time derivative of $\mathbf{R}_{10}$, one obtains

$$\dot{\mathbf{R}}_{10} = \dot{\varphi} \begin{bmatrix} -\sin\varphi & \cos\varphi & 0 \\ -\cos\varphi & -\sin\varphi & 0 \\ 0 & 0 & 0 \end{bmatrix} = \dot{\varphi} \begin{bmatrix} 0 & 1 & 0 \\ -1 & 0 & 0 \\ 0 & 0 & 0 \end{bmatrix} \begin{bmatrix} \cos\varphi & \sin\varphi & 0 \\ -\sin\varphi & \cos\varphi & 0 \\ 0 & 0 & 1 \end{bmatrix}.$$

Because the angular velocity of the body is given by

$$\omega = \begin{bmatrix} 0 & 0 & \omega_3 \end{bmatrix} = \begin{bmatrix} 0 & 0 & \dot{\varphi} \end{bmatrix},$$

$$\dot{\mathbf{R}}_{10} = \omega_3 \begin{bmatrix} 0 & 1 & 0 \\ -1 & 0 & 0 \\ 0 & 0 & 0 \end{bmatrix} \begin{bmatrix} \cos\varphi & \sin\varphi & 0 \\ -\sin\varphi & \cos\varphi & 0 \\ 0 & 0 & 1 \end{bmatrix} = \omega_3 \begin{bmatrix} 0 & 1 & 0 \\ -1 & 0 & 0 \\ 0 & 0 & 0 \end{bmatrix} \mathbf{R}_{10}.$$

Consider a body rotating with an angular velocity vector:

$$\omega = \begin{bmatrix} \omega_1 & \omega_2 & \omega_3 \end{bmatrix}.$$

Considering the sequence of rotations from the inertial to the body frame and the time derivative of the direction cosine matrix, it may be shown that

$$\left(\frac{d}{dt}\mathbf{T}_{BI} \right) \mathbf{T}_{IB} = \begin{bmatrix} 0 & \omega_3 & -\omega_2 \\ -\omega_3 & 0 & \omega_1 \\ \omega_2 & -\omega_1 & 0 \end{bmatrix}.$$

Hence it follows that

$$\frac{d}{dt}\mathbf{T}_{BI} = \begin{bmatrix} 0 & \omega_3 & -\omega_2 \\ -\omega_3 & 0 & \omega_1 \\ \omega_2 & -\omega_1 & 0 \end{bmatrix} \mathbf{T}_{BI}.$$

The relation is similar to the rate of change of a fixed vector $\mathbf{r}$ in a frame rotating with an angular velocity vector, $\omega = \begin{bmatrix} \omega_1 & \omega_2 & \omega_3 \end{bmatrix}$. In this case we have

$$\frac{d}{dt}\mathbf{r} = \begin{bmatrix} 0 & \omega_3 & -\omega_2 \\ -\omega_3 & 0 & \omega_1 \\ \omega_2 & -\omega_1 & 0 \end{bmatrix} \mathbf{r}.$$

Although the vector is fixed, the fact that the frame is rotating implies that the components of the vector in the rotating frame are continuously changing.

Further, because

$$\mathbf{T}_{IB}\frac{d}{dt}\mathbf{T}_{BI} + \left(\frac{d}{dt}\mathbf{T}_{IB}\right)\mathbf{T}_{BI} = 0,$$

$$\tfrac{d}{dt}\mathbf{T}_{IB} = -\mathbf{T}_{IB}\begin{bmatrix} 0 & \omega_3 & -\omega_2 \\ -\omega_3 & 0 & \omega_1 \\ \omega_2 & -\omega_1 & 0 \end{bmatrix}.$$

A.4 Four-Parameter Rotational Operators: Quaternions

Sir William Hamilton first introduced *quaternions* in 1843 to generalize complex numbers to three dimensions. The use of the quaternion in the solution of rigid body rotations as required by robot-manipulator dynamics, computer vision, flight simulation, and inertial guidance systems has proven to be efficient and accurate. Attitude parameterization by the quaternion is more compact than the direction cosine matrix and more computationally efficient than the Euler axis/angle and Euler angle methods. In addition, the Euler angle parameterization has singularities at certain angles, which limits the usefulness of the Euler angle representation.

A.4.1 Definition of the Quaternion

The Rodrigues quaternion or quaternion is a quadruple consisting of a real scalar part (q_4) and a hyperimaginary three-vector part (q_1, q_2, q_3), defined as follows:

$$\mathbf{q} = q_1\mathbf{i} + q_2\mathbf{j} + q_3\mathbf{k} + q_4,$$

where $\mathbf{i}^2 = \mathbf{j}^2 = \mathbf{k}^2 = -1$,

$$\mathbf{ij} = -\mathbf{ji} = \mathbf{k}, \ \mathbf{jk} = -\mathbf{jk} = \mathbf{i}, \quad \mathbf{ki} = -\mathbf{ik} = \mathbf{j}.$$

The literature contains a variety of definitions of the quaternion as having different orders and sign conventions. The components of the Rodrigues quaternion are often also called Euler symmetric parameters. The quaternion is also expressed in one of several representations such as

$$\mathbf{q} = (q_4, \bar{\mathbf{q}})$$

or as

$$\mathbf{q} = [q_1, \ q_2, \ q_3, \ q_4]^T.$$

Additional constraints are imposed when the quaternion is used to represent angular velocity; first, the quaternion should have a norm of unity. This prevents loss of precision in computers and simplifies the related vector transformations. Second, the quaternion is defined in terms of the Euler–Rodriques rotation parameters or equivalently Euler symmetric parameters, which allow rotations to be performed as quaternion products. Thus the components of the quaternion are defined as

$$q_4 = \cos\left(\frac{\phi}{2}\right), \quad q_1 = e_x\sin\left(\frac{\phi}{2}\right), \quad q_2 = e_y\sin\left(\frac{\phi}{2}\right), \quad q_3 = e_z\sin\left(\frac{\phi}{2}\right),$$

where $\mathbf{e} = [e_x \quad e_y \quad e_z]$ is a unit vector along the axis of rotation and ϕ is the total rotation angle. Further, the norm of this quaternion is unity, i.e.,

$$q_1^2 + q_2^2 + q_3^2 + q_4^2 = 1.$$

A.4.2 Defining the Axis of Rotation

The unit vector in the direction of the axis of rotation may be defined in a number of ways, e.g., by locating a point on the surface of a unit sphere, known as the pole point or pole.

When the rotation axis is located in terms of latitude and longitude (north latitude and east longitude are positive),

$$[e_x \quad e_y \quad e_z]^T = [\cos(\text{lat})\cos(\text{long}) \quad \cos(\text{lat})\sin(\text{long}) \quad \sin(\text{lat})]^T.$$

When the rotation axis is located in terms of spherical coordinates α and θ, where α is azimuth and θ is the angle from the z axis,

$$[e_x \quad e_y \quad e_z]^T = [\sin(\theta)\cos(\alpha) \quad \sin(\theta)\sin(\alpha) \quad \cos(\theta)]^T.$$

A.4.3 Conversion Between Euler Angles and a Quaternion

The quaternion equivalence to a set of Euler angles defined by a 3–2–1 (yaw–pitch–roll rotation) sequence may be established. Thus,

$$q_1 = \cos\left(\frac{\text{yaw}}{2}\right)\cos\left(\frac{\text{pitch}}{2}\right)\sin\left(\frac{\text{roll}}{2}\right) - \sin\left(\frac{\text{yaw}}{2}\right)\sin\left(\frac{\text{pitch}}{2}\right)\cos\left(\frac{\text{roll}}{2}\right),$$

$$q_2 = \cos\left(\frac{\text{yaw}}{2}\right)\sin\left(\frac{\text{pitch}}{2}\right)\cos\left(\frac{\text{roll}}{2}\right) + \sin\left(\frac{\text{yaw}}{2}\right)\cos\left(\frac{\text{pitch}}{2}\right)\sin\left(\frac{\text{roll}}{2}\right),$$

$$q_3 = \sin\left(\frac{\text{yaw}}{2}\right)\cos\left(\frac{\text{pitch}}{2}\right)\cos\left(\frac{\text{roll}}{2}\right) - \cos\left(\frac{\text{yaw}}{2}\right)\sin\left(\frac{\text{pitch}}{2}\right)\sin\left(\frac{\text{roll}}{2}\right),$$

$$q_4 = \cos\left(\frac{\text{yaw}}{2}\right)\cos\left(\frac{\text{pitch}}{2}\right)\cos\left(\frac{\text{roll}}{2}\right) + \sin\left(\frac{\text{yaw}}{2}\right)\sin\left(\frac{\text{pitch}}{2}\right)\sin\left(\frac{\text{roll}}{2}\right).$$

For small angles, the cosine terms in the preceding equations can be replaced with a value of unity. The quaternion is therefore approximately equal to

$$q_1 \sim \frac{\text{roll}}{2}, \quad q_2 \sim \frac{\text{pitch}}{2}, \quad q_3 \sim \frac{\text{yaw}}{2}, \quad \text{and } q_4 \sim 1.$$

Inverse relationships may also be established. Thus the Euler angles may be extracted from a quaternion as

$$\sin(\text{yaw})/\cos(\text{yaw}) = 2(q_1 q_2 + q_3 q_4)/(q_4^2 + q_1^2 - q_2^2 - q_3^2),$$

$$\sin(\text{pitch}) = -2(q_1 q_3 - q_2 q_4),$$

$$\sin(\text{roll})/\cos(\text{roll}) = 2(q_1 q_4 + q_3 q_2)/(q_4^2 - q_1^2 - q_2^2 + q_3^2).$$

The following relations are some representative Euler angles and their equivalent quaternions:

$$[\text{yaw, pitch, roll}] = [0^0 \quad 0^0 \quad 0^0] \rightarrow \mathbf{q} = [q_1\, q_2\, q_3\, q_4] = [0\, 0\, 0\, 1],$$

$$[\text{yaw, pitch, roll}] = [90^0 \quad 0^0 \quad 0^0] \rightarrow \mathbf{q} = [q_1\, q_2\, q_3\, q_4] = [0\, 0\, 0.707\, 0.707],$$

$$[\text{yaw, pitch, roll}] = [0^0 \quad 60^0 \quad 0^0] \rightarrow \mathbf{q} = [q_1\, q_2\, q_3\, q_4] = [0\, 0.5\, 0\, 0.866],$$

and

$$[\text{yaw, pitch, roll}] = \begin{bmatrix} 10^0\ 20^0\ 30^0 \end{bmatrix} \rightarrow$$
$$\mathbf{q} = [q_1\, q_2\, q_3\, q_4] = [0.239\, 0.189\, 0.038\, 0.951].$$

For the alternative Euler angle sequence or the 3–2–3 sequence, the quaternion equivalence relations are

$$q_1 = \sin(\beta/2)\cos[(\alpha - \gamma)/2], \quad q_2 = \cos(\beta/2)\sin[(\alpha + \gamma)/2],$$

$$q_3 = \cos(\beta/2)\cos[(\alpha + \gamma)/2], \quad q_4 = \sin(\beta/2)\sin[(\alpha - \gamma)/2].$$

The inverse relations are easily found from these relations.

A.4.4 Inverse of a Quaternion

The inverse of a quaternion is defined in terms of the complex conjugate. The complex conjugate is defined as

$$\mathbf{q}^* = (q_4, \bar{\mathbf{q}})^* = (q_4, -\bar{\mathbf{q}}) = \begin{bmatrix} -q_1 & -q_2 & -q_3 & q_4 \end{bmatrix}^T = [-\bar{\mathbf{q}} \quad q_4]^T.$$

The inverse of the quaternion is defined as

$$\mathbf{q}^{-1} = \frac{\mathbf{q}^*}{q_1^2 + q_2^2 + q_3^2 + q_4^2} = \frac{(q_4, -\bar{\mathbf{q}})}{q_1^2 + q_2^2 + q_3^2 + q_4^2},$$

and

$$\mathbf{q}^{-1} = \frac{1}{(q_1^2 + q_2^2 + q_3^2 + q_4^2)}[-q_1 \quad -q_2 \quad -q_3 \quad q_4]^T,$$

which simplifies to

$$\mathbf{q}^{-1} = \frac{1}{(q_1^2 + q_2^2 + q_3^2 + q_4^2)}\begin{bmatrix} -\bar{\mathbf{q}} \\ q_4 \end{bmatrix}.$$

When $q_1^2 + q_2^2 + q_3^2 + q_4^2 = 1$, which is normally assumed,

$$\mathbf{q}^{-1} = \mathbf{q}^* = (q_4, \bar{\mathbf{q}})^* = (q_4, -\bar{\mathbf{q}}) = [-q_1 \quad -q_2 \quad -q_3 \quad q_4]^T = [-\bar{\mathbf{q}} \quad q_4]^T.$$

A.4.5 Reversing the Direction of Rotation

We may reverse the direction of a rotation simply by taking the complex conjugate of the quaternion:

$$\mathbf{q}^* = (q_4, \bar{\mathbf{q}})^* = (q_4, -\bar{\mathbf{q}}) = [-q_1 \quad -q_2 \quad -q_3 \quad q_4]^T = [-\bar{\mathbf{q}} \quad q_4]^T.$$

A.4.6 Quaternion Normalization

The length or norm of a quaternion is defined as

$$|\mathbf{q}| = \sqrt{\mathbf{q}\mathbf{q}^*} = \sqrt{\mathbf{q}^*\mathbf{q}} = \sqrt{q_1^2 + q_2^2 + q_3^2 + q_4^2} = 1.$$

The norm of the Rodriques' quaternion should be 1. If the quaternion is integrated, the quaternion norm will no longer be equal to unity and generally diverges. The quaternion can be renormalized by dividing all its components by the quaternion norm.

A.4.7 Combined Rotations

A quaternion can be rotated into a new reference frame by a multiplication following the laws of quaternion algebra. For example, a set of quaternions that rotates A into B and B into C can be combined into a single quaternion that rotates A into C. Thus the composition of two quaternions is defined as

$$\mathbf{q}_c = \mathbf{q}_a \otimes \mathbf{q}_b = (q_{a4}, \bar{\mathbf{q}}_a) \otimes (q_{b4}, \bar{\mathbf{q}}_b) = \begin{bmatrix} q_{a4}\bar{\mathbf{q}}_b + q_{b4}\bar{\mathbf{q}}_a + \bar{\mathbf{q}}_a \times \bar{\mathbf{q}}_b \\ q_{a4}q_{b4} - \bar{\mathbf{q}}_a \cdot \bar{\mathbf{q}}_b \end{bmatrix}$$

and

$$|\mathbf{q}_a\mathbf{q}_b| = |\mathbf{q}_a|\,|\mathbf{q}_b| = 1.$$

Using the rules of vector algebra and the unit vector identities, we may write this in matrix form as

$$\begin{bmatrix} q_{c1} \\ q_{c2} \\ q_{c3} \\ q_{c4} \end{bmatrix} = \begin{bmatrix} q_{a4} & q_{a3} & -q_{a2} & q_{a1} \\ -q_{a3} & q_{a4} & q_{a1} & q_{a2} \\ q_{a2} & -q_{a1} & q_{a4} & q_{a3} \\ -q_{a1} & -q_{a2} & -q_{a3} & q_{a4} \end{bmatrix} \begin{bmatrix} q_{b1} \\ q_{b2} \\ q_{b3} \\ q_{b4} \end{bmatrix}.$$

After the vector–matrix product is formed,

$$\begin{bmatrix} q_{c1} \\ q_{c2} \\ q_{c3} \\ q_{c4} \end{bmatrix} = \begin{bmatrix} q_{a4}q_{b1} & +q_{a3}q_{b2} & -q_{a2}q_{b3} & +q_{a1}q_{b4} \\ -q_{a3}q_{b1} & +q_{a4}q_{b2} & +q_{a1}q_{b3} & +q_{a2}q_{b4} \\ q_{a2}q_{b1} & -q_{a1}q_{b2} & +q_{a4}q_{b3} & +q_{a3}q_{b4} \\ -q_{a1}q_{b1} & -q_{a2}q_{b2} & -q_{a3}q_{b3} & +q_{a4}q_{b4} \end{bmatrix},$$

which may also be expressed as

$$\begin{bmatrix} q_{c1} \\ q_{c2} \\ q_{c3} \\ q_{c4} \end{bmatrix} = \begin{bmatrix} q_{b4} & -q_{b3} & q_{b2} & q_{b1} \\ q_{b3} & q_{b4} & -q_{b1} & q_{b2} \\ -q_{b2} & q_{b1} & q_{b4} & q_{b3} \\ -q_{b1} & -q_{b2} & -q_{b3} & q_{b4} \end{bmatrix} \begin{bmatrix} q_{a1} \\ q_{a2} \\ q_{a3} \\ q_{a4} \end{bmatrix}.$$

A.4.8 Conversion of Latitude and Longitude to a Quaternion

The latitude and longitude positions on a planet can be converted to a quaternion by forming the quaternion product of the latitude and longitude quaternions. For example, a longitudinal rotation may be expressed as a quaternion:

$$q_4 = \cos\left(\frac{\phi}{2}\right), \quad q_1 = e_x \sin\left(\frac{\phi}{2}\right), \quad q_2 = e_y \sin\left(\frac{\phi}{2}\right), \quad q_3 = e_z \sin\left(\frac{\phi}{2}\right),$$

where $\mathbf{e} = [e_x \quad e_y \quad e_z]$ is a unit vector along the *N–S* axis of the Earth's rotation and ϕ is the total longitudinal rotation angle.

Often the direction of the axis of rotation is expressed in terms of the latitude and longitude. If we have two rotations, each represented as a quaternion, the result of both rotations is the product of their two quaternions.

A.4.9 Transformation of a Vector by a Quaternion

To rotate a vector by a quaternion, it should be recognized that a component of each of the reference axes in the direction of the quaternion axis does not undergo any change whereas the component perpendicular to it undergoes a two-dimensional rotation. Thus the direction cosine matrix may be expressed as

$$\mathbf{T}_{BI} = \begin{bmatrix} e_x \\ e_y \\ e_z \end{bmatrix} [e_x \quad e_y \quad e_z] + \left(\mathbf{I} - \begin{bmatrix} e_x \\ e_y \\ e_z \end{bmatrix} [e_x \quad e_y \quad e_z] \right) \cos\phi + \bar{Q}(\mathbf{e}_\phi) \sin\phi,$$

where ϕ is the rotation angle or Euler angle about the $\mathbf{e}_\phi$ axis, $\mathbf{e}_\phi = [e_x \quad e_y \quad e_z]$ is a unit vector along the rotation axis or Euler axis, and

$$\bar{Q}(\bar{\mathbf{e}}_\phi) = \begin{bmatrix} 0 & e_z & -e_y \\ -e_z & 0 & e_x \\ e_y & -e_x & 0 \end{bmatrix}.$$

When the transformation is applied to a vector in the direction of the Euler axis it remains invariant. Any three-element vector may be transformed by the direction cosine matrix or directly rotated by a quaternion in a manner similar to that used to rotate a quaternion. In this case the scalar part is zero. Thus,

$$\bar{\mathbf{v}}' = \mathbf{T}_{BI}\bar{\mathbf{v}} = \mathbf{q}^*\bar{\mathbf{v}}\mathbf{q} = (\mathbf{e}_\phi \cdot \bar{\mathbf{v}})\,\mathbf{e}_\phi + \cos\phi(\bar{\mathbf{v}} - (\mathbf{e}_\phi \cdot \bar{\mathbf{v}})\,\mathbf{e}_\phi) + \sin\phi(\mathbf{e}_\phi \times \bar{\mathbf{v}}),$$

which is equivalent to

$$\bar{\mathbf{v}}' = \left(2q_4^2 - 1\right)\bar{\mathbf{v}} + 2\bar{\mathbf{q}}\left(\bar{\mathbf{q}} \cdot \bar{\mathbf{v}}\right) + 2q_4\left(\bar{\mathbf{q}} \times \bar{\mathbf{v}}\right).$$

The preceding transformation correspond to a direct rotation of the vector by the quaternion and may be expressed as

$$\bar{\mathbf{v}}' = \left[\mathbf{I}\cos\phi + [e_x \quad e_y \quad e_z]^T [e_x \quad e_y \quad e_z](1 - \cos\phi) + \bar{Q}(\mathbf{e}_\phi)\sin\phi \right]\bar{\mathbf{v}},$$

where $\bar{\mathbf{v}}'$ is the output vector, $\bar{\mathbf{v}}$ is the input vector, and $\bar{\mathbf{q}}$ is the vector part of the quaternion, i.e., (q_1, q_2, q_3). The vector product $\mathbf{e}_\phi \times \bar{\mathbf{v}}$ is equivalent to

$$\mathbf{e}_\phi \times \bar{\mathbf{v}} = -\bar{\mathbf{v}} \times \mathbf{e}_\phi = -\bar{Q}(\bar{\mathbf{e}}_\phi)\,\bar{\mathbf{v}} = - \begin{bmatrix} 0 & e_z & -e_y \\ -e_z & 0 & e_x \\ e_y & -e_x & 0 \end{bmatrix} \bar{\mathbf{v}}.$$

A.4.10 The Direction Cosine Matrix

A direction cosine matrix can be expressed in terms of a quaternion that results in the following relationship:

$$\mathbf{T}_{BI}(\mathbf{q}) = (q_4^2 - \bar{\mathbf{q}}^2)\,\mathbf{I} + 2\bar{\mathbf{q}}\bar{\mathbf{q}}^T + 2q_4\,\bar{Q}(\bar{\mathbf{q}})$$

or

$$\mathbf{T}_{BI}(\mathbf{q}) = \begin{bmatrix} q_1^2 - q_2^2 - q_3^2 + q_4^2 & 2\,(q_1 q_2 + q_3 q_4) & 2\,(q_1 q_3 - q_2 q_4) \\ 2\,(q_1 q_2 - q_3 q_4) & -q_1^2 + q_2^2 - q_3^2 + q_4^2 & 2\,(q_2 q_3 + q_1 q_4) \\ 2\,(q_1 q_3 + q_2 q_4) & 2\,(q_2 q_3 - q_1 q_4) & -q_1^2 - q_2^2 + q_3^2 + q_4^2 \end{bmatrix},$$

where

$$\bar{Q}(\bar{\mathbf{q}}) = \begin{bmatrix} 0 & q_3 & -q_2 \\ -q_3 & 0 & q_1 \\ q_2 & -q_1 & 0 \end{bmatrix}.$$

The inverse relations are not unique and are given by

$$q_1 = \frac{1}{4q_4}(T_{23} - T_{32}) = \pm\frac{1}{2}\sqrt{1 + T_{11} - T_{22} - T_{33}},$$

$$q_2 = \frac{1}{4q_1}(T_{12} + T_{21}) = \frac{1}{4q_4}(T_{23} - T_{32}),$$

$$q_3 = \frac{1}{4q_4}(T_{12} - T_{21}) = \frac{1}{4q_1}(T_{31} + T_{13}),$$

$$q_4 = \frac{1}{4q_1}(T_{23} - T_{32}) = \pm\frac{1}{2}\sqrt{1 + T_{11} + T_{22} + T_{33}},$$

where $\mathbf{T}_{BI} = \{T_{ij}\}$.

In general, if a single vector needs to be transformed, the direct rotation of the vector by the quaternion is more efficient. If a large number of vectors need to be transformed, it is more efficient to first convert the quaternion into a direction cosine matrix and then use conventional matrix–vector products.

The product of two transformation matrices may be written in terms of the composition of the two quaternions:

$$\mathbf{T}_{BI}(\mathbf{q}_a)\,\mathbf{T}_{BI}(\mathbf{q}_b) = \mathbf{T}_{BI}(\mathbf{q}_a \otimes \mathbf{q}_b).$$

A.4.11 The Rate of Change of Attitude in Terms of Quaternions

If $\mathbf{q}$ is the attitude or orientation quaternion, then the rate of change of the quaternion of a body with an angular velocity vector,

$$\boldsymbol{\omega} = \boldsymbol{\omega}(t)^T = [\omega_1 \quad \omega_2 \quad \omega_3],$$

may be shown to be

$$d\mathbf{q}/dt = q' = (1/2)\,\text{quaternion}(\boldsymbol{\omega})\mathbf{q}.$$

The function quaternion $(\boldsymbol{\omega}) = \boldsymbol{\Omega}\,[\boldsymbol{\omega}(t)]$ converts the angular velocity vector $\boldsymbol{\omega}(t)$ into a quaternion with a zero scalar part and $\boldsymbol{\omega}$ as the vector part. Thus,

$$\boldsymbol{\Omega}\,[\boldsymbol{\omega}(t)] = \begin{bmatrix} 0 & \omega_3 & -\omega_2 & \omega_1 \\ -\omega_3 & 0 & \omega_1 & \omega_2 \\ \omega_2 & -\omega_1 & 0 & \omega_3 \\ -\omega_1 & -\omega_2 & -\omega_3 & 0 \end{bmatrix}.$$

$\mathbf{q}$ may be numerically integrated and updated to

$$\mathbf{q}^+ = \mathbf{q} + d\mathbf{q}/dt.$$

$\mathbf{q}^+$ should be normalized at the end of the numerical integration step. The inertial rate quaternion $(\mathbf{q}')$ is defined in terms of the inertial attitude quaternion and the body rates:

$$\mathbf{q}' = \begin{bmatrix} \dot{q}_1 \\ \dot{q}_2 \\ \dot{q}_3 \\ \dot{q}_4 \end{bmatrix} = \frac{d}{dt}\begin{bmatrix} q_1 \\ q_2 \\ q_3 \\ q_4 \end{bmatrix} = \frac{1}{2}\begin{bmatrix} 0 & \omega_3 & -\omega_2 & \omega_1 \\ -\omega_3 & 0 & \omega_1 & \omega_2 \\ \omega_2 & -\omega_1 & 0 & \omega_3 \\ -\omega_1 & -\omega_2 & -\omega_3 & 0 \end{bmatrix}\begin{bmatrix} q_1 \\ q_2 \\ q_3 \\ q_4 \end{bmatrix} = \frac{1}{2}\boldsymbol{\Omega}\,[\boldsymbol{\omega}(t)]\begin{bmatrix} q_1 \\ q_2 \\ q_3 \\ q_4 \end{bmatrix}.$$

In a typical vehicle dynamics model, the angular forces on the vehicle are converted into angular accelerations and then integrated in the vehicle reference frame to derive body-referenced rotation rates (yaw rate, pitch rate, and roll rate). The preceding relation is used to convert these rates into an inertial rate quaternion. The inertial rate quaternion is integrated to form an inertial attitude quaternion. The inertial attitude quaternion represents the vehicle attitude in inertial space. Thus

$$\mathbf{q}(t) = \exp\left\{\frac{1}{2}\int_0^t \boldsymbol{\Omega}\,[\boldsymbol{\omega}(t')]\,dt'\right\}\mathbf{q}(0).$$

These quaternion equations are most commonly implemented in a digital computer. An alternative approach suitable for an analog computer or operational amplifier implementation is also used in practice.

The equation for the inertial rate quaternion may be expanded and expressed as

$$\begin{bmatrix} \dot{q}_1 \\ \dot{q}_2 \\ \dot{q}_3 \\ \dot{q}_4 \end{bmatrix} = \frac{1}{2} \begin{bmatrix} \omega_3 q_2 & -\omega_2 q_3 & +\omega_1 q_4 \\ -\omega_3 q_1 & +\omega_1 q_3 & +\omega_2 q_4 \\ \omega_2 q_1 & -\omega_1 q_2 & +\omega_3 q_4 \\ -\omega_1 q_1 & -\omega_2 q_2 & -\omega_3 q_3 \end{bmatrix}.$$

To implement the equations on a computer, they are expressed as

$$\begin{bmatrix} \dot{q}_1 \\ \dot{q}_2 \\ \dot{q}_3 \\ \dot{q}_4 \end{bmatrix} = \frac{1}{2} \begin{bmatrix} \omega_3 q_2 & -\omega_2 q_3 & +\omega_1 q_4 \\ -\omega_3 q_1 & +\omega_1 q_3 & +\omega_2 q_4 \\ \omega_2 q_1 & -\omega_1 q_2 & +\omega_3 q_4 \\ -\omega_1 q_1 & -\omega_2 q_2 & -\omega_3 q_3 \end{bmatrix} + k\lambda \begin{bmatrix} q_1 \\ q_2 \\ q_3 \\ q_4 \end{bmatrix},$$

where $\lambda = 1 - (q_1^2 + q_2^2 + q_3^2 + q_4^2)$ and k is chosen to be less than $1/\Delta t$, where Δt is the integration step size. Initial values for the components of the attitude quaternion can be established by use of the relationships to the Euler angles ψ, θ, and ϕ, representing yaw, pitch, and roll.

When the direction of the angular velocity vector $\boldsymbol{\omega} = [\omega_1 \quad \omega_2 \quad \omega_3]^T$ is almost constant over an interval of interest or when the rotation vector defined by

$$\Delta\boldsymbol{\theta}(t) = \int_t^{t+\Delta t} \boldsymbol{\omega}(t')\,dt'$$

is small, then the attitude quaternion may be updated by use of

$$\mathbf{q}(t + \Delta t) = \mathbf{M}(\Delta\boldsymbol{\theta})\,\mathbf{q}(t),$$

where

$$\mathbf{M}(\Delta\boldsymbol{\theta}) = \cos(|\Delta\boldsymbol{\theta}|/2)\,\mathbf{I}_{4\times4} + \frac{\sin(|\Delta\boldsymbol{\theta}|/2)}{|\Delta\boldsymbol{\theta}|}\,\boldsymbol{\Omega}(\Delta\boldsymbol{\theta}).$$

Bibliography

Ayers, J., Davis, J. L., and Rudolph, A. (Eds.) (2002). *Neurotechnology for Biomimetic Robots* (1st ed.), MIT Press, Boston, MA.

Balafoutis, C. A. and Patel, R. V. (1991). *Dynamic Analysis of Robotic Manipulators: A Cartesian Tensor Approach*, Kluwer Academic, Boston, MA.

Bottema, O. and Roth, B. (1979). *Theoretical Kinematics*, North-Holland, New York (reprinted by Dover).

Boullart, L., Krijgsman, A., and Vingerhoeds, R. (Eds.) (1992). *Application of Artificial Intelligence Techniques in Process Control*, Pergamon, Oxford.

Burdea G. C. (1996). *Force and Touch Feedback for Virtual Reality*, Wiley, New York.

Canudas, C., Siciliano, B., and Bastin, G. (1996). *Theory of Robot Control*, Springer-Verlag, New York.

Craig, J. J. (1989). *Introduction to Robotics: Mechanics and Control*, Addison-Wesley, Reading, MA.

Denavit, J. and Hartenberg, R. S. (1955). A kinematic notation for lower pair mechanisms based on matrices, *J. Appl. Mech.* 2, 215–221.

Deo, N. (1974). *Graph Theory with Applications to Engineering and Computer Science*, Prentice-Hall, Englewood Cliffs, NJ.

Duffy, J. (1980). *Analysis of Mechanisms and Robot Manipulators*, Arnold, London.

Farin, G. (1988). *Curves and Surfaces for Computer Aided Geometric Design*, Academic, New York.

Featherstone, R. (1987). *Robot Dynamics Algorithms*, Kluwer Academic, Boston, MA.

Foley, J. D., van Dam, A., Feiner, S. K., and Hughes, J. K. (1996). *Computer Graphics: Principles and Practice* (2nd ed. in C), Addison-Wesley Longman, New York.

Garcia de Jalon, J. and Bayo, E. (1994). *Kinematic and Dynamic Simulation of Multibody Systems: The Real Time Challenge*, Springer-Verlag, New York.

Goldstein, H. (1980). *Classical Mechanics*, Addison-Wesley, Reading, MA.

Golub, G. H. and Van Loan, C. F. (1989). *Matrix Computations*, Johns Hopkins University Press, Baltimore, MD.

Greenwood, D. T. (1977). *Classical Dynamics*, Prentice-Hall, Englewood Cliffs, NJ.

Greenwood, D. T. (1988). *Principles of Dynamics*, Prentice-Hall, Englewood Cliffs, NJ.

Gupta, K. C. (1997). *Mechanics and Control of Robots*, Mechanical Engineering Series, Springer, Berlin.

Haug, E. J. (1989). *Computer-Aided Kinematics and Dynamics of Mechanical Systems*, Allyn & Bacon, Boston, MA, Vol. 1.

Hunt, K. H. (1978). *Kinematic Geometry of Mechanisms*, Clarendon, Oxford.

Ioannou P. and Fidan B. (2006). *Adaptive Control Tutorial*, Advances in Design and Control, Society for Industrial and Applied Mathematics, Philadelphia, PA.

Lanczos, C. (1986). *The Variational Principles of Mechanics*, Dover, New York.

Latombe, J. C. (1991). *Robot Motion Planning*, Kluwer Academic, Boston, MA.

Lewis, F. L., Abdallah, C. T., and Dawson, D. M. (1993). *Control of Robot Manipulators*, Macmillan, New York.

Mason, T. and Salisbury Jr., J. L. (1985). *Robot Hands and the Mechanics of Manipulation*, MIT Press, Cambridge, MA.

McCarthy, J. M. (1990). *An Introduction to Theoretical Kinematics*, MIT Press, Cambridge, MA.

McKerrow, P. J. (1991). *Introduction to Robotics*, Addison-Wesley, Sydney.

McMahon T. A. (1984). *Muscles, Reflexes, and Locomotion*, Princeton University Press, Princeton, NJ.

Meirovitch, L. (1970). *Methods of Analytical Dynamics*, McGraw-Hill, New York.

Mueller, T. J. (Ed.) (2002). *Fixed and Flapping Wing Aerodynamics for Micro Air Vehicle Applications*, Progress in Astronautics and Aeronautics Series, American Institute of Aeronautics and Astronautics, Washington, D.C.

Morecki, A. and Waldron, K. J. (Eds.) (1997). *Human and Machine Locomotion*, Springer-Verlag, New York.

Murray, R. M., Li, Z., and Sastry, S. S. (1993). *A Mathematical Introduction to Robot Manipulation*, CRC Press, Boca Raton, FL.

Nakamura, Y. (1991). *Theory of Robotics, Redundancy and Optimisation*, Addison-Wesley, Reading, MA.

Nigg B. M. and Herzog, W. (1999). *Biomechanics of Musculo-Skeletal System* (2nd ed.), Wiley, New York.

Nof, S. Y. (Ed.) (1999). *Handbook of Industrial Robotics*, Wiley, New York.

O'Shea, T. and Eisenstadt, M. (Eds.) (1984). *Artificial Intelligence: Tools, Techniques and Applications*, Harper & Row, New York.

Paul R. P. (1981). *Robot Manipulators, Mathematics, Programming and Control*, MIT Press, Cambridge, MA.

Rose J. and Gamble J. G. (1994). *Human Walking* (2nd ed.), Williams and Wilkins, Baltimore, MD.

Schilling, R. J. (1990). *Fundamentals of Robotics: Analysis and Control*, Prentice-Hall, Englewood Cliffs, NJ.

Selig, J. M. (1992). *Introductory Robotics*, Prentice-Hall International.

Selig, J. M. (1996). *Geometrical Methods in Robotics*, Springer, New York.

Shabana, A. A. (1998). *Dynamics of Multibody Systems* (2nd ed.), Wiley, New York.

Shyy W., Lian Y., Tang J., Viieru D., and Liu H. (2007). *Aerodynamics of Low Reynolds Number Flyers* (Cambridge Aerospace Series), Cambridge University Press, New York.

Slotine, J. J. E. and Li, W. (1991). *Applied Nonlinear Control*, Prentice-Hall, Englewood Cliffs, NJ.

Snyder, W. E. (1985). *Industrial Robots: Computer Interfacing and Control*, Prentice-Hall, Englewood Cliffs, NJ.

Spong, M. W. and Vidyasagar, M. (1989). *Robot Dynamics and Control*, Wiley, New York.

Spong, M. W. (1996). Motion control of robot manipulators, in *The Control Handbook*, W. S. Levine, Ed., CRC Press/IEEE, New York, pp. 1339–1350.

Srinivasan, A. V. and McFarland, D. M. (1999). *Smart Structures: Analysis and Design*, Cambridge University Press, Cambridge.

Stone, H. W. and Kanade, T. (Eds.) (1991). *Kinematic Modeling, Identification and Control of Robotic Manipulators*, Kluwer Academic, Dordrecht, The Netherlands.

Toko, K. (2000). *Biomimetic Sensor Technology*, Cambridge University Press, Cambridge.

Tsai, L. W. (1999). *Robotic Analysis*, Wiley, New York.

Tzafestas, S. G. (Ed.) (1992). *Robotic Systems: Advanced Techniques and Applications*, Kluwer Academic, Dordrecht, The Netherlands.

Vukobratovir, M., Frank, A. A., and Juricic, D. (1970). On the stability of biped locomotion, *IEEE Trans. Biomed. Eng.* BME-17, 25–36.

Wang, J. and Gosselin, C. M. (1998). A new approach for the dynamic analysis of parallel manipulators, in *Multibody System Dynamics*, Kluwer Academic, Dordrecht, The Netherlands, Vol. 2, pp. 317–334.

Yoshikawa, T. (1990). *Foundation of Robotics*, MIT Press, Cambridge, MA.

Index

A^* algorithm, 178–180
acceleration due to gravity, 151, 162, 190, 193, 197
accelerometers, 180, 181, 186, 188–197, 264, 265
Achilles tendon, 27
Ackermann steering gear, 16
action potential, 43, 45–50, 52–53
adaptation, 243–247, 251, 259
adaptive control, 242–249, 251–252, 255–258
adenosine triphosphate (ATP), 45
adjacency matrix, 10
adjoint, 208, 211
aerofoil, 25, 34, 285–287, 289–290, 293, 294–301, 303, 304, 306
afferent, 46–48, 50–53
alula, 34
anal fin, 38
analyte, 39–41, 43
angle-of-attack, 26, 34–36, 286–307
angular momentum, 283, 305, 306
angular velocity, 123, 124, 127–130, 140–141, 273, 278, 280, 283–285
anode, 41, 42
anthropomorphic, 22, 24, 76–79, 91, 220
area co-ordinates, 103, 104
Arnold–Kennedy theorem, 64
articulation, 33
artificial intelligence, 179–180, 248
associative memory, 35
attitude, 5, 318, 322–324, 326, 332–334
auditory nerve, 264, 265, 267, 268
austenite, 58, 59
axon, 47–50, 52, 54, 55

B spline, 168–169, 171, 193–194
β spline, 168
balance, 27, 28, 30, 38, 57, 58, 256, 258, 265, 272, 273, 277–279, 282
ball-and-socket joint, 12
ballistic, 27, 29
basal ganglia, 57, 58

Bernstein polynomials, 167, 170–171
Bézier curves, 167–169, 171
biomimesis, 25
biomimetic, 25–61, 76, 93, 228, 272–307
biomimicry, 30, 34
bionic limb, 46–48, 51, 55, 57, 61
biosensor, 39–41, 43, 61
biped, 27, 272, 279
blending function, 166, 168, 170, 171, 174
boundary layer, 37, 302

canonic form, 212, 217–219, 237
canonical equations, 200, 206
canonical form, 208
canonical transformation, 205, 227
capulla, 265
Cartesian control, 174
Cartesian, 5, 22, 86, 104, 273–274, 280
catheter, 59, 60
cathode, 41
caudal fin, 38, 303
caudal peduncle, 38
cells, 32, 39, 40, 42–50, 57, 304
center of gravity, 258
center of mass, 143, 147, 148, 154, 158, 160–162, 273–274, 276, 278, 280
center of rotation, 64, 258
centrifugal force, 151–154, 269
cerebellum, 57–58, 256–257, 265
chain, 6, 8–13, 24
Chasles' Theorem, 68
Chebyshev polynomial, 167
chromosome, 180, 259
circulation, 34
clap, 25, 26, 34–36
cochlea, 264, 265
compliance, 113, 115–117
composition, 89, 99, 101, 102, 108, 109
computed torque control, 220, 221, 241, 248, 260, 269
connectivity, 133

constraint 6, 11–13, 111, 116–117, 119, 133, 134, 137
contact transformation, 204, 205
control surface, 244, 248
control, 202, 203, 206–207, 210–214, 217–225, 229, 233, 235, 237–239, 241, 242–271, 272, 279, 281, 282–287, 297, 299–302, 304
controllability, 211, 214, 217–219
controller partioning, 246, 259
Coriolis force, 152–154, 190, 269, 283
cortex, 32, 57, 58, 257
Corvid, 34
counterstroking, 35
coupler curve, 8
coupler, 8, 11
crank and slotted lever, 15
crank, 8, 11, 13–16, 24
cruciate, 27
C-space, 177, 178
cubic spline, 167, 172, 174, 194
cuff electrodes, 47
cutaneous, 50, 268
cylindrical pair, 12, 13

da Vinci™ manipulator, 20, 21, 23
degrees of freedom (DOF), 5, 6, 8, 11–13, 18, 20, 21, 74, 76, 81, 88, 94–106, 120–141, 164, 207, 217, 229, 237, 257, 272–278, 280, 281, 287
Denavit–Hartenberg convention, 74–88, 99
Denavit–Hartenberg decomposition, 66–67, 69, 71
Denavit–Hartenberg parameters, 74
denticles, 37
DH convention, 74–88
DH parameters, 75–88
direct kinematics, 74–88
direction cosine matrix (DCM) 185, 325–326, 330–331
direction cosine, 185, 318–320, 325–326, 330–331
dissipative, 203, 223
dorsal fin, 38
drag, 34–38, 285, 286, 289, 301–304
dragonfly, 35, 36, 286
dual quaternions, 103, 188
duck, 167
Dynamic programming, 173
dynamically tuned gyro, 190
dynamics, 28, 38, 55, 59, 128, 142–163, 172, 189–241, 242, 246, 248, 249, 253, 260, 266, 269, 272, 279–282, 305, 317, 322, 326, 332

efferent, 46–55
electrochemical, 39, 40, 43
electrode, 40–42, 46, 47, 50–52
electrolyte, 42
emarginations, 36
end-effector, 7, 74, 77–79, 84, 86–119, 164, 165, 194, 220, 242, 253, 257, 258, 260, 261, 283, 284
endolymph, 265

endoscopy, 56, 59
energy, 25, 28, 29, 37, 38, 152–154, 158–160, 199–203, 207, 225, 232, 274, 285, 286, 287, 294, 297–300, 306
enzymes, 39, 42
epicyclic gear train, 17
equilibrium potential, 43–45
Euler angle, 67, 72, 80, 87, 121, 127, 134, 140, 322–324, 326–328, 330, 333
Euler axis, 67, 72, 323, 326, 330
Euler–Lagrange equations, 152–155, 160, 225, 226, 232, 234, 275
Euler's equations, 145

falcon, 35
feathering, 35
feathers, 33, 34, 36
feedback linearization, 198, 210–219, 237–239, 260
feedback, 36, 47–51, 53–58, 198, 206–208, 210–217, 219, 221, 224, 225, 236, 237–239, 242, 243, 245, 246, 253, 254, 257, 260, 265, 267, 268, 284
feedforward, 36, 58, 245–246, 257, 259, 260, 261, 271
femur, 27
fiber-optic gyro, 190
field-effect transistor, 39
fin, 38, 303
finlets, 38
fish, 33, 37–39
flapping–translation, 25, 34–36, 282
flapping, 25–26, 34–36, 272, 282, 285–287, 294, 299–301, 303, 304, 306
flight simulators, 26, 101–102
fling, 25–26, 34–36
flow separation, 34
footstep planning, 31
force-control, 111–117, 252, 258, 260
forward dynamics, 154–156
forward kinematics, 74–88
four-bar mechanism 7, 8, 11, 16, 24, 161–163
frontal cortex, 32
functional electrical stimulation, 49, 52
fuzzy-logic, 180, 249–251, 258, 259, 269, 270

gain scheduling, 244
gear train, 13, 17
genetic algorithm, 179, 259
gimbal, 17, 18, 80
gluconic acid, 42
glucose oxidase, 42
glucose, 41–43
GOD, 42
Goldman–Hodgkin–Katz equation, 44, 45
graph, 7
graph-theoretic, 7
grasp, 111–113, 117
Gray's Paradox, 302
group, 205, 208–209
Grübler mobility criterion, 133, 134

gyroscope, 3, 13, 17–19, 180, 181, 188–193, 195, 197, 264, 283

Hamiltonian equations, 200–201, 203–207, 210, 227, 237, 239
Hamiltonian, 198, 200–210, 223–225, 237, 239
Hamilton–Jacobi, 205–207
Hamilton's Principle, 203, 206, 226
haptic, 50, 52, 55–58, 267, 268
heave, 18
Hermite spline, 168–169, 193
higher kinematic pair, 12
high-lift, 34, 36
Hobby, 35
holonomic, 207, 228
homogeneous coordinate, 65, 188
homogeneous transformation, 62, 66–72, 74–85, 89–90, 114, 122, 123, 127, 128, 133
Hooke's joint, 17, 18, 102, 141
humerus, 33
hydrogen peroxide, 42

imitation, 30–32
immobilization, 39, 40, 42, 43, 61
inchworm, 29
independent joint control, 260
inelastic collision, 28
inertial measuring unit (IMU), 181, 188
inertial navigation, 180, 189
inertial reference, 181–184, 228, 230, 317, 321–323
inferior frontal cortex, 32
interchange graph, 7
inverse dynamics, 59, 154–156
inverse Jacobian, 131, 132, 134
inverse kinematics, 89, 91, 92, 94, 98, 99, 101, 102, 172–174
inverse model, 246, 249, 251, 269, 270
involutive, 208, 214, 215, 218, 219
ion, 40, 43–46
isomorphic, 10
isomorphism, 10
Isaac Asimov's laws, 24
IUPAC, 39

Jacobi identity, 208, 211
Jacobian, 120–141, 204, 205, 215, 285
joint angle, 66, 75, 129
joint interpolated control, 171, 174
joint, 5, 6, 11–13, 15, 17, 18, 20–24, 27, 28, 30, 33, 37, 46, 50–55, 74–78, 80–81, 87–110, 120–124, 128–141, 242, 243, 249, 257–262, 268, 269

Kalman filter, 190
kinematic chain 6–12, 24, 80–83, 90, 95, 99, 102, 133, 135
kinematic constraints 6, 116, 117
kinematic geometry, 10, 24, 91
kinematic inversion, 11
kinematic pairs, 12, 13, 27, 99

kinematic structure, 6, 7
kinematic system, 5–7, 11, 12, 117
kinesthetic, 50, 268
kinetic energy, 29, 37, 152–154, 159–160, 199–203, 225, 232, 274, 294, 297, 306
kinetics, 50, 142, 145
knee joint, 27

Lagrange interpolation, 166–168
Lagrangian, 142, 152–155, 198–201, 203, 204, 223, 226, 227, 237, 238, 306
leading-edge suction, 33, 35
leading-edge vortex, 35, 36
leading edge, 33–37
learning control, 255
learning, 248, 253, 255, 259, 271
Lie algebra, 208–209, 211
Lie bracket, 208, 214–217
Lie derivative, 208, 211, 218
Lie group, 208–209
lift, 25, 26, 33–38, 281, 285–286, 289, 292, 296, 298, 301, 304, 307
lifting surface, 33, 289
ligament, 27
limit cycle, 28
link length, 66, 75, 80
link offset, 66, 75, 80
link parameters, 76–88
link twist, 66, 75, 80
link, 5–13, 15–18, 24, 74–88, 242, 248, 258, 260, 261, 263, 269
locomotion, 25, 27, 29 31, 38, 281
lower kinematic pairs, 12, 13, 27
Luh, Walker and Paul algorithm, 150
Lyapunov approach, 104, 137, 138, 141, 207, 243

manipulability, 135–137
manipulator, 2–9, 11, 13, 17–24, 25, 46, 56–60, 71, 74–110, 142, 149–160, 164–166, 171–176, 194, 207, 213, 220–222, 225, 237, 239–241, 272, 281–285
mass, 23, 105, 124, 143, 146–148, 153–162, 189, 195, 218, 220–222, 227, 228, 230, 231, 235, 237–240, 248, 273, 274, 276, 277, 280, 288, 294, 297, 304, 306, 317
matensite, 58, 59
mechanism, 6–20, 23, 24, 25–29, 32–40, 48, 49, 59, 60, 64, 74–76, 81, 87, 91, 99, 106, 109, 110, 118, 130–138, 142, 161, 162, 209, 242–247, 251, 258, 263, 272, 285, 286, 294, 299, 303, 304, 306
median fins, 38, 303
membrane potential, 43–46, 49–50
membrane, 39–46, 49–50
mimicking, 25, 30–32
mirror neurons, 32
mobile robot, 164, 172, 174, 176, 177, 181, 185
mobility, 133–135
model following control, 245, 246, 266
model predictive control, 249, 252, 255

model reference adaptive control, 246
model-based control, 242, 251, 255, 259, 260, 270
moment of force, 111–119, 142, 150–152
moment of inertia, 145–147, 161–162, 273, 278
moment of momentum, 144, 146, 278, 283
moment, 111–119, 142–162
momentum, 142–144, 146, 198, 199, 225, 236, 276, 278, 282, 283, 285, 303–306
motion planning, 172, 174, 177, 178, 180, 225
motor cortex, 32, 57, 58, 257
muscles, 33, 39, 46, 51–55, 55, 57
myelin, 50, 54

natural selection, 179, 259
navigation, 164, 176, 180, 181, 189, 191–194, 197
Nernst equation, 40, 44, 45
neural network, 32, 51, 248, 249, 259, 271
neural, 30, 32, 51, 248, 249, 259, 265, 266, 271
neuroanatomy, 1, 30, 31, 54
neurobiology, 32
neuromuscular 33, 52, 53, 55
neuron, 32, 45–50, 53–55, 248, 249, 259
neurophysiology, 1, 32
neurotransmitter, 48, 49, 52, 58
Newton–Euler equations, 145, 148–150, 15153, 155, 158, 160
Newton's laws, 142, 145, 149
Nitinol, 58, 60
non-holonomic, 165, 178, 207, 208, 225–228, 230, 233, 283
novel mechanisms, 13–17

obstacle avoidance, 164, 165, 172, 174–178, 180
ossicles, 264
optic flow, 26, 175–176, 264
optimal, 34, 37, 54, 206, 209, 223–224, 243, 248, 252, 254, 258, 260, 267, 301
otolith organs, 265
oxygen, 40–43

Padé approximant, 194
pantograph, 15
parallel manipulator, 99–110, 155
passivity, 36, 202, 203, 207, 254
path following, 164
path planner, 164
path planning, 164, 165, 172–175, 177, 179, 180
path tracker, 164
path tracking, 164, 172
pawl-and-ratchet mechanism, 14
pectoral fin, 38
pelvic fin, 38
perception, 26, 45, 46, 253, 254, 263, 265, 267, 268
pitch, 5, 13, 18, 68–72, 112
planar, 6, 8, 11, 12, 64, 76–78, 80, 86–88, 91, 93, 94, 96, 99–101, 108–109
planning, 31, 164, 165, 169, 172–180, 225, 254, 257, 258, 263
planograph, 15

platinum, 41
Plücker coordinates, 103
Poisson brackets, 205, 208
Poisson constraint equations, 205
polynomial, 166–171, 194
pose, 64, 66
posture, 89, 91, 94, 95, 106
potential energy, 152, 154, 160, 198–203, 206, 223, 226, 232, 238, 274, 287–290
potential flow, 26, 34, 175
potential, 26, 29, 34, 39–53, 60, 152, 154, 159–161, 198, 199, 203, 206, 223, 226, 232, 238, 274, 287–290
premotor cortex, 32
pressure, 287–290, 292, 294, 295, 298, 303, 307
principle of least action, 203, 225, 226
principle of virtual work, 156–158, 163
prismatic joint, 13, 18, 20, 22, 23
product of exponentials, 70
propatagium, 33
proprioceptive, 46, 50–52, 268
propulsion, 37, 38, 272–307
propylene, 41
prosthesis, 46, 48, 50–52, 55
proximity, 264, 268
PUMA, 20–22, 81–84, 95, 134, 141
Purkinje cell, 256

quadruped, 279–281, 306
quartz crystal microbalance, 39
quaternion, 69, 80, 188, 198, 317, 324, 326–333
quick-return mechanisms, 15, 16

rack-and-pinion mechanism, 14
radius, 33
reachability, 105–107, 109, 110
real time 260–262
receptors, 39, 42, 43, 46, 55
recognition element, 39, 40, 42, 43
redundant manipulator, 94, 136
reference frame, 180–185, 192, 317–324, 329, 332
reference model, 243–247, 251, 252, 263
regenerative electrode, 47
relative degree, 211, 213, 214, 239
remiges, 33, 34
resting potential, 43–45, 48, 50
revolute joint, 12, 13, 17, 18, 20–23
Reynolds number, 34
riblets, 34, 37
ring LASER gyro, 190
robot manipulator, 3, 4, 5–11, 13, 17, 21, 24, 74–119, 136, 150, 160, 164, 166, 172, 176, 281
roll, 5, 18

saccule, 265, 266
SCARA manipulator, 20, 22, 23, 88, 109, 160
scheduling, 244, 255, 258, 262
Schuler frequency, 193, 197
screw joint, 13, 128

screw motion, 62, 67–69, 124, 127–130, 133, 140, 145, 149–151, 158
screw vector, 127–130, 140
screw, 1, 3, 13, 62, 67–72, 124, 127–131, 134, 140
search, 178–180
semicircular canals, 265, 266
semipermeable membrane, 39–44, 49
separation bubble, 36
separation, 34, 36
serial manipulator, 74, 99, 101, 149, 155
shape memory alloy, 56, 59
sidereal rate, 183, 184, 187
Simon filter, 59
singularity, 106, 134–138, 140, 141
skin friction, 37
slider–crank mechanism, 11, 13–16, 24
slider, 11, 16
sliding joint, 11
spanning, 35
sparrowhawk, 35
spatial manipulator, 18–20
spherical pair, 12
spherical wrist, 80, 81, 84, 90, 91, 95, 96, 107
spline, 167–169, 171, 172, 174, 193–194
stance leg, 27, 272, 275, 276
Stanford manipulator, 84, 86, 97, 107
static equilibrium, 113–114
statically determinate, 114, 115
statically indeterminate, 113–115
Stewart platform, 18–20, 101–104
strapped-down, 180, 190
Strouhal number, 293
subsumption, 255
Sun and planet gear train, 17
superior parietal lobule, 32
support, 168
surface acoustic wave, 39
surge, 18
sway, 18
swimming, 37–39
swing leg, 27, 272, 275, 276
symplectic, 200, 202, 204, 223
synapse, 48, 49, 54
systolic array, 76

tactile, 50, 51, 57, 264, 268
tendon, 27, 52–54, 268
Theodorsen function, 285, 286, 293, 294
three-centers-in-line theorem, 64

thrust effectors, 38
thrust, 26, 34, 37–39, 285–287, 294, 298–304, 306
tibia, 27
tool center point, 105, 106, 110, 117, 165, 242
torque, 111–119
touch, 267, 268
trajectory control, 164
trajectory following, 164, 257
trajectory, 164–166, 171, 173, 246, 257–259, 264
transducer, 39, 43, 53, 61
transmission zeros, 212
tremor, 36, 56–58
triangular coordinates, 103
trigonometric formulae, 91, 92, 103
twist coordinates, 69–70
twist, 66, 69–70

ulna, 33
Unimation Puma 560, 20, 21
universal joint, 17, 18
utricle, 265, 266

vector field, 208–211, 214
velocity graph, 10, 11
velocity potential, 287–290
vestibular system, 256, 265, 266
vestibulo-ocular reflex (VOR), 265
via points, 164, 166–168
virtual displacement, 156–157
virtual work, 156–158, 163
visual cues, 266–268
Voronoi diagram, 177
Voronoi region, 177
vortex drag, 37
vortex flow, 35, 36, 287–304
vortex lift, 35, 36

Wagner effect, 25, 34–36
walking, 27–31, 272–281, 304
Whitworth quick-return mechanism, 16
wingbeat, 35
winglets, 34
workspace, 6, 8, 22, 94, 101, 105–107, 109, 110
wrench, 111–119, 145
wrist, 80, 81, 83, 84, 87, 88, 90, 91, 95, 96, 98, 107

yaw, 5, 18

zero dynamics, 213, 235, 236, 238, 239
zero moment point, 277–279

rinted in the United States
Baker & Taylor Publisher Services

Printed in the United States
by Baker & Taylor Publisher Services